T0133259

Aquatic and Standing Water Plants
of the Central Midwest

Books in the Aquatic and Standing Water Plants
of the Central Midwest Series by Robert H. Mohlenbrock
Cyperaceae: Sedges
Filicineae, Gymnospermae, and Other Monocots, Excluding Cyperaceae:
 Ferns, Conifers, and Other Monocots, Excluding Sedges
Acanthaceae to Myricaceae: Water Willows to Wax Myrtles
Nelumbonaceae to Vitaceae: Water Lotuses to Grapes

Other Southern Illinois University Press Books
by Robert H. Mohlenbrock
Guide to the Vascular Flora of Illinois, revised and enlarged edition
Distribution of Illinois Vascular Plants, with Douglas M. Ladd
A Flora of Southern Illinois, with John W. Voigt

In the Illustrated Flora of Illinois Series
Ferns, 2nd edition
Flowering Plants: Basswoods to Spurges
Flowering Plants: Flowering Rush to Rushes
Flowering Plants: Hollies to Loasas
Flowering Plants: Lilies to Orchids
Flowering Plants: Magnolias to Pitcher Plants
Flowering Plants: Nightshades to Mistletoe
Flowering Plants: Pokeweeds, Four-o'clocks, Carpetweeds,
 Cacti, Purslanes, Goosefoots, Pigweeds, and Pinks
Flowering Plants: Smartweeds to Hazelnuts
Flowering Plants: Willows to Mustards
Grasses: Bromus to Paspalum, 2nd edition
Grasses: Panicum to Danthonia, 2nd edition
Sedges: Carex
Sedges: Cyperus to Scleria, 2nd edition

Acanthaceae to Myricaceae
Water Willows to Wax Myrtles

Robert H. Mohlenbrock

Southern
Illinois
University Press

Carbondale

Cover Illustration by Mark W. Mohlenbrock

Library of Congress Cataloging-in-Publication Data

Mohlenbrock, Robert H., 1931–

  Acanthaceae to Myricaceae : water willows to wax myrtles / Robert H. Mohlenbrock.

    p. cm. — (Aquatic and standing water plants of the central Midwest)

  Includes index.

  ISBN-13: 978-0-8093-2790-4 (alk. paper)

  ISBN-10: 0-8093-2790-2 (alk. paper)

  1. Angiosperms—Middle West—Identification. 2. Angiosperms—Middle West—Pictorial
works. I. Title.

QK128.M64 2007

581.977—dc22   2007021714

Printed on recycled paper. ♻

The paper used in this publication

meets the minimum requirements

of American National Standard

for Information Sciences—

Permanence of Paper for

Printed Library Materials,

ANSI Z39.48-1992. ⊚

This book is dedicated to the late Robert E. Woodson, my doctoral advisor at Washington University in St. Louis, who provided me with my taxonomic philosophy and taught me so many things about taxonomy that no one else could have done. It was his dedication to teaching that instilled within me the passion I have for teaching.

Contents

Illustrations

Series Preface

The purpose of the four books in the Aquatic and Standing Water Plants of the Central Midwest series is to provide illustrated guides to the plants of the central Midwest that may live in standing or running water at least three months a year, though a particular species may not necessarily live in standing or running water during a given year. The states covered by these guides include Iowa, Illinois, Indiana, Ohio, Kansas, Kentucky, Missouri, and Nebraska, except for the Cumberland Mountain region of eastern Kentucky, which is in a different biological province. Since 1990, I have taught week-long wetland plant identification courses in all of these states on several occasions.

The most difficult task has been to decide what plants to include and what plants to exclude from these books. Three groups of plants are within the guidelines of the manuals. One group includes those aquatic plants that spend their entire life with their vegetative parts either completely submerged or at least floating on the water's surface.

This group includes obvious submerged aquatics such as *Ceratophyllum*, the Najadaceae, the Potamogetonaceae, *Elodea, Cabomba, Brasenia, Numphaea*, some species of *Ranunculus, Utricularia*, and a few others.

Plants in a second group are called emergents. These plants typically are rooted under water, with their vegetative parts standing above the water surface. Many of these plants can live for a long period of time, even their entire life, out of the water. Included in this group are *Sagittaria, Alisma, Peltandra, Pontederia, Saururus, Justicia*, and several others.

The most difficult group of plants that I had to consider is made up of those wetland plants that live most or all of their lives out of the water, but which on occasion can live at least three months in water. I concluded that I would include within these books only those species that I personally have observed in standing water during the year, or which have been reported in the literature as living in water. In this last group, for example, I have included *Poa annua*, since Yatskievich, in his *Steyermark's Flora of Missouri* (1999), indicates that this species may occur in standing water, even though I have not observed this myself.

In these books, I have included most plants that live in lakes, ponds, rivers, and streams, as well as plants that live in marshes, bogs, fens, wet meadows, sedge meadows, wet prairies, swampy woods, and temporary depressions in woods, on cliffs, and in barrens.

Swink and Wilhelm (1999) consider a marsh as a transition between aquatic communities and drier communities, or in large flats which are regularly inundated by shallow surface water for much of the growing season. Cattails are often frequent in marshes. Trees are generally uncommon in a typical marsh.

Wet meadows, as treated here, are similar to marshes, but without pools of water. Woody plants are generally absent, and cattails do not dominate and are often not present.

Sedge meadows are similar, except that the overwhelming majority of species is composed of members of the sedge family (Cyperaceae). Wet prairies are also similar, but the dominant vegetation consists of grasses. Ladd (1995) recognizes dry, moist, and wet prairies, and only the last of these is included in these manuals.

Fens are wetlands where the underlying groundwater is rich with calcium or magnesium carbonates. Fens may or may not have woody plants present. Swink and Wilhelm (1999) also recognize marly fens that occur on open prairie slopes and hillside fens that are wooded seeps on steep bluffs.

Bogs are habitats typified by acidic, usually organic, substrates. However, bogs that have been influenced by carbonate-rich water have been called alkaline bogs by Swink and Wilhelm. Where minerotrophic water is insignificant, acid bogs develop. Floating sedge mats are acidic and often develop in sand flats or basins that rise and fall with the water table.

Swampy woods are forested wetlands in poorly drained flats or basins. In areas around Lake Michigan, swamps occur in wet sandy flats and on the moraine in wet depressions and in large flats behind the high dunes. In the southern part of the central Midwest, swampy woods may have standing water throughout the year.

Occasional depressions that fill with water harbor species that are included in these books. These depressions may occur in various kinds of woodlands, on exposed sandstone blufftops, and, in Kentucky, even on rocky barrens. Seeps may occur on the faces of cliffs, and the plants that live in this constantly wet habitat are also included in these books.

It is likely that I failed to include a few plants that should have been included, but that I had not observed myself.

The nomenclature that I have used in these manuals reflects my own opinion as to what I believe the scientific names should be. If these names differ from those used by the U.S. Fish and Wildlife Service, I have indicated this. A partial list of synonymy is included for each species, particularly accounting for synonyms that have been in use for several decades.

After the description of each plant, I have indicated the habitats in which the plant may be found, followed by the states in which the plant occurs. I have indicated the U. S. Fish and Wildlife Service wetland designation for each species for the states in which each occurs. In 1988, the National Wetland Inventory Section of the U.S. Fish and Wildlife Service attempted to give a wetland designation for every plant occurring in the wild in the United States. The states covered by these aquatic manuals occur in three regions of the Fish and Wildlife Service. Kentucky and Ohio are in region 1; Illinois, Indiana, Iowa, and Missouri are in region 3; and Kansas and Nebraska are in region 5. Definitions of the Fish and Wildlife Service wetland categories are:

OBL (Obligate Wetland). Occur almost always under natural conditions in wetlands, at least 99% of the time.

FACW (Facultative Wetland). Usually occur in wetlands 67%–99% of the time, but occasionally found in non-wetlands.

FAC (Facultative). Equally likely to occur in wetlands or non-wetlands 34%–66% of the time.

FACU (Facultative Upland). Usually occur in non-wetlands 67%–99% of the time, but occasionally found in wetlands.

UPL (Upland). Occur in uplands at least 99% of the time, but under natural conditions not found in wetlands.

NI (Not Indicated). Due to insufficient information.

A plus or minus sign (+ or −) may appear after FACW, FAC, and FACU. The plus means leaning toward a wetter condition; the minus means leaning toward a drier condition.

Although the Fish and Wildlife Service made changes to the wetland status of several species in an updated version in 1997, this later list has never been approved by the Congress of the United States.

Following this is one or more common names currently employed in the central Midwest. A brief discussion of distinguishing characteristics and nomenclatural notes is often included. Illustrations accompany each species, showing the diagnostic characteristics. In some of the illustrations, a gap in the stem signifies that a portion of the stem has been omitted due to space limitations.

The sequence of families in these aquatic manuals is as follows:

1. Azollaceae	28. Potamogetonaceae	55. Caprifoliaceae
2. Blechnaceae	29. Ruppiaceae	56. Caryophyllaceae
3. Equisetaceae	30. Scheuchzeriaceae	57. Ceratophyllaceae
4. Isoetaceae	31. Sparganiaceae	58. Convolvulaceae
5. Lycopodiaceae	32. Typhaceae	59. Cornaceae
6. Marsileaceae	33. Xyridaceae	60. Corylaceae
7. Onocleaceae	34. Zannichelliaceae	61. Cuscutaceae
8. Osmundaceae	35. Acanthaceae	62. Droseraceae
9. Thelypteridaceae	36. Aceraceae	63. Elatinaceae
10. Pinaceae	37. Amaranthaceae	64. Ericaceae
11. Taxodiaceae	38. Anacardiaceae	65. Escalloniaceae
12. Acoraceae	39. Apiaceae	66. Euphorbiaceae
13. Alismataceae	40. Apocynaceae	67. Fabaceae
14. Araceae	41. Aquifoliaceae	68. Fagaceae
15. Butomaceae	42. Araliaceae	69. Gentianaceae
16. Cyperaceae	43. Aristolochiaceae	70. Grossulariaceae
17. Eriocaulaceae	44. Asclepiadaceae	71. Haloragidaceae
18. Hydrocharitaceae	45. Asteraceae	72. Hamamelidaceae
19. Iridaceae	46. Balsaminaceae	73. Hippuridaceae
20. Juncaceae	47. Betulaceae	74. Hydrophyllaceae
21. Juncaginaceae	48. Bignoniaceae	75. Hypericaceae
22. Lemnaceae	49. Boraginaceae	76. Juglandaceae
23. Maranthaceae	50. Brassicaceae	77. Lamiaceae
24. Najadaceae	51. Cabombaceae	78. Lauraceae
25. Orchidaceae	52. Caesalpiniaceae	79. Leitneriaceae
26. Poaceae	53. Callitrichaceae	80. Lentibulariaceae
27. Pontederiaceae	54. Campanulaceae	81. Limnanthaceae

82. Linaceae
83. Lythraceae
84. Malvaceae
85. Melastomaceae
86. Menyanthaceae
87. Molluginaceae
88. Myricaceae
89. Nelumbonaceae
90. Nymphaeaceae
91. Nyssaceae
92. Oleaceae
93. Onagraceae
94. Parnassiaceae

95. Plantaginaceae
96. Platanaceae
97. Podostemaceae
98. Polemoniaceae
99. Polygalaceae
100. Polygonaceae
101. Primulaceae
102. Ranunculaceae
103. Rhamnaceae
104. Rosaceae
105. Rubiaceae
106. Salicaceae

107. Sarraceniaceae
108. Saururaceae
109. Saxifragaceae
110. Scrophulariaceae
111. Solanaceae
112. Styracaceae
113. Ulmaceae
114. Urticaceae
115. Valerianaceae
116. Verbenaceae
117. Violaceae
118. Vitaceae

This volume consists of families 35–88. Volume 4 will have the remaining dicots, from families 89 to 118. The first volume in the series was devoted exclusively to the Cyperaceae, family 16. The second volume included the ferns, gymnosperms, and monocots, families 1–15 and 17–34.

Aquatic and Standing Water Plants
of the Central Midwest

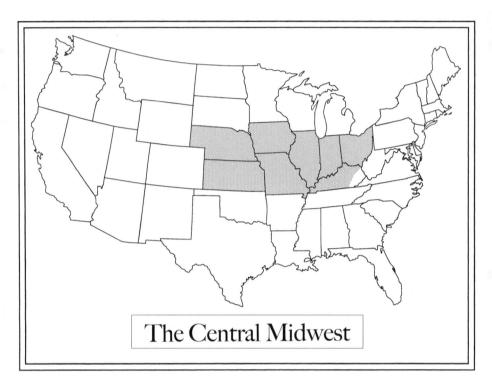

The Central Midwest

Descriptions and Illustrations

General Key to Groups of Aquatic and Wetland Dicots

Group 1. Plants aquatic, living in standing water the entire year.

1. Plants attached to rocks or wood in flowing water..................................97. Podostemaceae*
1. Plants not attached to rocks or wood in flowing water.
  2. Some of the leaves deeply divided or compound.
    3. Flowers borne together in a solitary, yellow head, each flower sharing a common receptacle.. *Megalodonta* in 45. Asteraceae

* Family names marked with an asterisk are included in volume 4.

3. Flowers not crowded, each with its own receptacle.
    4. Leaves trifoliolate .... *Menyanthes* in 86. Menyanthaceae
    4. Leaves 5- to 9-parted or pinnatisect.
        5. Leaves 5- to 9-parted .... *Nasturtium* in 50. Brassicaceae
        5. Leaves pinnatisect.
            6. Underwater structures bearing small bladders; flowers zygomorphic ....
            .... 80. Lentibulariaceae
            6. Underwater structures not bearing bladders; flowers actinomorphic.
                7. Some or all the flowers unisexual.
                    8. Calyx (or involucre) 8- to 12-cleft; stamens 10–20; ovary unlobed ....
                    .... 57. Ceratophyllaceae
                    8. Calyx 3- or 4-cleft; stamens 3–8; ovary 2- to 4-lobed ..... 71. Haloragidaceae
                7. All flowers perfect.
                    9. Pistils 2–several, free, or pistil 1 but deeply 2- to 4-lobed.
                        10. Sepals 3; petals 3; stamens 3 or 6 .... 51. Cabombaceae
                        10. Sepals 4–5; petals 0, 4, or 5; stamens 3, 4, 8, or numerous.
                            11. Sepals 5; petals 5; stamens numerous ....
                          ....*Ranunculus* in 102. Ranunculaceae*
                            11. Sepals 4; petals 0 or 4; stamens 3, 4, or 8 .... 71. Haloragidaceae
                  9. Pistil 1, unlobed.
                      12. Sepals 5; petals 5, united at base; stamens 5 ....
                      .... *Hottonia* in 101. Primulaceae*
                      12. Sepals 4; petals 4, free; stamens 6 .... *Neobeckia* in 50. Brassicaceae
2. Leaves simple, none deeply divided.
    13. Leaves with conspicuous sheathing stipules (ocreae) at base .... 100. Polygonaceae*
    13. Leaves without sheathing stipules at base.
        14. Flowers unisexual; petals absent.
            15. Calyx absent; fruit heart-shaped .... 53. Callitrichaceae
            15. Sepals 3; fruit ellipsoid ....*Proserpinaca* in 71. Haloragidaceae
        14. Flowers perfect; perianth, or at least the calyx, present.
            16. Leaves peltate.
                17. Sepals, petals, and stamens each 5 .... *Hydrocotyle* in 39. Apiaceae
                17. Sepals, petals, and stamens some number other than 5.
                    18. Sepals 3–4; petals 3–4; stamens 12–18; leaves up to 6 cm across, covered
                    with a gelatinous material ....*Brasenia* in 51. Cabombaceae
                  18. Sepals and petals indistinguishable, together totaling more than 8 seg-
                    ments; stamens more than 20; leaves more than 6 cm across, not gelati-
                    nous .... 89. Nelumbonaceae*
            16. Leaves not peltate.
                19. Leaves opposite or whorled.
                  20. Stamen 1; leaves linear ....73. Hippuridaceae
                  20. Stamens 3–5; leaves not linear.
                    21. Sepals and petals usually 3 each .... *Elatine* in 63. Elatinaceae
                    21. Sepals and petals (if present) usually 4 or 5 each.
                      22. Corolla absent ....*Didiplis* in 83. Lythraceae
                      22. Corolla present.
                        23. Stamens 2; leaves lanceolate or ovate ....
                      ....*Justicia* in 35. Acanthaceae
                      23. Stamens 4; leaves more or less orbicular to oblong ....
                      ....*Bacopa* in 110. Scrophulariaceae*
                19. Leaves alternate or basal.
                24. Petals absent (sepals appearing petaloid in *Caltha*).

25. Flowers white; pistil 1..108. Saururaceae*

25. Flowers yellow; pistils 4–12..............*Caltha* in 102. Ranunculaceae*

24. Petals present (what appear to be petals in *Caltha* are actually sepals).

26. Petals united .... *Nymphoides* in 86. Menyanthaceae

26. Petals free.

27. Flowers borne in umbels; stamens 5 ....
....*Hydrocotyle* in 39. Apiaceae

27. Flowers not borne in umbels; stamens 8–numerous.

28. Petals 4–6; stamens 8–11 .... *Ludwigia* in 93. Onagraceae*

28. Petals more than 6; stamens more than 12...........................
.... 90. Nymphaeaceae*

Group 2. Plants terrestrial, at least part of the year; plants with latex.

1. Many ray flowers crowded into heads, each sharing the same receptacle............................
.... *Prenanthes* in 45. Asteraceae

1. Flowers not in heads and not sharing the same receptacle.

2. Leaves opposite.

3. Pistils 2; fruit in pairs; stems often pink or purple .... 40. Apocynaceae

3. Pistil 1; fruit solitary; stems not pink or purple .... 44. Asclepiadaceae

2. Leaves alternate.

4. Leaves toothed.. 54. Campanulaceae

4. Leaves entire.

5. Plants scabrous; leaves up to 5 mm wide .... 54. Campanulaceae

5. Plants not scabrous; leaves more than 5 mm wide.

6. Pistils 2; fruit in pairs; stems often pink or purple......................... 40. Apocynaceae

6. Pistil 1; fruit solitary; stems not pink or purple .... 44. Asclepiadaceae

Group 3. Plants terrestrial, at least part of the year; some part of plant prickly or spiny.

1. Plants woody.

2. Leaves compound.

3. Leaves twice-compound .... *Gleditsia* in 52. Caesalpiniaceae

3. Leaves once-compound or trifoliolate.

4. Leaves trifoliolate.

5. Base of petiole dilated... *Rosa* in 104. Rosaceae*

5. Base of petiole not dilated .... *Rubus* in 104. Rosaceae*

4. Leaves once-pinnate, with 5 or more leaflets.

6. Leaflets more than 5.

7. Stamens more than 10; shrubs ....104. Rosaceae*

7. Stamens 5; trees.. *Gleditsia* in 52. Caesalpiniaceae

6. Leaflets 5.

8. Leaves pinnately compound... *Rosa* in 104. Rosaceae*

8. Leaves palmately compound .... *Rubus* in 104. Rosaceae*

2. Leaves simple.

9. Stamens 15 or more ....104. Rosaceae*

9. Stamens 5.

10. Woody vines .... *Solanum* in 111. Solanaceae*

10. Shrubs or trees.

11. Leaves usually lobed... 70. Grossulariaceae

11. Leaves not lobed .... 103. Rhamnaceae*

1. Plants herbaceous.

12. Flowers crowded into heads, each flower sharing the same receptacle ....
.... *Cirsium* in 45. Asteraceae

12. Flowers not crowded into heads and not sharing the same receptacle.
  13. Petals absent; leaves with a sheathing stipule .... 100. Polygonaceae*
  13. Petals present; leaves without a sheathing stipule .... 74. Hydrophyllaceae

Group 4. Plants terrestrial, at least part of the year; plants climbing or twining.

1. Plants woody.
  2. Leaves simple.
    3. Leaves opposite .... *Lonicera* in 55. Caprifoliaceae
    3. Leaves alternate.
      4. Some of the leaves lobed.
        5. Plants without tendrils .... 111. Solanaceae*
        5. Plants with tendrils .... 118. Vitaceae*
      4. None of the leaves lobed.
        8. Leaves toothed .... 118. Vitaceae*
        8. Leaves entire .... 43. Aristolochiaceae
  2. Leaves compound.
    9. Leaves opposite .... 48. Bignoniaceae
    9. Leaves alternate.
      10. Leaves bipinnate or tripinnate .... 118. Vitaceae*
      10. Leaves once-pinnate or trifoliolate.
        11. Leaves pinnate, with 5 or more leaflets .... 67. Fabaceae
        11. Leaves trifoliolate .... 38. Anacardiaceae
1. Plants herbaceous.
  12. Leaves opposite.
    13. Leaves compound .... *Clematis* in 102. Ranunculaceae*
    13. Leaves simple.
      14. Plants with latex; leaves entire .... *Trachelospermum* in 40. Apocynaceae
      14. Plants without latex; leaves toothed .... *Mikania* in 45. Asteraceae
  12. Leaves alternate.
    15. Leaves compound .... 67. Fabaceae
    15. Leaves simple, sometimes with a pair of tiny lobes at base.
      16. Tendrils present; stamens 8 .... *Brunnichia* in 48. Polygonaceae*
      16. Tendrils absent; stamens 5.
        17. Some part of plant prickly .... *Tracaulon* in 48. Polygonaceae*
        17. Plants not prickly.
          18. Leaves entire .... 58. Convolvulaceae
          18. Some leaves with a pair of basal lobes .... 111. Solanaceae*

Group 5. Plants terrestrial, at least part of the year; plants woody (excluding vines); leaves opposite or whorled.

1. Leaves simple.
  2. Some or all the leaves whorled.
    3. Flowers white; petals 4, united; stamens 4; ovary inferior; flowers and fruits in globose clusters .... *Cephalanthus* in 105. Rubiaceae*
    3. Flowers pink; petals 5, free; stamens 10; ovary superior; flowers and fruits in axillary whorled .... *Decodon* in 83. Lythraceae
  2. All leaves opposite.
    4. Leaves entire.
      5. Sepals 5; petals 5 .... 55. Caprifoliaceae
      5. Sepals 4; petals 4.
        6. Flowers in globose heads, white .... *Cephalanthus* in 105. Rubiaceae*
        6. Flowers not in globose heads, yellow or white.

7. Petals united at base, yellow; stamens 2 .... *Forestiera* in 92. Oleaceae*
    7. Petals free, white; stamens 4 .... 59. Cornaceae
  4. Leaves toothed or lobed.
    8. Leaves lobed.
      9. Petals small and inconspicuous or absent; fruit a samara .... 36. Aceraceae
      9. Petals white, conspicuous; fruit fleshy .... *Viburnum* in 55. Caprifoliaceae
    8. Leaves toothed.
      10. Sepals 4; petals 0; stamens 2 .... *Forestiera* in 92. Oleaceae*
      10. Sepals 5; petals 5; stamens 5 ....55. Caprifoliaceae
1. Leaves compound.
  11. Stamens 2; sepals 4 ....*Fraxinus* in 92. Oleaceae*
  11. Stamens 5; sepals 5.
    12. Petals white; fruit a berry .... *Sambucus* in 55. Caprifoliaceae
    12. Petals yellow or green or absent; fruit a samara.................................. 36. Aceraceae

Group 6. Plants terrestrial, at least part of the year; plants woody (excluding vines); leaves alternate, simple, entire, toothed, or lobed.

1. Leaves entire.
  2. Fruit an acorn... 68. Fagaceae
  2. Fruit not an acorn.
    3. Petals absent.
      4. Sepals present.
        5. Sepals 6; stamens 9; plants aromatic .... 78. Lauraceae
        5. Sepals 5; stamens 5 or 10; plants not aromatic.
          6. Leaves 3-veined from the base; stamens 5......................*Celtis* in 113. Ulmaceae*
          6. Leaves not 3-veined from the base; stamens 10 ....91. Nyssaceae*
      4. Sepals absent.
        7. Leaves glandular, aromatic .... 88. Myricaceae
        7. Leaves eglandular, not aromatic.
          8. Fruit drupelike; leaves 10–15 cm long .... 79. Leitneriaceae
          8. Fruit a capsule; leaves up to 8 cm long .... *Salix* in 106. Salicaceae*
    3. Petals present.
      9. Petals free.
        10. Ovary superior; hypanthium absent.
          11. Sepals absent; petals yellow......................... *Nemopanthus* in 41. Aquifoliaceae
          11. Sepals present; petals usually not yellow ....64. Ericaceae
        10. Ovary inferior; hypanthium present.
          12. Stamens numerous ....104. Rosaceae*
          12. Stamens 4, 5, or 10... 53. Ericaceae
      9. Petals united .... 53. Ericaceae
1. Leaves toothed or lobed.
  13. Leaves toothed.
    14. Stems spiny...*Crataegus* in 104. Rosaceae*
    14. Stems not spiny.
      15. Flowers unisexual.
        16. Petals present...41. Aquifoliaceae
        16. Petals absent.
          17. Leaves aromatic, glandular .... 88. Myricaceae
          17. Leaves not aromatic, not glandular.
            18. Fruit an acorn.. 68. Fagaceae
            18. Fruit not an acorn.
              19. Sepals 5; fruit fleshy or a samara............................. 113. Ulmaceae*

19. Sepals absent (at least in the staminate flowers); fruit dry, but not a samara.
  20. Carpels 2; seeds with a tuft of hairs.................106. Salicaceae*
  20. Carpel 1; seeds without a tuft of hairs.
    21. Staminate flowers without a calyx; pistillate flowers not in catkins ....60. Corylaceae
    21. Staminate flowers with a calyx; pistillate flowers in catkins..
    .... 47. Betulaceae
15. Flowers perfect.
  22. Petals absent.. 103. Rhamnaceae*
  22. Petals present.
    23. Petals 4 ....64. Ericaceae
    23. Petals 5.
      24. Petals united.
        25. Pubescence stellate... 112. Styracaceae*
        25. Pubescence not stellate, or absent.............................64. Ericaceae
      24. Petals free.
        26. Stamens 5.
          27. Flowers white, in racemes ....65. Escalloniaceae
          27. Flowers not white, not in racemes.............. 103. Rhamnaceae*
        26. Stamens 15 or more ....104. Rosaceae*
13. Leaves lobed.
  28. Leaves star-shaped... 72. Hamamelidaceae
  28. Leaves not star-shaped.
    29. Flowers unisexual.
      30. Petals present; fruit dry, globose ....96. Platanaceae*
      30. Petals absent; fruit an acorn.. 68. Fagaceae
    29. Flowers perfect.
      31. Stamens 5 .... 70. Grossulariaceae
      31. Stamens 10 or more...104. Rosaceae*

Group 7. Plants woody (at least part of the year); leaves alternate, compound.

1. Leaves palmately compound, with at least 5 leaflets.
  2. Plants prickly; flowers white .... *Rubus* in 104. Rosaceae*
  2. Plants not prickly; flowers yellow .... *Pentaphylloides* in 104. Rosaceae*
1. Leaves pinnately compound.
  3. Plants prickly or spiny.
    4. Leaves bipinnately or tripinnately compound .... 52. Caesalpiniaceae
    4. Leaves once pinnately compound ....104. Rosaceae*
  3. Plants neither prickly nor spiny.
    5. Flowers, or some of them, unisexual; stamen 1 .... 76. Juglandaceae
    5. Flowers perfect.
      6. Flowers zygomorphic; stamens 9 or 10.................................... 67. Fabaceae
      6. Flowers actinomorphic; stamens 15 or more.......................104. Rosaceae*

Group 8. Plants terrestrial, at least part of the year; herbaceous; leaves all basal.

1. Leaves compound.
  2. Leaves ternately decompound.
    3. Sepals 5; petals 5; stamens 5.. 42. Araliaceae
    3. Sepals 4 or more; petals absent; stamens numerous.................... 102. Ranunculaceae*
  2. Leaves trifoliolate .... *Coptis* in 102. Ranunculaceae*
1. Leaves simple.

4. Leaves or some of them lobed.
  5. Leaves peltate .... *Hydrocotyle* in 39. Apiaceae
  5. Leaves not peltate.
    6. Flowers in umbels; plants usually aquatic .... 39. *Hydrocotyle* in 39. Apiaceae
    6. Flowers not in umbels; plants rarely aquatic .... 109. Saxifragaceae*
4. Leaves not lobed.
  7. Leaves toothed.
    8. Petals united, at least at base .... 101. Primulaceae*
    8. Petals free.
      9. Stamens numerous .... *Dalibarda* in 104. Rosaceae*
      9. Stamens 5 .... 117. Violaceae*
  7. Leaves entire.
    10. Leaves with sticky hairs .... 62. Droseraceae
    10. Leaves without sticky hairs.
      11. Sepals 4; petals 4.
        12. Petals white or yellow, free; stamens 6; ovary superior ....50. Brassicaceae
        12. Petals translucent, united; stamens 2–4; ovary inferior ....
        ....95. Plantaginaceae*
      11. Sepals or calyx lobes 5; petals or corolla lobes 5.
        13. Leaves linear; sepals spurred; pistils numerous ....
        .... *Myosurus* in 102. Ranunculaceae*
        13. Leaves not linear; sepals not spurred; pistil 1.
          14. Petals united, at least at base .... 101. Primulaceae*
          14. Petals free .... 94. Parnassiaceae*

Group 9. Plants terrestrial, at least part of the year; herbaceous; leaves whorled.

1. Flowers crowded in heads, each sharing the same receptacle .... 45. Asteraceae
1. Flowers not crowded in heads, at least not sharing the same receptacle.
  2. Leaves compound .... *Hottonia* in 101. Primulaceae*
  2. Leaves simple.
    3. Leaves toothed ....*Veronicastrum* in 110. Scrophulariaceae*
    3. Leaves entire.
      4. Flowers zygomorphic; petals 3; stamens 8 ....99. Polygalaceae*
      4. Flowers actinomorphic; petals 4–5 (rarely 3 in some species of *Galium*); stamens 3,
        4, 5, or 10.
        5. Petals 4 .... 105. Rubiaceae*
        5. Petals 3 or 5.
          6. Petals free; ovary superior .... 56. Caryophyllaceae
          6. Petals united; ovary inferior .... 105. Rubiaceae*

Group 10. Plants terrestrial, at least part of the year; herbaceous; leaves, or some of them, opposite, simple, entire.

1. Flowers crowded into heads, each sharing the same receptacle .... 45. Asteraceae
1. Flowers not crowded into heads, at least not sharing the same receptacle.
  2. Latex present.
    3. Pistils 2; fruits borne in pairs .... 40. Apocynaceae
    3. Pistil 1; fruits borne singly .... 44. Asclepiadaceae
  2. Latex absent.
    4. Petals absent.
      5. Stamen 1; dwarf plant; fruits heart-shaped .... 53. Callitrichaceae
      5. Stamens 3–5, 8 or 10 or 20; fruits not heart-shaped.
        6. Stamens 8 or 10 or 20.

7. Plants lying on the ground .... *Glinus* in 87. Molluginaceae
7. Plants ascending to erect .... 56. Caryophyllaceae
  6. Stamens 3–5.
    8. Sepals 4; stamens 4.
      9. Plants prostrate.
        10. Ovary inferior...*Ludwigia* in 93. Onagraceae*
        10. Ovary superior .... *Didiplis* in 83. Lythraceae
      9. Plants erect .... 83. Lythraceae
    8. Sepals 3 or 5; stamens 3 or 5.
      11. Stipules present.. 56. Caryophyllaceae
      11. Stipules absent .... 37. Amaranthaceae
4. Petals present.
  12. Petals free.
    13. Stamens 8–numerous.
      14. Stamens numerous; leaves often punctate .... 75. Hypericaceae
      14. Stamens 8–12.
        15. Stamens 8... 85. Melastomaceae
        15. Stamens 9–12.
          16. Stamens 9; flowers maroon or pink........................ 75. Hypericaceae
          16. Stamens 10–12.; flowers purple or white.
            17. Stipules present... 56. Caryophyllaceae
            17. Stipules absent .... 83. Lythraceae
    13. Stamens 4–5.
      18. Sepals 4; petals 4.....................................*Rotala* in 83. Lythraceae
      18. Sepals 5; petals 5.
        19. Stipules present; flowers white................................ 56. Caryophyllaceae
        19. Stipules absent; flowers yellow.................................82. Linaceae
  12. Petals united.
    20. Corolla 4-lobed.
      21. Ovary superior...77. Lamiaceae
      21. Ovary inferior .... 105. Rubiaceae*
    20. Corolla 5-lobed, or petals 5.
      22. Stamens 5; flowers actinomorphic.
        23. Corolla rotate .... 101. Primulaceae*
        23. Corolla tubular.
          24. Ovary 1-locular ....69. Gentianaceae
          24. Ovary 3-locular .... 98. Polemoniaceae*
      22. Stamens 2 or 4; flowers zygomorphic.
        25. Ovary 4-parted; fruit separating into 4 nutlets ....77. Lamiaceae
        25. Ovary not 4-parted; fruit not separating into 4 nutlets.
          26. Flowers nearly actinomorphic ....35. Acanthaceae
          26. Flowers zygomorphic..................................... 110. Scrophulariaceae*

Group 11. Plants terrestrial, at least part of the year; herbaceous; leaves opposite, toothed, or lobed.

1. Flowers crowded into heads, each sharing the same receptacle..................... 45. Asteraceae
1. Flowers not crowded into heads, at least not sharing the same receptacle.
  2. Leaves toothed.
    3. Petals absent... 114. Urticaceae*
    3. Petals present.
      4. Petals 4; sepals 4...77. Lamiaceae
      4. Petals 5; sepals 5, or absent.

5. Petals free; leaves glandular-toothed; flowers actinomorphic ....
.... *Bergia* in 63. Elatinaceae
5. Petals united; leaves not glandular-toothed; flowers zygomorphic.
  6. Calyx absent; ovary inferior .... 115. Valerianaceae*
  6. Calyx present; ovary superior.
    7. Ovary 4-parted.
      8. Ovary deeply 4-lobed, the style arising from between the lobes ....
      .... 77. Lamiaceae
      8. Ovary shallowly 4-lobed or, if more deeply lobed, the style arising from the
      top of the ovary .... 116. Verbenaceae*
    7. Ovary single, not 4-parted .... 116. Verbenaceae*
2. Leaves lobed.
  9. Ovary 4-parted; fruit separating into 4 nutlets .... 77. Lamiaceae
  9. Ovary not 4-parted; fruit solitary .... 110. Scrophulariaceae*

Group 12. Plants terrestrial, at least part of the year; herbaceous; leaves opposite, compound.

1. Flowers crowded into heads, each sharing the same receptacle .... 45. Asteraceae
1. Flowers not crowded into heads, at least not sharing the same receptacle ....
.... 115. Valerianaceae*

Group 13. Plants terrestrial, at least part of the year; herbaceous; leaves alternate, simple, entire.

1. Flowers crowded into heads, each sharing the same receptacle .... 45. Asteraceae
1. Flowers not crowded into heads, at least not sharing the same receptacle.
  2. Latex present.
    3. Flowers zygomorphic; corolla split nearly to base .... 4. Campanulaceae
    3. Flowers actinomorphic; corolla not split nearly to base.
      4. Pistils 2; fruit borne in pairs; flowers deep blue .... *Amsonia* in 40. Apocynaceae
      4. Pistil 1; fruit not borne in pairs; flowers not deep blue .... 54. Campanulaceae
  2. Latex absent.
    5. Flowers unisexual.
      6. Perianth and bracts scarious .... 37. Amaranthaceae
      6. Perianth and bracts not scarious .... 66. Euphorbiaceae
    5. Flowers perfect.
      7. Petals absent.
        8. Flowers zygomorphic; stamens 6 .... 43. Aristolochiaceae
        8. Flowers actinomorphic; stamens 5 or 8, less commonly 6.
          9. Leaves with sheaths at base, not cordate .... 100. Polygonaceae*
          9. Leaves without sheaths at base, cordate .... 108. Saururaceae*
      7. Petals present.
        10. Flowers zygomorphic; stamens 8; petals 3 .... 99. Polygalaceae*
        10. Flowers actinomorphic; stamens 4, 5, 8, or 10; petals absent or other than 3.
          11. Petals united.
            12. Ovary distinctly 4-lobed .... 49. Boraginaceae
            12. Ovary not distinctly 4-lobed.
              13. Flowers in a scorpioid cyme .... *Heliotropium* in 49. Boraginaceae
              13. Flowers not in a scorpioid cyme.
                14. Ovary inferior ....*Samolus* in 101. Primulaceae*
                14. Ovary superior.
                  15. Stigmas 3 .... 98. Polemoniaceae*
                  15. Stigma 1 .... 101. Primulaceae*

11. Petals free.
    16. Petals 6...83. Lythraceae
    16. Petals 4 or 5.
        17. Petals 5.
            18. Stamens united into a tube ....82. Linaceae
            18. Stamens tree ....*Ludwigia* in 93. Onagraceae*
        17. Petals 4.
            19. Stamens 6; ovary superior ....50. Brassicaceae
            19. Stamens 8; ovary inferior......................................93. Onagraceae*

Group 14. Plants terrestrial, at least for part of the year; herbaceous; leaves simple, alternate, toothed or lobed.

1. Flowers crowded into heads, each sharing the same receptacle......................45. Asteraceae
1. Flowers not in heads, at least not sharing the same receptacle.
  2. Leaves toothed.
    3. Petals absent.
      4. Sepals 5–several, petaloid.
        5. Leaves cordate; stamens numerous; pistils 4–12......*Caltha* in 102. Ranunculaceae*
        5. Leaves not cordate; stamens 10; pistils 5 ....*Penthorum* in 109. Saxifragaceae*
      4. Sepals 3, not petaloid, green...................................*Proserpinaca* in 71. Haloragidaceae
    3. Petals present.
      6. Flowers zygomorphic.
        7. Petals free, one of them spurred...46. Balsaminaceae
        7. Petals united, not spurred.
            8. Corolla tube split; stamens 5 ....*Lobelia* in 54. Campanulaceae
            8. Corolla tube not split; stamens 2 or 4 ....110. Scrophulariaceae*
      6. Flowers actinomorphic.
        9. Stamens and pistils attached to a central column; stamens numerous...................
        ....84. Malvaceae
        9. Stamens and pistils not attached to a central column; stamens 5, 6, or 8.
            10. Ovary inferior; stamens 8 ....93. Onagraceae*
            10. Ovary superior; stamens 5 or 6.
                11. Sepals 4; petals 4; stamens 6................................50. Brassicaceae
                11. Sepals 5; petals 5; stamens 5 ....*Eryngium* in 39. Apiaceae
  2. Leaves lobed.
    12. Petals absent; sepals 3; stamens 3 ....*Proserpinaca* in 71. Haloragidaceae
    12. Petals present; sepals 4 or more; stamens 2 or 4 or more.
      13. Petals free.
        14. Stamens and pistils borne on a central column ....84. Malvaceae
        14. Stamens and pistils not borne on a central column.
            15. Plants trailing; flowers borne in a small purple head ....
            ....*Eryngium* in 39. Apiaceae
            15. Plants upright; flowers not borne in a head, yellow or white...... 50. Brassicaceae
      13. Petals united.
        16. Flowers zygomorphic; stamens 4................................. 110. Scrophulariaceae*
        16. Flowers actinomorphic; stamens 5...111. Solanaceae*

Group 15. Plants terrestrial; herbaceous; leaves alternate, trifoliolate or ternately compound.

1. Leaves trifoliolate.
  2. Flowers zygomorpic; fruit a legume ....67. Fabaceae
  2. Flowers actinomorphic; fruit not a legume.

3. Stamens 10 or more ....104. Rosaceae*
3. Stamens 5.
    4. Petals free; flowers in umbels; ovary inferior .... 39. Apiaceae
    4. Petals united; flowers not in umbels; ovary superior ....
    ....*Menyanthes* in 86. Menyanthaceae
1. Leaves ternately compound.
    5. Stamens and pistils numerous; ovary superior............................... 102. Ranunculaceae*
    5. Stamens 5; pistil 1, ovary inferior.. 39. Apiaceae

Group 16. Plants terrestrial, at least part of the year; herbaceous; leaves alternate, pinnately compound.
1. Flowers crowded into heads, each sharing the same receptacle...................... 45. Asteraceae
1. Flowers not crowded into heads, at least not sharing the same receptacle.
    2. Leaves bipinnately or tripinnately compound.
        3. Petals absent; ovary or ovaries superior .... 102. Ranunculaceae*
        3. Petals present; ovary inferior... 39. Apiaceae
    2. Leaves once pinnately compound.
        4. Petals absent; stamens 4 ....*Sanguisorba* in 104. Rosaceae*
        4. Petals present; stamens some number other than 4.
        5. Petals free.
            6. Sepals 3; petals 3 ....81. Limnanthaceae
            6. Sepals 4 or more; petals 4 or more.
                7. Sepals 4; petals 4.
                    8. Stamens 6; ovary superior ....50. Brassicaceae
                    8. Stamens 8; ovary inferior..................................... 93. Onagraceae*
                7. Sepals 5 (–7); petals 5 (–7).
                  9. Stamens usually more than 10 (sometimes as few as 5 in *Agrimonia*); pistils 2–numerous ....104. Rosaceae*
                  9. Stamens 5, pistil 1 .... 39. Apiaceae
        5. Petals united ....111. Solanaceae*

Group 17. Plants terrestrial, at least part of the year; herbaceous; leaves alternate, palmately compound.
1. Stamens 10 or more; pistils numerous ....104. Rosaceae*
1. Stamens 5; pistil 1 .... *Cynosciadium* in 39. Apiaceae

35. ACANTHACEAE—ACANTHUS FAMILY

Annual or perennial herbs; leaves opposite, simple, entire, without stipules; flowers perfect, actinomorphic or zygomorphic; calyx usually 5-parted, united at least at the base; corolla usually 5-parted, united, tubular; stamens 2 or 4; ovary superior, bilocular; fruit a capsule, usually exploding when dehiscing.

This family consists of 250 genera and about twenty-five hundred species worldwide. In addition to the genus described below, *Dicliptera* and *Ruellia* also occur in the central Midwest.

1. **Justicia** L.—Water Willow

Perennial herbs (in our area); leaves opposite, simple, entire, sessile or short-petiolate; flowers zygomorphic; sepals 5, linear, united at base; corolla 2-lipped, the upper lip more or less notched, the lower lip 3-lobed, violet or white, marked with purple; stamens 2; ovary superior; fruit a capsule, elevated on a stipe.

This mostly tropical and subtropical genus consists of approximately three hundred species. The following often grow in standing water in the central Midwest.

1. Leaves linear to lanceolate; flowers densely clustered in a headlike spike ....1. *J. americana*
1. Leaves elliptic to oblong; flowers loosely scattered along the peduncle ....2. *J. ovata*

1. **Justicia americana** (L.) Vahl, Symb. Bot. 2:15. 1791. Fig. 1.
Dianthera americana L. Sp. Pl. 27. 1753.

Glabrous perennial; stems grooved, erect, unbranched, to nearly 1 m tall (occasionally taller); leaves simple, opposite, linear to lanceolate, acute to acuminate, tapering to the nearly sessile base, to 15 cm long, to 2.5 cm wide, often much narrower, entire; flowers violet or white with purple markings, clustered in headlike spikes; sepals narrowly lanceolate, 4–8 mm long; corolla 8–12 mm long, the lobes as long as or a little longer than the tube; capsule to 1 cm long, the stipe about as long as the body; seeds deeply verrucose, about 3 mm across. May–October.

Shallow water in streams and at the edge of lakes and ponds; also in mud.
IA, IL, IN, KS, KY, MO, OH (OBL).
American water willow.
The strongly grooved stems are a good field character for this species.

Fig. 1. *Justicia americana* (American water willow). Habit (center). Flower (upper right). Fruit with calyx (lower right). Fruit (lower left).

2. **Justicia ovata** (Walt.) Lindau in Urb. Symb. Antill. 2:237. 1895. Fig. 2.
Dianthera ovata Walt. Fl. Carol. 63. 1788.
Dianthera ovata Walt. var. *lanceolata* Chapm. Fl. S. States 304. 1860.

Glabrous perennial; stems ascending to erect, usually unbranched, to 50 cm tall; leaves simple, opposite, elliptic to oblong, acute to acuminate, tapering to the short-petiolate base, to 10 cm long, to 3.5 cm wide, entire; flowers usually pale purple, loosely scattered along the peduncle; sepals linear, 7–10 mm long; corolla 6–10 mm long; capsule to 1 cm long; seeds smooth, about 3 mm broad. May–June.

Shallow water of swamps; bottomland woods.

IL, KY, MO (OBL).

Broad-leaved water willow; coastal plain water willow.

Some botanists place this and the preceding species in the genus *Dianthera*, but I prefer to consider *Dianthera* and *Justicia* to be one and the same.

36. ACERACEAE—MAPLE FAMILY

Deciduous trees; leaves opposite, simple or compound; flowers actinomorphic, perfect or more commonly unisexual; sepals usually 5, united at base; petals usually 5, sometimes absent, free from each other; stamens usually 8, usually associated with a nectary disk; ovary superior or half-inferior, bilocular; fruit usually a pair of samaras.

Fig. 2. *Justicia ovata* (Broad-leaved water willow).
Habit (center). Sepals (upper left). Fruits (lower right).
Fruit (lower left).

There are two genera and 112 species in this family. All but two species of the Chinese genus *Dipteronia* are in the genus *Acer.*

1. **Acer** L.—Maple

Deciduous trees; leaves opposite, simple and lobed or compound; flowers appearing before or with the leaves; fruit a pair of samaras.

There are 110 species of *Acer* found throughout much of the world. Several species occur in the central Midwest, with the following sometimes occurring in standing water.

1. Leaves compound; flowers in hanging racemes .... 2. *A. negundo*
1. Leaves simple; flowers in cymes or corymbs.
  2. Leaves cleft more than halfway to the middle .... 4. *A. saccharinum*
  2. Leaves cleft less than halfway to the middle.
    3. Leaves white-tomentose beneath; petioles tomentose; samaras 3.5–6.0 cm long............
    .... 1. *A. drummondii*
    3. Leaves glabrous or sparsely pubescent beneath, not white-tomentose; petioles glabrous
    or sparsely pubescent, but not tomentose; samaras up to 3 cm long........... 3. *A. rubrum*

1. **Acer drummondii** H. & A. in Hook. Journ. Bot. 1:200. 1814. Fig. 3.
Acer rubrum L. var. *drummondii* (H. & A.) Sarg. 10th Census U.S. 9:50. 1884.

Tree to 13 m tall with a trunk diameter up to 1 m; bark gray and smooth when young, becoming darker and scaly; twigs slender, white-hairy when young, usually becoming glabrous or nearly so, reddish, with pale lenticels; buds reddish, rounded, pubescent, up to 5 mm long; leaves simple, opposite, to 15 cm long, nearly as broad, palmately 5-lobed, the lobes cut less than halfway to the middle, each lobe acute,

Fig. 3. *Acer drummondii*
(Swamp red maple). Leaves
(center). Fruit (lower right).

the margins serrate, green and somewhat pubescent on the upper surface, densely covered on the lower surface with white tomentum; petioles tomentose, rarely becoming glabrous, up to 8 cm long; staminate and pistillate flowers borne separately but sometimes on the same tree, in dense red clusters, opening before the leaves begin to unfold; samaras borne in pairs, each member composed of an erect wing with a seed at the base, bright red, 3.5–6.0 cm long. February–March.

Wooded swamps, often in standing water.

IL, IN, MO (OBL), KY (FACW+).

Swamp red maple; Drummond's red maple.

This species differs from *Acer rubrum* by the larger fruits and the dense white tomentum on the lower surface of the leaves and often on the petioles. Many botanists consider this plant to be a variety of *A. rubrum*, but the larger fruits, dense white tomentum, and deeper aquatic habitats have convinced me to recognize it as a separate species.

2. **Acer negundo** L. Sp. Pl. 1056. 1753. Fig. 4.

Tree to 20 m tall, with a trunk diameter up to 1.3 m; bark light brown, ridged when young, becoming deeply furrowed; twigs glabrous or puberulent, green, glaucous, or rarely purplish, shiny, usually with white lenticels; buds rounded, white-pubescent, up to 3 mm long; leaves opposite, pinnately compound, with 3, 5, or 7 leaflets, the leaflets elliptic to ovate, up to 10 cm long, about half as broad, pointed at the tip, tapering or rounded at the sometimes asymmetrical base, shallowly or usually coarsely toothed along the margins or even shallowly lobed, light green and glabrous on the upper surface, paler and glaucous or pubescent on the lower

Fig. 4. *Acer negundo* (Box elder). Leaves (center). Cluster of flowers (lower left). Fruit (lower center). Twig (lower right).

surface; staminate and pistillate flowers borne on separate trees, several in drooping racemes, greenish yellow, appearing as the leaves begin to expand; fruits borne in pairs, in drooping racemes, composed of a curved wing with a seed at the base, greenish yellow, up to 4 cm long. March–April.

Moist or swampy woods, often along rivers and streams.

IA, IL, IN, KS, MO, NE, (FACW–), KY, OH (FAC+).

Box elder.

The leaves of this species resemble those of ash trees but box elders differ by their green or glaucous twigs, their larger teeth on the leaflets, and their paired samaras. There is considerable variation in the color of the twigs, with the twigs sometimes being purplish.

Twigs that are glaucous may be known as var. *violaceum*. Plants with puberulent twigs and shallowly toothed leaflets have been called var. *texanum* and occur in the southern part of the central Midwest. Plants with puberulent twigs and coarsely lobed leaflets are known as var. *interior*. These plants occur in the western part of our range.

3. Acer rubrum L. Sp. Pl. 1055. 1753.

Tree to 27 m tall with a trunk diameter up to 1 m; bark gray and smooth when young, becoming darker and scaly; twigs slender, mostly glabrous, more or less reddish, usually with pale lenticels; buds rounded, reddish, usually pubescent, up to 6 mm long; leaves simple, opposite, to 15 cm long, nearly as broad, 3- to 5-lobed, the lobes cut less than half-way to the middle, each lobe acute, the margins serrate, pale green and glabrous on the upper surface, paler and glabrous to sparsely pubescent on the lower surface; petioles glabrous or finely pubescent, up to 4.5 cm long; staminate and pistillate flowers borne separately, but sometimes on the same tree, in dense clusters, bright red or yellow, opening before the leaves begin to unfold; fruits borne in pairs, each member composed of an erect wing with a seed at the base, red or yellow, up to 3 cm long.

Two variations of *Acer rubrum* occur in the central Midwest.

a. Leaves 3- and 5-lobed, cordate at base, the margins serrate .... 3a. *A. rubrum* var. *rubrum*
a. Leaves 3-lobed, rounded at base, the margins usually entire .... 3b. *A. rubrum* var. *trilobum*

3a. Acer rubrum L. var. rubrum Fig. 5.

Leaves 3- or 5-lobed, the lobes serrate, cordate at the base, glabrous or sparsely pubescent on the lower surface; petioles glabrous; samaras 1.5–2.5 cm long. March–April.

Rocky woods, wooded slopes, moist woods, swamps (in the northern part of the Midwest).

IA, IL, IN, KY, MO, OH (FAC).

Red maple.

Red maple occurs in almost every woodland habitat, from the driest woods on ridgetops to swampy woods. In the northern part of the central Midwest, red maple may actually occur in standing water. The leaves of this species usually turn yellow or red in the autumn.

3b. **Acer rubrum** L. var. **trilobum** K. Koch, Hort. Dendrol. 80. 1853. Not illustrated.

Leaves 3-lobed, the lobes entire, rounded at the base, glabrous or sparsely pubescent on the lower surface; petioles glabrous; samaras 1.5–2.5 cm long. March–April. Wooded swamps, often in standing water.

IL, IN, MO (OBL), KY (FACW+). This variety was not recognized as distinct from typical *A. rubrum* by the U.S. Fish and Wildlife Service in their 1987 determination. Tri-lobed red maple.

This variety differs from typical *A. rubrum* by its 3-lobed leaves with entire lobes.

4. **Acer saccharinum** L. Sp. Pl. 1055. 1753. Fig. 6.

Tree to 35 m tall, with a trunk diameter up to 1.7 m; bark gray or silvery, glabrous at first, becoming loose and scaly or even somewhat shaggy when old; twigs slender, reddish brown, glabrous, often curving upward; buds more or less rounded, reddish brown, glabrous to finely pubescent, up to 3 mm long; leaves opposite, simple, up to 20 cm long, nearly as broad, deeply palmately 5-lobed, the margins of the

Fig. 5. *Acer rubrum* (Red maple). Leaves and fruits (center). Cluster of flowers (right).

leaves sharply toothed, pale green and glabrous on the upper surface, silvery-white and usually glabrous on the lower surface, except in the leaf axils; petioles glabrous, up to 12 cm long, often reddish; staminate and pistillate flowers borne separately, but sometimes on the same tree, in dense clusters, greenish yellow, opening before the leaves begin to unfold; fruits borne in pairs, each member composed of a curved wing with a seed at the base, green or yellow, up to 5.5 cm long. February–March.

Floodplain woods, swampy woods, river banks.

IA, IL, IN, KY, MO, OH (FACW), KS, NE (FAC).

Silver maple.

This is a tree primarily of river and stream banks and floodplain forests. On occasion it may occur in swampy woods. The leaves turn yellow in autumn.

37. AMARANTHACEAE—PIGWEED FAMILY

Annual or perennial herbs (in the central Midwest); leaves alternate or opposite, simple, entire; flowers perfect or, if unisexual, the plants monoecious or dioecious, arranged in spikes or glomerules, the spikes often aggregated into panicles; calyx (3-) 5-parted, free or united below, sometimes united into a cup or tube, or calyx absent in some pistillate flowers; petals absent; stamens (1–) 5, the filaments sometimes united; styles 1–3; ovary superior, 1-locular; fruit dehiscent or indehiscent; seed 1.

There are about sixty-five genera and nine hundred species in this family, primarily in the tropics and subtropics. Five genera occur in the central Midwest, but only three have species that may grow in water.

1. All or most of the leaves opposite.
    2. Flowers perfect; inflorescence of capitate spikes; leaves more or less fleshy, sessile ....
    .... 1. *Alternanthera*
    2. Flowers unisexual, the plants dioecious; inflorescence a broad panicle of spikes; leaves
       not fleshy, petiolate .... 3. *Iresine*
1. Leaves alternate.. 2. *Amaranthus*

1. **Alternanthera** Forsk.—Alligator Weed

Annual or perennial (ours) herbs; leaves opposite, simple; flowers bisexual, crowded into terminal and axillary, white or silvery, capitate spikes; sepals 5, free, white; petals absent; stamens usually 5, united near base into a short tube, with 5 staminodia; style 1; ovary superior, 1-locular; fruit a 1-seeded, indehiscent, compressed utricle; seed 1, lenticular.

Approximately 170 species comprise this genus. Most of them occur naturally in tropical America. Only the following barely enters the central Midwest:

1. **Alternanthera philoxeroides** Griseb. Goett. Abh. 24:36. 1789. Fig. 7.

Perennial emergent aquatic or semiterrestrial herb; stems prostrate to decumbent, mat-forming, branched or unbranched, stoloniferous, to 1 m long, glabrous or with a pair of puberulent lines; leaves opposite, simple, more or less fleshy, linear to narrowly obovate, acute and sometimes mucronate at the apex, cuneate at the base, to 10 cm long, to 2 cm wide, entire, glabrous, sessile; flowers perfect, borne

Fig. 6. *Acer saccharinum* (Silver maple). Leaves and fruits (center). Fruits (lower left). Twig (lower right).

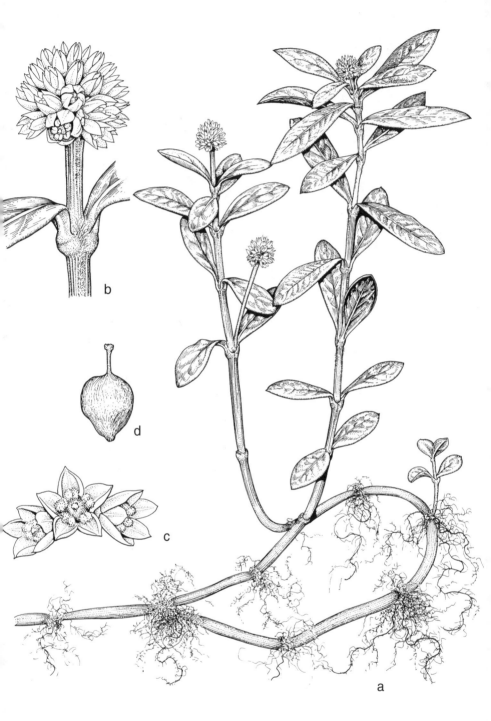

Fig. 7. *Alternanthera philoxeroides*
(Alligator weed).

a. Habit.
b. Flowering head.

c. Cluster of flowers
d. Ovary.

in terminal and axillary capitate spikes; sepals 5, free, silvery white, 5–6 mm long, lanceolate to narrowly ovate, glabrous; petals absent; stamens 5; style 1; fruit a 1-seeded, indehiscent, compressed utricle. June–August.

Along rivers; introduced from the southeastern United States.

IL, KY (OBL).

Alligator weed.

This plant is very aggressive in the southeastern United States, but the cold winters in the central Midwest prevent its spread in our area.

The silvery white calyx distinguishes this species from any other aquatic plant in the central Midwest.

2. Amaranthus L.—Amaranth; Pigweed

Annual or perennial monoecious or dioecious herbs from taproots; stems erect, ascending, or prostrate, branched or unbranched; leaves alternate, simple, entire, petiolate; flowers unisexual, less commonly perfect, borne in spikes or glomerules, the spikes often arranged in panicles, terminal or axillary; bracts subtending at least some of the flowers sometimes rigid-tipped, longer or shorter than the sepals and fruits; staminate flowers with (3–) 5 sepals; pistillate flowers with (3–) 5 sepals, or sometimes reduced to 1 sepal and 1 rudimentary sepal; petals absent; stamens (3–) 5; styles 3; ovary superior, 1-locular; fruit a 1-seeded circumscissile utricle; seed 1, lenticular to orbicular, black or dark red-brown.

Amaranthus is a genus of about sixty species found worldwide. In tropical regions particularly, some species are grown for their seeds that have value as food.

Some botanists segregate the genus *Acnida* from *Amaranthus*, based on the reduced sepals in the pistillate flowers.

Two species of *Amaranthus* may be found in standing water, although they are mostly terrestrial species.

1. Fruits regularly dehiscent, circumscissile; sepals of staminate flowers usually awn-tipped ... . 1. *A. rudis*
1. Fruits bursting irregularly upon dehiscence, not circumscissile; sepals of staminate flowers acute, rarely awn-tipped .... 2. *A. tuberculatus*

1. Amaranthus rudis J. Sauer, Madrono 21:428. 1972. Fig. 8.

Amaranthus tamariscinus Nutt. Trans. Am. Phil. Soc. II 5:165. 1837, misapplied.
Acnida tamariscina (Nutt.) Wood, Bot. & Fl. 289. 1873.

Annual dioecious herb from an elongated taproot; stems slender or more commonly stout, erect, to 2 m tall, glabrous, branched or unbranched; leaves lanceolate to oblong to narrowly ovate, obtuse and sometimes emarginate at the apex, cuneate at the base, to 10 cm long, to 3.5 cm wide, glabrous, entire, with slender petioles up to half as long as the blade; flowers unisexual, the staminate crowded into slender terminal spikes up to 20 cm long and up to 1 cm thick, or crowded into a short panicle of spikes, the pistillate crowded into a solitary or paniculate group of very slender spikes; bracts 1.5–2.0 mm long, excurrent, shorter than the staminate sepals but longer than the fruits; staminate flowers with 5 sepals, the sepals 2–3 mm long, acuminate with a stiff subulate tip; pistillate flowers with 1–2 sepals, if 2, then

Fig. 8. *Amaranthus rudis*
(Water hemp).

a. Habit of staminate plant.
b. Staminate flowers with leaf.
c. Habit of pistillate plant.
d. Pistillate flower.

e. Staminate flowers.
f. Bract.
g. Fruit.
h. Seed.

1 of them very rudimentary; stamens 5; styles 3; fruit a dehiscent, circumscissile utricle, ovoid, 1.3–1.5 mm long; seeds lenticular, about 1 mm long, dark red-brown. August–October.

Moist disturbed soil, not commonly in standing water.

IA, IL, IN, KS, MO, NE (FACW), KY, OH (FACW–).

Water hemp.

This species is distinguished from the very similar appearing *A. tuberculatus* by its regularly dehiscent circumscissile fruits and its awn-tipped staminate sepals. It is sometimes segregated into the genus *Acnida*.

2. **Amaranthus tuberculatus** (Moq.) J. Sauer, Madrono 13:18. 1955. Fig. 9.

Acnida tuberculata Moq. in DC. Prodr. 13 (2):277. 1849.

Acnida altissima Riddell ex Moq. in DC. Prodr. 13 (2):278. 1849.

Acnida tuberculata Moq.var. *subnuda* S. Wats. in Gray, Man. ed. 6, 429. 1889.

Acnida tamariscina Riddell var. *prostrata* Uline & Bray, Bot. Gaz. 20:158. 1895.

Acnida tuberculata Moq. var. *prostrata* (Uline & Bray) B. L. Robins. Rhodora 10:32. 1908.

Acnida altissima Riddell var. *subnuda* (S. Wats.) Fern. Rhodora 43:288. 1941.

Annual dioecious herb from a slender taproot; stems slender to stout, prostrate to ascending to erect, to 2.5 m tall, glabrous, branched or unbranched; leaves oblanceolate to lanceolate to obovate, obtuse at apex, less commonly acute, cuneate at base, to 15 cm long, to 8 cm wide, entire, glabrous, with slender petioles up to half as long as the blades; flowers unisexual, the staminate in continuous, slender, axillary spikes, or the spikes arranged in a terminal panicle, the pistillate in continuous or interrupted, slender or globose spikes, both axillary and terminal; bracts lanceolate and sharp-tipped, 3.5–4.5 mm long, longer than the sepals and the fruits; staminate flowers with 5 sepals, the sepals 2–3 mm long, acute to acuminate, often cuspidate; pistillate flowers with 1–2 sepals, if 2, then one of them very rudimentary; stamens 5; styles 3; fruit a 1-seeded utricle that bursts irregularly, not circumscissile; seed lenticular, 0.7–1.0 mm long, dark red-brown. August–October.

Moist soil, particularly stream banks, pond margins, and ditches, sandbars, moist disturbed soil; less commonly in shallow standing water.

IA, IL, IN, MO, NE (OBL), KY, OH (FACW).

Water hemp.

Because of the reduced number of sepals in the pistillate flower, this species is sometimes segregated into the genus *Acnida*.

Amaranthus tuberculatus is a variable species, and three entities may be recognized. Typical variety *tuberculatus* is erect with leaves that may be narrowly ovate and reach a length of 15 cm, and with pistillate flowers crowded into slender spikes. Variety *subnuda* is ascending or prostrate with leaves that are narrower and only as much as 8 cm long, and with pistillate flowers aggregated into dense, distinct, globular glomerules. Variety *prostrata* is prostrate, with often somewhat pointed leaves, and with pistillate flowers arranged in loose, few-flowered glomerules.

Fig. 9. *Amaranthus tuberculatus* (Water hemp).

a. Pistillate plant.
b, c. Ovary with sepals.
d. Seed.

e. Staminate plant.
f. Staminate flower.
g. Bract.

3. Iresine P. Br.—Bloodleaf

Annual or perennial (in the central Midwest), monoecious or dioecious herbs; leaves opposite, simple, petiolate; flowers perfect or unisexual, crowded into spikes arranged in panicles; bracts present; calyx deeply 5-parted; petals 0; stamens 5, the filaments connate at base; styles 2–3, very short; ovary superior, 1-locular, compressed; fruit an indehiscent, 1-seeded utricle; seed inverted.

There are about eighty species in this genus, primarily in North America, Africa, and tropical America.

Only the following species occurs in the central Midwest:

1. **Iresine rhizomatosa** Standl. Proc. Biol. Soc. Wash. 28:172. 1915. Fig. 10.

Perennial dioecious herb from slender rhizomes; stems erect, to 1.5 m tall, usually unbranched, glabrous or sometimes pilose at the nodes; leaves opposite, simple, thin, lance-ovate to ovate, acute to acuminate at the apex, cuneate at the base, to 15 cm long, to 7 cm wide, entire, glabrous or sparsely pubescent on the veins above and on the surface beneath, bright green, decurrent onto a short petiole; flowers unisexual, borne in spikes aggregated into panicles, the panicles terminal and in the uppermost leaf axils, to 30 cm long; bracts ovate, shorter than the calyx; calyx 5-parted, the lobes 1.2–1.5 mm long, silvery-white, ovate-lanceolate, subtended by long hairs in the pistillate flowers; petals 0; stamens 5; fruit a 1-seeded, indehiscent utricle, globose, 2.0–2.5 mm in diameter; seed 1, suborbicular, about 0.5 mm in diameter, shiny, dark red. August–October.

Swampy woods.

IL, IN, KY, MO, OH (FACW–), KS (FACW).

Bloodleaf.

This species occurs in swampy woods, but it is usually rather rare. It is distinguished by its opposite leaves, its silvery-white unisexual inflorescences, and the long white hairs that subtend the sepals in the pistillate flowers.

38. ANACARDIACEAE—SUMAC FAMILY

Trees, shrubs, or woody vines; leaves alternate, usually trifoliolate or pinnately compound, without stipules; flowers actinomorphic, perfect or unisexual; sepals 5, connate at the base; petals 5, free; stamens 5 (or 10); nectary disk present; ovary superior, tricarpellate, each with one ovule; fruit a drupe or berry.

This family consists of sixty to eighty genera and about six hundred species. Most of the species are tropical and subtropical.

In addition to the genus listed below, *Rhus* also occurs in the central Midwest.

1. Toxicodendron Mill.

Woody vines, shrubs, or trees; leaves alternate, trifoliolate or pinnately compound, poisonous to the touch; flowers perfect; sepals 5, connate below; petals 5, free; stamens 5; ovary superior; drupe white, 1-seeded.

There are ten species in this genus. Only the following may sometimes live in standing water:

1. Leaves with 3 leaflets; vines or small shrubs.. 1. *T. radicans*
1. Leaves with 7–13 leaflets; tall shrub or tree... 2. *T. vernix*

Fig. 10. *Iresine rhizomatosa*
(Bloodleaf).

a. Upper part of plant.
b. Leaves.
c. Inflorescence.

d. Flower.
e. Seed.

1. **Toxicodendron radicans** (L.) Kuntze, Rev. Gen. 153. 1891. Fig. 11.
Rhus radicans L. Sp. Pl. 266. 1753.

Woody vines, or sometimes erect shrubs, with aerial roots; leaves alternate, trifoliolate, the leaflets membranaceous to subcoriaceous, with entire, serrate, or dentate leaflets, the terminal leaflet 3–20 cm long, up to 12 cm wide, larger than the lateral leaflets, narrowly ovate to ovate, acute at the apex, mostly rounded at the base, glabrous or pubescent, petiolate, the petioles up to 20 cm long, glabrous or pubescent; flowers in panicles, actinomorphic; sepals 5, united at base; petals 5, free, yellow-green; stamens 5; ovary superior; fruit a drupe, globose, up 3–4 mm in diameter, white to cream, glabrous. May–July.

A variety of habitats including swampy woods.

IA, IL, IN, MO (FAC+), KS, KY, NE, OH (FAC).

Poison ivy.

All parts of this plant may be poisonous to the touch.

The leaves and growth form of this species are extremely variable. Plants that live in swampy woods usually have membranaceous leaves and are high-climbing vines, although erect shrubs sometimes occur. In drier habitats, the leaves usually are more firm in texture. Plants in the western part of our range with stout erect stems without aerial roots may be a different species known as *Toxicodendron rydbergii*.

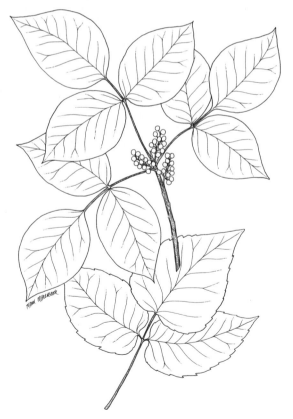

Fig. 11. *Toxicodendon radicans* (Poison ivy). Leaves with fruits (above). Leaf (below).

2. **Toxicodendron vernix** (L.) Kuntze, Rev. Gen. Pl. 153. 1891. Fig. 12.
Rhus vernix L. Sp. Pl. 265. 1753.

Shrubs or trees to 10 m, with a trunk diameter up to 20 cm; leaves pinnately compound, with 7–13 leaflets, the leaflets oval to ovate to obovate, acute to acuminate at the apex, cuneate at the base, entire, rather thin, glabrous or puberulent, poisonous to the touch; flowers borne in loose, axillary panicles; sepals 5, united at the base; petals 5, free, greenish; stamens 5; ovary superior; drupes globose or a little longer than broad, gray-white, 3–4 mm in diameter, glabrous. May–July.

Bogs, marshes, swamps.

IL, IN, KY (OBL).

Poison sumac.

The leaves of this species often turn reddish in the autumn. Touching any part of the plant may result in a severe skin rash.

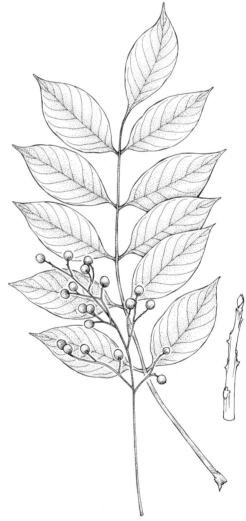

Fig. 12. *Toxicodendron vernix*
(Poison sumac). Leaf and fruits
(center). Twig (right).

39. APIACEAE—CARROT FAMILY

Herbaceous annuals, biennials, or perennials; leaves alternate, often compound, frequently with a dilated petiole base; flowers usually arranged in umbels, actinomorphic, perfect; sepals none or minute and 5-parted; petals 5, free; stamens 5; ovary inferior, bilocular; fruit of 2 dried seedlike carpels, called mericarps. The family consists of about three hundred genera and approximately three thousand species, most of them in the northern hemisphere. Several genera in the central Midwest have species that may grow regularly in water.

1. Leaves all simple.
  2. Leaves reniform or orbicular; flowers in umbels ....7. *Hydrocotyle*
  2. Leaves ovate or lanceolate; flowers in capitate clusters ....6. *Eryngium*
1. Leaves pinnately compound, or basal leaves simple in *Cynosciadium*.
  3. Leaves once compound, either pinnate, palmate, trifoliolate, or ternate.
    4. Leaves trifoliolate ....14. *Zizia*
    4. Leaves pinnately or palmately divided, with more than 3 leaflets.
      5. Leaves palmately divided; basal leaves simple ....5. *Cynosciadium*
      5. Leaves all pinnately divided.
        6. Leaflets filiform, at most 1 mm wide ....11. *Ptilimnium*
        6. Leaflets broader, always more than 1 mm wide.
          7. Bulblets present in some of the upper leaf axils...................... 3. *Cicuta*
          7. Bulblets absent.
            8. Margins of leaflets entire.
              9. Leaflets cross-septate; bracts at base of inflorescence many ....
                  .... 8. *Limnosciadium*
              9. Leaflets not cross-septate; bracts at base of inflorescence 0–2 ....
                  ....10. *Oxypolis*
            8. Margins of leaflets serrate.
              10. Margins of leaflets with up to 8 teeth per side.........................10. *Oxypolis*
              10. Margins of leaflets with more than 8 teeth per side.
                11. Leaflets regularly toothed .... 12. *Sium*
                11. Leaflets unequally incised......................................2. *Berula*
3. Leaves more than once compound.
  12. Ultimate leaflet segments of nonsubmersed leaves filiform.
    13. Leaflet divisions capillary; fruit linear-oblong, 8–10 mm long .... 13. *Trepocarpus*
    13. Leaflet divisions filiform; fruit oval to elliptic to suborbicular, 1.5–3.5 mm long.......
    ....11. *Ptilimnium*
  12. Ultimate leaflet segments of nonsubmersed leaves broader than filiform, more than 1 mm wide.
    14. Bulblets present in the axils of some of the upper leaves .... 3. *Cicuta*
    14. Bulblets absent.
      15. Stems pubescent; flowers yellow... 1. *Angelica*
      15. Stems glabrous; flowers white (except in *Zizia*).
        16. Leaves 2- to 3-ternate.
          17. Flowers yellow; some leaflets more than 2 cm wide ....14. *Zizia*
          17. Flowers white; none of the leaflets more than 1.5 cm wide....................
          .... 4. *Conioselinum*
        16. Leaves 2- to 3-pinnate.
          18. Plants creeping or floating; submersed leaves often present and filiform.
          ....9. *Oenanthe*
          18. Plants erect; submersed leaves absent.

19. Stems often purple-speckled; fruits 2–4 mm long, flattened laterally . ... 4. *Cicuta*
19. Stems not purple-speckled; fruits 4.0–5.5 mm long, flattened dorsally .... 4. *Conioselinum*

1. **Angelica** L.—Angelica

Robust perennials; leaves alternate, 2- to 3-pinnate; flowers in large umbels, subtended by minute or no bracts; sepals 5, minute or absent; petals 5, free, inflexed at tip; stamens 5; ovary inferior; fruits ovoid, compressed.

This genus consists of approximately forty species found in the northern hemisphere and New Zealand. Two species occur in the central Midwest, with the following sometimes growing in standing water.

1. **Angelica atropurpurea** L. Sp. Pl. 251. 1753. Fig. 13.

Robust perennials to 4 m tall; stems erect, glabrous or pubescent in the upper part, usually purple-speckled; lower leaves up to 1 m across, 2- to 3-pinnate, the upper ones progressively smaller, alternate, glabrous, on widely dilated petioles, the leaflets more or less ovate, sharply serrate to shallowly lobed; umbels up to 25 cm across, with up to 15 rays, the rays to 10 cm long; flowers yellow, on slender pedicels to 15 mm long; fruit oval, to 8 mm long, glabrous. May–August.

Fens, calcareous meadows, damp thickets.

IA, IL, IN, KY, OH (OBL).

Great angelica.

This is the tallest species of the Apiaceae in the central Midwest, sometimes attaining a height of 4 meters. The dilated leaf sheaths, often purple-striated, are remarkable in their size and color.

2. **Berula** Besser—Cut-leaved Water Parsnip

Perennial herbs; leaves alternate, once-pinnate, with serrate leaflets; umbels compound, subtended by several bracts; sepals minute; petals 5, free, white; stamens 5; ovary inferior; fruit flattened laterally with corky-thickened outer walls.

Only the following species comprises the genus.

1. **Berula erecta** (Huds.) Coville, Contr. Nat. Herb. 4:115. 1893. Fig.14.
Sium erectum Huds. Fl. Angl. 103. 1762.
Sium incisum Torr. Rep. Exped. Rocky Mts. 90. 1845.
Berula erecta (Huds.) Coville var. *incisa* (Torr.) Cronq. Vasc. Pl. Pacific N.W. 3:519. 1961.

Glabrous, rather stout perennials, with fibrous roots; leaves alternate, once-pinnate, the leaflets 7–19, oval to narrowly oblong, irregularly incised, up to 8 cm long; umbels compound, subtended by several narrow bracts; calyx teeth minute; petals 5, white; stamens 5; ovary inferior, bilocular; fruit ovate to orbicular, 1.5–2.0 mm across, flattened laterally, the outer wall corky-thickened. June–September.

Bogs, marshes, sometimes in flowing water.

IA, IL, KS, MO, NE (OBL).

Cut-leaved water parsnip.

The seeds of this species rarely mature.

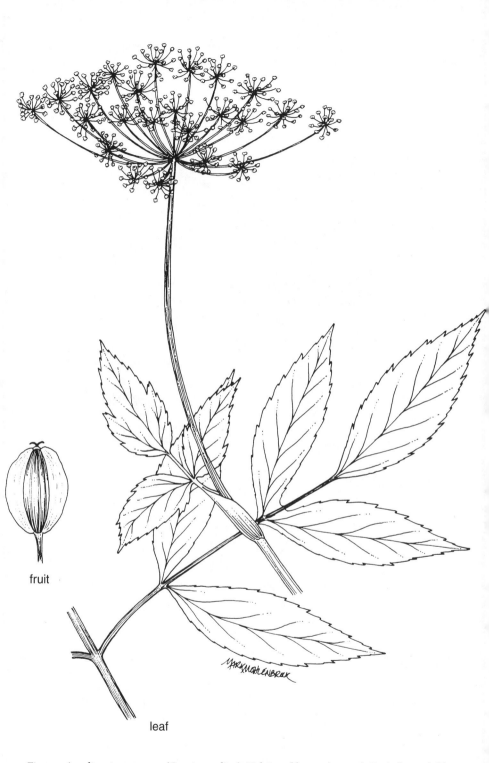

fruit

leaf

Fig. 13. *Angelica atropurpurea* (Great angelica). Habit and leaves (center). Fruit (lower left).

This species is similar to *Sium suave* in appearance but has irregular teeth. It does not form finely dissected leaves under water as *Sium suave* does. On rare occasions, this species may be totally submerged in flowing water. As long as it is submerged, it does not flower.

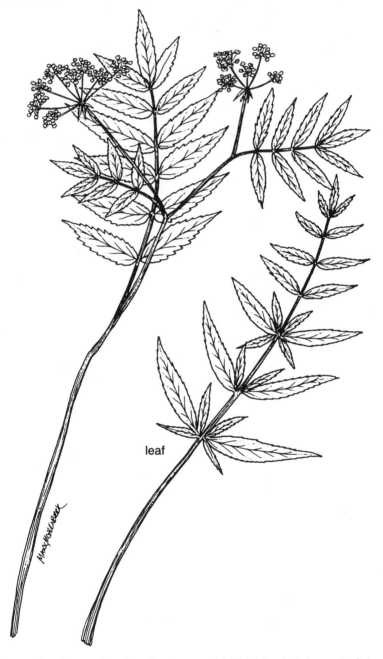

leaf

Fig. 14. *Berula erecta* (Cut-leaved water parsnip). Habit (center). Leaves (right).

3. Cicuta L.—Water Hemlock

Slender or stout perennials from thickened roots; leaves alternate, 1- to 3-pinnate, the leaflets serrate; umbels compound, with no involucre of bracts, although bractlets are present beneath each ultimate umbel; calyx teeth 5; petals 5, white, free; stamens 5; ovary inferior, bilocular; fruit with flat corky ribs.

There are four species in the genus, with two of them found in the central Midwest. These two are sometimes found in shallow water.

1. Upper leaves bearing bulblets in some of their axils; leaflets linear .... 1. *C. bulbifera*
1. None of the leaves bearing bulblets; leaflets narrowly lanceolate or broader ... 2. *C. maculata*

1. **Cicuta bulbifera** L. Sp. Pl. 255. 1753. Fig. 15.

Slender, erect, glabrous perennials to 1 m tall, with fleshy tuberous roots; leaves 1- to 3-pinnate, the leaflets linear, sparsely serrate, to 3 cm long, to 5 mm wide; bulblets present in the axils of some of the upper leaves; umbels often not formed, when present, up to 5 cm across; petals white; stamens 5; ovary inferior; fruit broadly ovate, 2.0–2.5 mm long. July–September.

Swamps, sometimes in shallow standing water.

IA, IL, IN, OH (OBL), KS, NE (NI).

Bulblet water hemlock.

This species is poisonous if eaten. Flower-bearing umbels are sometimes not present, but when they are present, the fruits seldom mature. Reproduction is done almost exclusively through the bulblets.

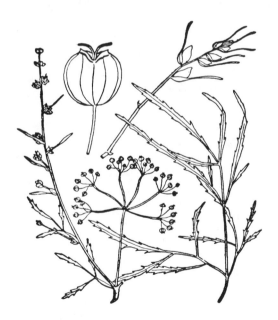

Fig. 15. *Cicuta bulbifera* (Bulblet water hemlock).
Inflorescence (left). Fruit (upper left). Branch with bulblets
(upper right). Leaves (lower right).

2. Cicuta maculata L. Sp. Pl. 256. 1753. Fig. 16.

Stout perennials from a cluster of fleshy roots; stems erect, glabrous, usually purple-speckled, hollow, to 2 m tall; lower leaves bi- or tri-pinnately compound, up to 35 cm long, up to 25 cm wide, on long petioles; upper leaves once-pinnate to ternate; leaflets glabrous, lanceolate, up to 3 cm broad, serrate; inflorescence umbellate, terminal and/or axillary, compound, subtended by an involucre of bracts, the bracts unequal, linear, ternate to simple, glabrous, up to 2 cm long; sepals 5, green, deltoid, up to 1.5 mm long; petals 5, white, inflexed, clawed at base; stamens 5; ovary inferior; fruit ovoid to ellipsoid, 2.5–3.0 mm long, 1.5–2.0 mm wide. June–September.

Marshes, wet prairies, wet ditches, moist woods, occasionally in shallow water. IA, IL, IN, KS, KY, MO, NE, OH (OBL).

Water hemlock.

This species most closely resembles *Conium maculatum*, from which it differs by its broader ultimate leaflets. All parts of this plant are poisonous if eaten.

Fig. 16. *Cicuta maculata* (Water hemlock). Habit (center). Fruit (lower left).

4. Conioselinum Hoffm.—Hemlock Parsley

Perennials from thickened roots; stems erect; leaves alternate, 2- to 3-pinnate; umbels compound, with 0–several bracts and with bractlets present beneath each ultimate umbel; calyx absent or only rudimentary; petals 5, white, free; stamens 5; ovary inferior, bilocular; fruit dorsally flattened, broadly winged.

This genus consists of ten species in temperate North America and Eurasia. Only the following species occurs in the central Midwest.

1. **Conioselinum chinense** (L.) BSP. Prel. Cat. N. Y. 22. 1888. Fig. 17.
Athamanta chinensis L. Sp. Pl. 245. 1753.

Slender to rather stout, erect perennials; stems up to 1.5 m tall, striate, glabrous; leaves 2- to 3-pinnate, the ultimate segments linear to oblong, subacute, to 1 cm across, the petioles often broadly winged; inflorescence umbellate, the rays usually puberulent, terminal and/or axillary, compound, subtended by a few narrow bracts, or bracts absent, but bractlets present at base of ultimate umbels; sepals absent or nearly so; petals 5, white, inflexed; stamens 5; ovary inferior; fruit ellipsoid to oblongoid, 2.5–4.5 mm long, distinctly winged. April–September.

Swamps, bogs, wet meadows.

IA, IL, IN, MO (OBL), OH (FACW); NE (not listed by the U.S. Fish and Wildlife Service).

Hemlock parsley.

This species is distinguished from all other similar erect species of white-flowered Apiaceae by its subacute leaflets and its distinctly winged fruits.

Despite the specific epithet *chinense*, this species does not occur in China. Linnaeus, who originally named this species, apparently was confused concerning the origin of the type specimen.

Fig. 17. *Conioselinum chinense* (Hemlock parsley). Leaves and inflorescence (center). Fruit (lower right). Cross-section of fruit (lower left).

5. Cynosciadium DC.—Dogshade

Glabrous annual herbs; basal leaves simple, cross-septate, cauline leaves palmately divided, with narrow, entire leaflets; inflorescence of axillary and terminal umbels, subtended by a few undivided bracts and a pair of bractlets; sepals 5, minute; petals 5, free, white; stamens 5; ovary inferior; fruits flattened or subterete, with narrow wings and prominent corky wings.

Only the following species comprises the genus.

1. Cynosciadium digitatum DC. Mem. Umb. 44. 1829. Fig. 18.

Annual herbs with fibrous roots; stems erect, to 50 cm tall, glabrous, sparsely branched; basal leaves simple, linear to narrowly lanceolate, cross-septate, acute at the apex, tapering to the base; cauline leaves palmately compound, with 3 or 5 leaflets, the leaflets linear to narrowly lanceolate, to 4 cm long, to 2 cm wide, glabrous; flowers in compound axillary and terminal umbels on rays up to 8 cm long, subtended by a few linear, undivided unequal bracts and 2 bractlets; flowers actinomorphic, up to 3 mm across; sepals minute; petals 5, free, white, 2–3 mm long; stamens 5; ovary inferior; fruit ovoid, 2–3 mm long, flattened or subterete, with narrow ribs and prominent corky wings. May–June.

Swamps, wet woods, along streams and bayous.

IL, MO (FACW).

Finger dogshade.

This uncommon species is recognized by its palmately divided leaves with very narrow, entire leaflets.

6. Eryngium L.—Eryngo

Biennial or perennial herbs; leaves alternate, simple, sometimes spiny along the margins, petiolate; flowers crowded into heads, the heads borne in the axils of the leaves in cymes or racemes, subtended by several bracts, the bracts often spinulose; sepals 5, more or less free, persistent; petals 5, free, white, blue, or purple; stamens 5; ovary inferior; fruit without ribs but usually papillose or tuberculate.

This genus consists of about two hundred species. The following may occur in habitats where water stands for a portion of the year.

1. Plants erect; heads in cymes..1. *E. integrifolium*
1. Plants prostrate to ascending; head solitary in leaf axils ....2. *E. prostratum*

1. Eryngium integrifolium Walt. Fl. Carol. 112. 1788. Fig. 19.

Perennial herbs from narrow roots; stems erect, slender, glabrous, to 75 cm tall, branched in the inflorescence; leaves alternate, simple, ovate to oblong to narrowly lanceolate, up to 8 cm long, up to 4 cm wide, entire or variously toothed, sometimes with spinulose teeth, glabrous, the lowermost petiolate and usually absent at flowering time, the middle and uppermost sessile; flowers in heads, borne in cymes, the heads spherical or occasionally hemispherical, 1.0–1.5 cm across, subtended by 6–10 bracts, the bracts linear, with spinulose teeth and a spiny apex; sepals 5, free, green, narrowly ovate, acuminate, 1.5–2.0 mm long; petals 5, free,

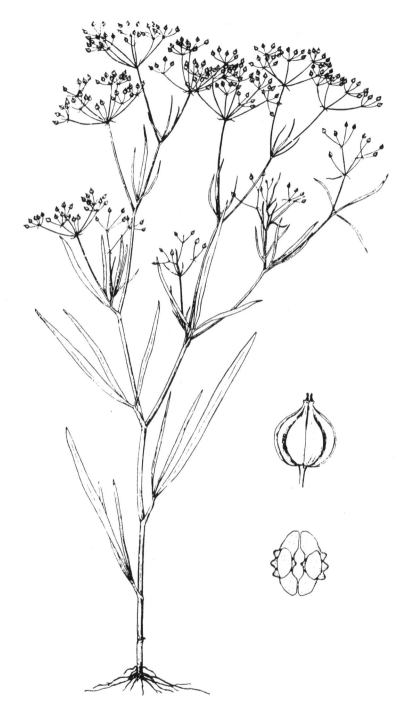

Fig. 18. *Cynosciadium digitatum* (Finger dogshade). Leaves and inflorescence (left). Fruits (right).

bluish; stamens 5; ovary inferior; fruit obpyramidal, 1.5–2.0 mm long, papillose. July–September.

Swampy woods, borders of ponds.

KY (NI).

Single-leaf eryngo.

This is the only erect, simple-leaved species of the Apiaceae in wet areas of the central Midwest in which all of the leaves are simple.

2. **Eryngium prostratum** Nutt. in DC. Prodr. 4:92. 1830. Fig. 20.

Biennial or perennial herbs from fibrous roots; stems freely branched from the base, prostrate to weakly ascending, glabrous, up to 30 cm long; leaves alternate, simple, lanceolate to ovate, acute to subacute at the apex, tapering to the base, entire or more commonly dentate or rarely pinnatifid, glabrous, the lower leaves to 6.5 cm long, to 2.5 cm wide, petiolate, the upper leaves smaller and sessile; flowers crowded into heads, the heads solitary in the axils of the leaves, oblong to oblong-cylindric, to 8 mm long, to 4 mm wide, on peduncles up to 3 cm long; bracts 5–10 per head, up to 6 mm long, linear, entire, glabrous, more or less reflexed; bractlets linear; sepals 5, free, green; petals 5, free, blue, 2–3 mm long; stamens 5; ovary inferior; fruit globose or slightly broader than long, about 2 mm in diameter, short-papillose. May–November.

Fig. 19. *Eryngium integrifolium* (Single-leaf eryngo). Habit. Leaf (center).

Wet areas that are temporarily flooded, particularly around lakes and ponds, swampy woods, wet ditches.

IL, KY, MO (OBL), KS (FACW+).

Creeping eryngo; spreading eryngo.

This small prostrate species is easily overlooked unless it has its small, blue, axillary flowering heads. Usually only one flower head is formed per leaf axil.

7. **Hydrocotyle** L.—Water Pennywort

Slender, creeping, glabrous perennials; leaves alternate, simple, sometimes

Fig. 20. *Eryngium prostratum* (Creeping eryngo). Habit. Flower (above center).

peltate; umbels solitary or appearing one above another; calyx teeth minute; petals 5, white; stamens 5; mericarps with slender ribs.

There are about seventy-five species in this genus, with most of them living in subtropical or tropical regions of the world. All the species in the central Midwest live in wetlands.

1. Leaves peltate.
   2. Inflorescence composed of a single umbel..4. *H. umbellata*
   2. Inflorescence composed of two or more umbels...5. *H. verticillata*
1. Leaves not peltate.
   3. Leaves at least 1 cm wide; inflorescence shorter than the petiole; flowers and fruit pedicellate.
      4. Leaves wider than high, lobed to about the middle............................2. *H. ranunculoides*
      4. Leaves more or less orbicular, shallowly lobed...................................... 1. *H. americana*
   3. Leaves up to 1 cm wide; inflorescence longer than the petiole; flowers and fruits sessile.....
     ....3. *H. sibthorpoides*

1. **Hydrocotyle americana** L. Sp. Pl. 234. 1753. Fig. 21.

Stems very slender, creeping or more rarely floating; leaves suborbicular or nearly so, with 6–10 shallow lobes, up to 5 cm wide; peduncles shorter than the leaves; rays 2–7 per umbel, nearly sessile; fruit suborbicular, about 1.5 mm across, with sharp ribs. June–September.
Shallow water and in mud.
IN, KY, OH (OBL).
Marsh pennywort.
This species differs from the other aquatic species of *Hydrocotyle* by its leaves that are not peltate, by its shallowly lobed leaves, its nearly sessile flowers, and its sharply ribbed fruits.

2. **Hydrocotyle ranunculoides** L. f. Suppl. 177. 1781. Fig. 22.

Stems floating or creeping; leaves often a little wider than long, with 5–6 deep crenate lobes, up to 6 cm wide; peduncles shorter than the leaves; rays 5–10 per umbel, up to 3 mm long; fruit suborbicular, 2–3 mm across, with obscure ribs. May–August.
Shallow water and in mud.
IL, KS, KY, OH (OBL).
Buttercup pennywort.
May–August.
This species is readily recognized by its deeply lobed leaves that are wider than they are high.

3. **Hydrocolyle sibthorpoides** Lam. Encycl. 3:153. 1789. Fig. 23.
Hydrocotyle rotundifolia Roxb. Hort. Beng. 21. 1814, non Wallich (1812).

Stems filiform, creeping or less commonly floating; leaves not peltate, reniform to orbicular, up to 1 cm across, shallowly 7-lobed, glabrous or hispidulous on the lower surface, the petioles 1–2 cm long; inflorescence longer than the petiole of the subtending leaf, consisting of a single umbel with 3–10 sessile flowers; fruits sessile, nearly orbicular, 1.0–1.5 mm wide. April–September.

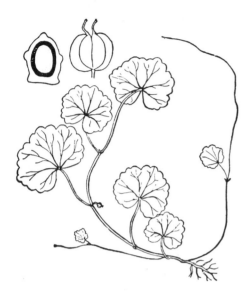

Fig. 21. *Hydrocotyle americana* (Marsh pennywort). Habit. Section of fruit (upper left). Fruit (upper center).

Fig. 22. *Hydrocotyle ranunculoides* (Buttercup pennywort). Habit. Fruit (right).

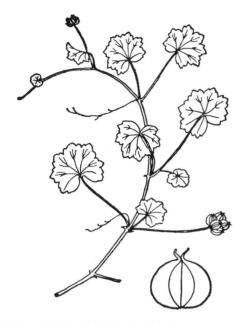

Fig. 23. *Hydrocotyle sibthorpoides* (Lawn pennywort). Habit. Fruit (lower right).

In mud or shallow water.

IN, KY (not listed by the U.S. Fish and Wildlife Service).

Lawn pennywort; Asiatic pennywort.

This species rarely escapes from greenhouses.

Hydrocotyle sibthorpoides differs from all other species of *Hydrocotyle* in the central Midwest by its sessile flowers and fruits and its smaller leaves.

4. **Hydrocotyle umbellata** L. Sp. Pl. 1:234. 1753. Fig. 24.

Stems creeping or less commonly floating; leaves suborbicular, peltate, crenate, up to 4 cm across; inflorescence shorter than the subtending leaf, consisting of a single umbel with 10–25 flowers on pedicels up to 2.2 cm long; fruits reniform, 2–3 mm wide. June–September.

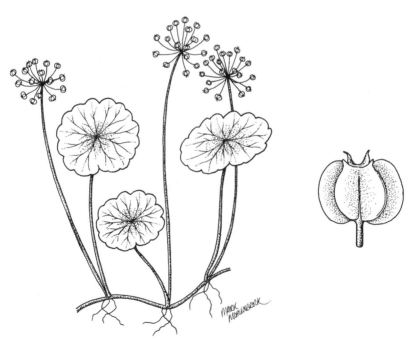

Fig. 24. *Hydrocotyle umbellata* (Common pennywort). Habit (left). Fruit (right).

In water or in mud.

IL, IN, OH (OBL).

Common pennywort.

This and *H. verticillata* are the only species of *Hydrocotyle* with peltate leaves. *Hydrocotyle umbellata* differs by having just one umbel of flowers per plant.

5. **Hydrocotyle verticillata** Thunb. Dis. 2:415, pl. 3. 1798. Fig. 25.
Hydrocotyle interrupta Muhl. Cat. 30. 1813.

Stems creeping or floating; leaves suborbicular, peltate, crenate, up to 4 cm across; inflorescence shorter than the subtending leaf, consisting of 2–3 separated whorls of 10–15 flowers on pedicels up to 1.8 cm long; fruits reniform, 1.8–2.2 mm wide. June–September.

In water or in mud.

MO (OBL).

Whorled pennywort.

Although very similar in appearance to *H. umbellata*, *Hydrocotyle verticillata* differs by having two or more whorls of flowers per plant.

8. **Limnosciadium** Math. & Const.

Annual herbs; stems erect; leaves basal and alternate, the basal ones simple or pinnately divided, the upper ones alternate, pinnately divided; inflorescence umbel-

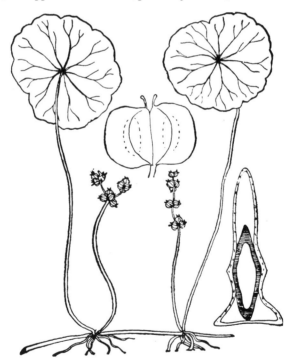

Fig. 25. *Hydrocotyle verticillata* (Whorled pennywort). Habit. Fruit (above center). Section of fruit (lower right).

Fig. 27. *Oenanthe aquatica* (Water fennel). Habit.

Fig. 28. *Oxypolis rigidior* (Cowbane). Leaves and inflorescence (center). Fruit (lower left).

in the central Midwest with once-pinnate leaves with entire leaflets or with leaflets with not more than 8 teeth per side.

11. **Ptilimnium** Raf.—Mock Bishop's-weed

Annual herbs from fibrous roots; leaves alternate, opposite, or whorled, 2- to 3-pinnate, with the leaflets very narrow; inflorescence a series of compound umbels subtended by small bracts; sepals more or less absent; petals 5, white, the tip inflexed; stamens 5; ovary inferior; fruit ovoid, more or less compressed.

The genus consists of about six species, most of them North American except for one in the East Indies.

1. Bracts deeply divided .... 1. *P. capillaceum*
1. Bracts undivided.
    2. Division of leaves whorled; fruit 3.0–3.5 mm long .... 2. *P. costatum*
    2. Division of leaves alternate or opposite; fruit 1.5–2.0 mm long .... 3. *P. nuttallii*

1. **Ptilimnium capillaceum** (Michx.) Raf. Bull. Bot. Geneve 1:217. 1830. Fig. 29.
Ammi majus Walt. Fl. Car. 113. 1788, non L. (1753).
Ammi capillaceum Michx. Fl. Bor. Am. 1:164. 1803.

Annual herbs from fibrous roots; stems slender, glabrous, to 0.5 m tall, with spreading to ascending branches; leaves alternate, 2- to 3-pinnate, the leaflets with filiform segments, the segments up to 0.5 mm wide, glabrous; uppermost leaves sessile or nearly so, the lower ones petiolate; inflorescence of compound umbels up to 6 cm across, each with 6–10 rays, subtended by deeply divided linear bracts; sepals absent; petals 5, free, white, 1.5–3.0 mm long, incurved at the tip; stamens 5; ovary inferior; fruit ovoid, acute, up to 3 mm long.
June–October.

Swamps, wet ground, sometimes in standing water.

KS (FACW), KY, MO (OBL).

Atlantic mock bishop's-weed.

The bracts that subtend the umbels are deeply divided, separating this species from the other species of *Ptilimnium* in the central Midwest.

2. **Ptilimnium costatum** (Ell.) Raf. Neogenyton 2. 1825. Fig. 30.
Ammi costatum Ell. Sketch Fl. Carol. 1:350. 1816.

Fig. 29. *Ptilimnium capillaceum* (Atlantic mock bishop's-weed). Habit. Fruit (lower left). Flower (above center). Cross-section of fruit (lower right).

Fig. 30. *Ptilimnium costatum*
(Ribbed mock bishop's-weed).

a. Flowering branch.
b. Leaves.
c. Lower part of stem.
d. Leaf.

e. Flower.
f. Fruit on pedicel.
g. Fruit.

Annual herbs from fibrous roots; stems more or less erect, branched, glabrous, to 1.5 m tall; leaves appearing whorled, deeply divided into numerous segments, the segments filiform, glabrous, up to 0.5 mm wide, the petiole dilated at base and hyaline; flowers in once compound umbels, the primary umbels on glabrous peduncles up to 10 cm long, subtended by several linear, entire bracts up to 10 mm long and bearing up to 20 secondary umbels; secondary umbels subtended by linear, undivided bracts up to 4 mm long, and with up to 20 flowers per umbel, the flowers borne on glabrous pedicels up to 5 mm long; sepals 5, free, deltoid, acute at the apex, persistent, up to 1 mm long; petals 5, free, white, 1–2 mm long, curved inward at the apex; stamens 5; ovary inferior; fruit ovoid, 3.0–3.5 mm long, with corky ribs. July–September.

Swampy woods.

IL, MO (OBL), KY (NI).

Ribbed mock bishop's-weed.

This species has the largest fruits of any species of *Ptilimnium* in the central Midwest. The leaves appear whorled on the stem.

3. **Ptilimnium nuttallii** (DC.) Britt. Mem. Torrey Club 5:244. 1894. Fig. 31.
Discopleura nuttallii DC. Mem. Umb. 38. 1829.

Annual herbs from fibrous roots; stems more or less erect, branched, glabrous, to

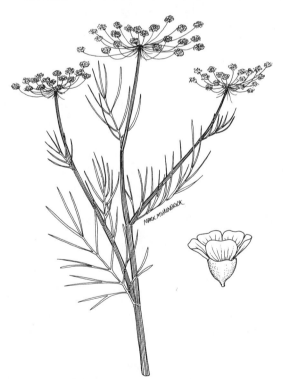

75 cm tall; leaves appearing opposite or alternate, deeply divided into several segments, the segments filiform, glabrous, to 0.5 mm wide, the petioles dilated at the base and hyaline; flowers in once compound umbels, the primary umbel on a peduncle up to 10 cm long, subtended by several filiform, entire bracts up to 10 mm long and bearing up to 30 secondary umbels; secondary umbels subtended by several filiform entire bractlets up to 10 mm long, and with up to 30 flowers per head, the flowers borne on glabrous pedicels up to 0.8 mm long; sepals 5, free, deltoid, about 0.5 mm long; petals 5, free, white, 1–2 mm long; stamens 5; ovary inferior; fruit ovoid, 1.0–1.5 mm long, with corky ribs. June–August.

Fig. 31. *Ptilimnium nuttallii* (Nuttall's mock bishop's-weed). Habit (left). Flower (right).

Swampy woods.

IL, MO (FACW+), KS (FAC), KY (FACW).

Nuttall's mock bishop's-weed.

This species is distinguished by its entire bracts, alternate or opposite leaves, and smaller fruits that the other species of *Ptilimnium*.

12. **Sium** L.—Water Parsnip

Glabrous perennials; leaves alternate, once-pinnate, often dimorphic, the leaflets serrate or less commonly pinnatifid; umbels compound, subtended by several bracts; calyx teeth minute; petals 5, free; stamens 5; ovary inferior, bilocular; fruit with conspicuous corky ribs.

There are eight species in this genus including a few in South Africa. Only the following occurs in the central Midwest.

1. **Sium suave** Walt. Fl. Carol. 115. 1788. Fig. 32.

Robust perennials to 2.5 m tall, from fibrous roots; stems strongly grooved; leaves dimorphic, those submerged divided into filiform segments, those above water once-pinnate with 7–17 leaflets, the leaflets up to 12 cm long, up to 5 cm wide, serrate, glabrous, gradually becoming reduced upward; umbels up to 10 cm across, with up to 25 rays; petals white; stamens 5; ovary inferior; fruits ovoid to ellipsoid, 2.5–3.0 mm long. June–September.

Swamps, roadside ditches, wet meadows.

IA, IL, IN, KS, KY, MO, NE, OH (OBL).

Water parsnip.

This is the only member of the Apiaceae in the central Midwest that has once-pinnate leaves that are regularly serrate with more than 8 teeth per side. Leaves that live under water for a period of time may be divided into filiform segments.

13. **Trepocarpus** Nutt.

Annual herbs; leaves alternate, doubly or triply pinnately compound, with filiform segments; flowers in compound umbels, the primary and secondary umbels subtended by 1–several unequal bracts; sepals 5, free, subulate; petals 5, free, white; stamens 5; ovary inferior; fruit elongated, strongly ribbed.

Only the following species comprises the genus.

1. **Trepocarpus aethusae** Nutt. Mem. Umbel. 56. 1829. Fig. 33.

Annual herbs from a slender taproot; stems erect, branched, glabrous, to 80 cm tall; leaves alternate, doubly or triply pinnately compound, the segments flat, linear, up to 1 mm wide, glabrous, on petioles 1–2 cm long; flowers in axillary and terminal umbels, the primary umbel subtended by 1–several linear, entire, unequal bracts up to 10 mm long, on peduncles up to 6 cm long; secondary umbels subtended by 1–several linear, entire, unequal bracteoles up to 10 mm long, 2- to 8-flowered, on peduncles up to 15 mm long; sepals 5, free, subulate, up to 2 mm long; petals 5, free, white, 1.5–2.5 mm long, incurved at the apex; stamens 5; ovary inferior; fruits linear-oblong, to 10 mm long, glabrous, with several strong ribs. May–June.

Fig. 32. *Sium suave* (Water parsnip). Terrestrial leaf (left). Habit (next to left). Aquatic leaf (next to right). Fruit (right).

Swampy woods.

IL, MO (not indicated for this region), KY (NI), but should undoubtedly be OBL.

Whitenymph.

This species has flat filiform leaflets and elongated, strongly ribbed fruits. The leaves have a fernlike appearance.

14. Zizia Koch—Golden Alexanders

Perennial herbs with thickened roots; leaves basal and alternate, simple or ternately compound; flowers in compound umbels, the primary umbel without bracts, the secondary umbels subtended by several bractlets, the central flower of each umbel sessile; sepals 5, free, small; petals 5, free, yellow, incurved at the apex; stamens 5; ovary inferior; fruit more or less flattened, with filiform ribs.

Fig. 33. *Trepocarpus aethusae* (Whitenymph). Habit. Fruit (lower right).

Zizia consists of three species, all in the eastern and central United States.

Only the following species sometimes occurs in wetland habitats in the central Midwest.

1. Zizia aurea (L.) Koch, Nov. Act. Caes. Leop. 12:129. 1825. Fig. 34.
Smyrnium aureum L. Sp. Pl. 262. 1753.

Perennial herbs from thickened roots; stems erect, branched or unbranched, glabrous, to 80 cm tall; leaves alternate, 2- to 3-ternate, on petioles up to 10 cm long, the leaflets ovate to ovate-lanceolate, acute to acuminate at the apex, rounded or tapering to the base, serrate, glabrous, to 4 cm long, to 2.5 cm wide; flowers borne in compound umbels, the primary umbel without bacts and with up to 18 rays, the secondary umbels subtended by a few linear bractlets, the central flower sessile, each flower on a slender pedicel; sepals 5, free, deltoid, 1–2 mm long; petals 5, free, yellow, 1.0–2.5 mm long; stamens 5; ovary inferior; fruit usually slightly flattened, oblongoid, to 4 mm long, with filiform ribs. April–June.

Fens, wet meadows, swampy woods, edge of woods, ditches.

Fig. 34. *Zizia aurea* (Golden Alexanders). Leaves and inflorescence (center). Fruit (right). Cross-section of fruit (left).

IA, IL, IN, MO (FAC+), KS, NE (FAC−), KY, OH (FAC).

Golden Alexanders.

This species is readily recognized by its yellow flowers borne in umbels and its ternately compound leaves. Although this species does not have a particularly wet designation by the U.S. Fish and Wildlife Service, it is often found in fens, wet meadows, and swampy woods.

40. APOCYNACEAE—DOGBANE FAMILY

Perennial herbs or vines (some trees in the tropics) with latex; leaves simple, opposite or occasionally alternate, without stipules; flowers perfect, in cymes or panicles; calyx 5-parted; corolla 5-parted; stamens 5, inserted on the corolla tube; ovaries usually 2, superior, with numerous ovules; fruit a pair of follicles (in our area); seeds sometimes comose.

The family consists of about two hundred genera and a little more than two thousand species. In addition to the two genera described below, *Apocynum* and *Vinca* occur in the central Midwest.

1. Leaves alternate; plants erect .... 1. *Amsonia*
1. Leaves opposite; plants viny.. 2. *Trachelospermum*

1. Amsonia Walt.—Blue Star

Perennial herbs with latex; leaves alternate, simple, entire; flowers borne in cymes, perfect; calyx deeply 5-parted; corolla salverform, 5-lobed, villous within; stamens 5; ovaries 2, superior, with many ovules; fruit a pair of elongated follicles; seeds not comose.

There are twenty-five species in this genus, all occurring either in North America or in Japan. The following wetland species sometimes occurs for a time in shallow water.

1. Amsonia tabernaemontana Walt. Fl. Carol. 98. 1788. Fig. 35.

Amsonia salicifolia Pursh, Fl. Am. Sept. 184. 1814.
Amsonia tabernaemontana Walt. var. *salicifolia* (Pursh) Woodson, Ann. Mo. Bot. Gard. 115:406. 1928.

Perennial herbs; stems erect, glabrous or nearly so, branched or unbranched, to 1 m tall; leaves alternate, lanceolate to ovate, acuminate at the apex, tapering or slightly rounded at the base, entire, glabrous or occasionally slightly pubescent beneath, green or glaucous beneath, to 12 cm long, to 4 (–6) cm wide, the petioles to 8 mm long; flowers several in cymes; calyx 5-parted, the segments subulate, up to 2 mm long; corolla salverform, blue, the tube to 1 cm long, villous within; stamens 5; ovaries 2; fruit a pair of slender, elongated follicles up to 10 cm long, long-tapering at the apex, glabrous; seeds papillose. April–June.

Moist or wet woods, wet prairies, rarely in standing water.

IL, IN, KY, MO (FACW), KS (FACW−).

Blue star.

This is primarily a wetland species, but it may on occasion grow in shallow standing water for short periods of time. Specimens with the leaves glaucous on the lower surface may be known as var. *salicifolia*.

Fig. 35. *Amsonia tabernaemontana* (Blue star). Habit (right). Twin fruits
(upper left). Flower, cut open (lower left).

2. **Trachelospermum** Lemaire—Climbing Dogbane

Vines, with latex; leaves opposite, entire; flowers in terminal and axillary cymes, perfect, small; calyx deeply 5-parted; corolla funnelform to salverform, 5-lobed, with glands within; stamens 5, included; ovaries 2, superior, with numerous ovules; fruit a pair of elongated follicles; seeds comose at the apex.

Most of the thirty species in this genus occur in Asia. The following sometimes grows in shallow standing water of swamps.

1. **Trachelospermum difforme** (Walt.) Gray, Syn. Fl. 2: part 1, 85. 1878. Fig. 36. *Echites difformis* Walt. Fl. Carol. 98. 1788.

High-climbing vines; stems glabrous or sparsely pubescent; leaves opposite, very narrowly lanceolate to oval to ovate, acute to acuminate at the apex, tapering or rarely rounded at the base, entire, glabrous or sparsely pubescent, 5–10 cm long, on petioles to 1.5 cm long; flowers in terminal or axillary cymes, on bracteolate pedi-

cels; calyx 5-parted, the segments to 2 mm long; corolla yellowish to cream, the tube 5–6 mm long, the lobes 3–4 mm long; stamens 5; fruit a pair of very slender, elongated follicles up to 25 cm long, glabrous; seeds comose. May–July.

Swamps, wet woods, occasionally in shallow standing water.

IL, IN, KY, MO (FACW).

Climbing dogbane.

There is considerable variation in the shape of the leaves. Plants with leaves that are very narrowly lanceolate and only 5–6 mm wide look very different from broader leaved plants.

Fig. 36. *Trachelospermum difforme* (Climbing dogbane). Habit (left). Leaf (bottom center). Twin fruits (right).

41. AQUIFOLIACEAE—HOLLY FAMILY

Trees or shrubs; leaves simple, alternate, entire or toothed, often coriaceous, sometimes spinulose, sometimes evergreen, with minute or no stipules; flowers solitary or in cymes, axillary, actinomorphic, perfect or unisexual (often on the same plant); calyx small, 4- to several-lobed, more or less persistent; petals 4–9, free or united at the base; stamens as many as the petals, alternate with the petals, free or attached to the base of the petals; ovary 1, superior, 4- to several-locular, with 1–2 ovules per locule; style very short or absent; fruit drupaceous, with 4–several nutlets.

There are three genera and about four hundred species in the temperate and tropical regions of the world. The two genera described below have species that sometimes occur in habitats that have standing water;

1. Petals white, oblong, united at base; stamens attached to base of the petals; leaves toothed, not mucronate..1. *Ilex*
1. Petals yellow, linear, free; stamens free; leaves entire, mucronate..................2. *Nemopanthus*

1. Ilex L.—Holly

Trees or shrubs; leaves simple, alternate, entire, toothed, or spinulose-toothed, evergreen or deciduous; flowers perfect or unisexual, axillary, the pistillate flowers solitary, the staminate and perfect flowers in crowded cymes; calyx 4- to 9-lobed, small; petals 4–9, free or united at the base; stamens 4–9, united to the base of the corolla; ovary superior, 4- to 9-locular, with 1 ovule per locule; fruit drupaceous, with 4–9 small nutlets.

Staminate, pistillate, and perfect flowers sometimes occur on the same plant. The staminate flowers have a rudimentary pistil, while the pistillate flowers often have stamens with tiny anthers.

This genus is worldwide in distribution and contains between three and four hundred species.

1. Leaves evergreen, with spiny teeth ....2. *I. opaca*
1. Leaves deciduous, without spiny teeth.
  2. Teeth of calyx acute, glabrous; nutlets ribbed; petals usually 4–5, at least in the pistillate flowers; apex of leaves obtuse ....1. *I. decidua*
  2. Teeth of calyx obtuse, ciliate; nutlets not ribbed; petals usually 6, at least in the pistillate flowers; apex of leaves more or less acute to acuminate..............................3. *I. verticillata*

1. **Ilex decidua** Walt. Fl. Car. 241. 1788. Fig. 37.

Ilex decidua Walt. f. *aurantiaca* Mohl. & Ozment, Trans. Ill. Acad. Sci. 60:187. 1967.

Shrubs or small trees to about 10 m tall, the twigs gray, glabrous, often bearing short shoots; leaves deciduous, obovate to oblong, obtuse at the apex, cuneate at the base, crenate, dark green and glabrous on the upper surface, paler and pubescent on the lower surface, at least on the main nerve, to 6 cm long, to 15 mm wide; petiole glabrous or pubescent, up to 1/6 as long as the blade; flowers 1–several from the axils of the leaves, on slender pedicels; staminate flowers with 4–5 ovate, acute, glabrous calyx lobes, 4–5 oblong, obtuse, nearly free, greenish or greenish white petals, and 4–5 stamens adnate to the base of the petals; fertile flowers similar except for the presence of a single, superior ovary and slightly shorter peduncles; fruit a drupe, red or orange, globose, up to 7 mm in diameter, containing 1 or 2 ribbed nutlets. April–May.

Low ground, frequently in swamps but occasionally growing above mid-slope on low hills.

IL, IN, KY, MO (FACW), KS, NE (FACW–).

Swamp holly; possum haw; deciduous holly.

This species is similar in appearance to *I. verticillata*, differing most clearly in characters of the flower and the nutlets. The two species are very much alike vegetatively, although the leaves of *I. decidua* are usually obtuse at the apex and crenate along the margins, while those of *I. verticillata* are usually acute at the apex and serrate along the margins. The leaves of *I. verticillata* are sometimes rugose.

Occasionally plants occur with orange fruits and have been called f. *aurantiaca*.

Fig. 37. *Ilex decidua* (Swamp holly). Habit (center). Twig (far left). Fruits (lower left). Leaf (right).

2. Ilex opaca Ait. Hort. Kew. 1:169. 1789. Fig. 38.

Trees to about 20 m tall, with the young branchlets sparsely pubescent; leaves coriaceous, evergreen, elliptic to obovate, coarsely spinulose-toothed, to 8.5 cm long, usually less than half as wide; petioles up to 6 mm long, pubescent; staminate flowers 3–10 in axillary cymes, pistillate and perfect flowers usually solitary in the axils; calyx 4-lobed, the lobes acute, ciliate; petals usually 4, free, suborbicular to oblong, white or greenish, up to 6 mm long; stamens 4, united to the base of the petals; ovary superior; style usually absent; stigma capitate; fruit a drupe, globose, usually red, up to 8 mm in diameter; nutlets grooved on the back. May–June.

Mesic woods, swampy woods.

IL, KY, MO, OH (FACU+).

American holly.

Although this species usually occurs in mesic woods, it is sometimes found in swampy woods in Kentucky and Ohio. It is readily identified by its spinulose, coriaceous, evergreen leaves.

3. Ilex verticillata (L.) Gray, Man., ed. 2:264. 1856. Fig. 39.
Prinos verticillatus L. Sp. Pl. 330. 1753.
Prinos padifolia Willd. Enum. Hort. Berol. 394. 1805.
Ilex verticillata (L.) Gray var. *padifolia* (Willd.) Torr. & Gray ex S. Wats. Bibliogr. Ind. N. Am. Bot. 160. 1878.

Shrubs or small trees to 5 m, the twigs usually brown, smooth; leaves deciduous, oval to lanceolate to oblong, more or less acute to acuminate at the apex, rounded at the base, serrate, dark green and usually glabrous on the upper surface, paler and pubescent on the veins below or rarely over the entire lower surface, to 7 cm long, to 2 cm wide; petiole glabrous or pubescent, up to 1/5 as long as the blade; flowers 1–several from the axils of the leaves, on short pedicels; staminate flowers with 4–6, ovate, obtuse, ciliate calyx lobes, 4–6 oblong, obtuse, nearly free, greenish or greenish white petals, and 4–6 stamens adnate to the base of the petals; fertile flowers similar, except for usually 6 calyx lobes, 6 petals, and a single, superior ovary; fruit a drupe, red, up to 7 mm in diameter, containing 1–2 unribbed nutlets. June–August.

Swamps, bogs, along streams, less commonly on exposed sandstone cliffs.

IA, IL, IN, KY, MO, OH (FACW+)

Winterberry.

This species may grow in standing water, particularly in bogs. Its brilliant red drupes make it a good candidate as an ornamental.

2. Nemopanthus Raf.—Mountain Holly

Shrubs; leaves alternate, simple, entire or toothed, petiolate, without stipules; flowers 1–few in the axils of the leaves, polygamodioecious, on long pedicels; staminate flowers: calyx 4- to 5-toothed, minute; petals 4–5, free, yellow; stamens 4–5, free; pistillate flowers: calyx absent; petals 4–5, free, yellow; ovary superior; fruit a red drupe, with 4–6 nutletes.

Two species comprise this genus, which differs from *Ilex* by its linear, yellow petals and its free stamens.

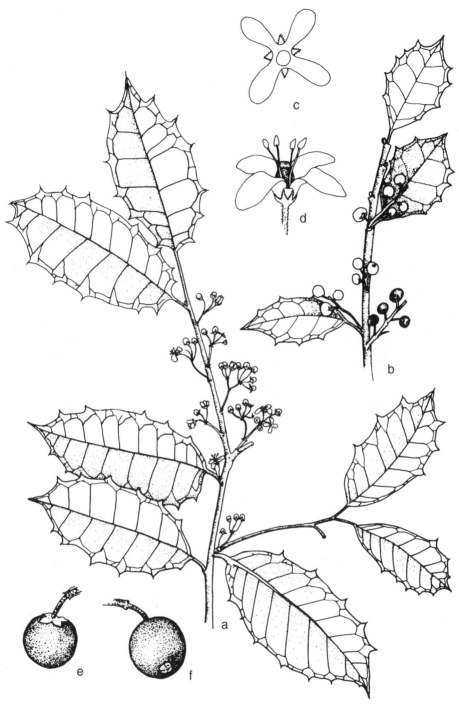

Fig. 38. *Ilex opaca*
(American holly).

a. Habit.
b. Habit, in fruit.

c, d. Flowers.
e, f. Fruits.

Fig. 39. *Ilex verticillata* (Winterberry). Habit (right). Flower (lower left).

1. **Nemopanthus mucronatus** (L.) Trel. Trans. Acad. Sci. Louis 5:349. 1889. Fig. 40.
Vaccinium mucronatum L. Sp. Pl. 350. 1753.

Much branched shrub to 3 m tall, with gray bark and twigs; leaves alternate, simple, elliptic to oblong, obtuse to acute at the apex, mucronate, rounded at the base, entire or obscurely serrulate, glabrous, green on the upper surface, paler on the lower surface, to 5 cm long, to 3 cm wide, on often reddisih petioles up to 1 cm long; calyx 4- to 5-toothed in the staminate floweres, minute, absent in the pistillate flowers; petals 4–5, free, linear, yellow, 2.0–2.5 mm long; stamens 4–5, free; ovary superior; drupes eglobose, red, 5–6 mm in diameter, with 4–5 smooth or ribbed nutlets. May–June.

Swampy woods.

IL, IN, OH (OBL).

Mountain holly.

This species is distinguished from the deciduous hollies in the central Midwest by its usually entire leaves, usually reddish petioles, and yellow flowers with linear petals.

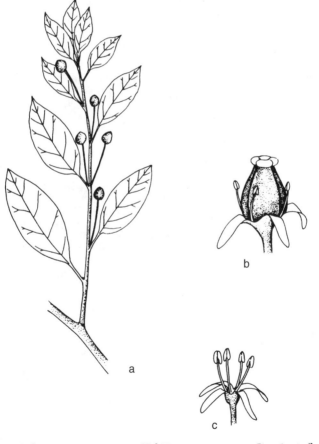

Fig. 40. *Nemopanthus mucronatus* a. Habit.   c. Staminate flower.
(Mountain holly).   b. Pistillate flower.

42. ARALIACEAE—SPIKENARD FAMILY

1. Aralia L.

Herbs, shrubs, or trees, sometimes prickly; leaves alternate, once- or several-pinnate; inflorescence umbellate; flowers several to numerous; calyx 5-lobed, minute; petals 5, free; stamens 5; ovary inferior; fruit a drupe.

Aralia consists of thirty-five species, most of them in Indonesia, Asia, and North America. Only the following may be found in a habitat where standing water is present for some time during the year.

1. **Aralia nudicaulis** L. Sp. Pl. 274. 1753. Fig. 41.

Perennial herbs from rhizomes; leaves arising directly from the underground rhizome, ternately compound, on a glabrous petioles to 0.5 m long, each part of the ternate leaf 3- or 5-pinnate, the leaflets oblong to obovate, acute to acuminate at the apex, more or less rounded at the base, serrate, glabrous, to 15 cm long, to 10 cm wide; umbels 2–7, borne on a leafless scape, the peduncle up to 0.3 m tall; calyx 5-toothed; petals 5, free, greenish; stamens 5; ovary inferior; drupes globose, black, 2–4 mm in diameter. May–July.

Tamarack bogs, mesic woods.

IA, IL, IN, KY, MO, OH (FACU), NE (NI).

Wild sarsaparilla.

Although primarily a species of mesic woods, *A. nudicaulis* is also found in tamarack bogs in the northern part of the central Midwest.

Fig. 41. *Aralia nudicaulis* (Wild sarsaparilla). Leaves and inflorescence (center). Section of ovary (lower left). Fruit (upper right). Flower (lower right).

43. ARISTOLOCIACEAE—BIRTHWORT FAMILY

Herbs or shrubs, sometimes climbing; leaves basal or alternate; stipules absent; flowers perfect, actinomorphic or zygomorphic; calyx tubular, united at least at the base, 3-lobed; petals absent; stamens 6 or 12, closely appressed to the styles; ovary inferior or subinferior, 6-locular; fruit a capsule, many-seeded.

Eight genera and about four hundred species make up this family, most of them occurring in the tropics.

1. **Aristolochia** L. Birthwort

Perennial herbs or high climbing vines; leaves alternate, entire; flower solitary (in the central Midwest species), zygomorphic, perfect; calyx tubular, the tube curved, 3-lobed, adnate to the ovary, at least at the base; stamens 6, the anthers adnate to the stigmas; ovary partly inferior, 6-locular; fruit a capsule; seeds many.

Approximately three hundred species, mostly tropical, comprise this genus.

1. Erect herbs, lower surface of leaves pubescent, not white-woolly; calyx dark purple, to 15 mm long; roots with odor of turpentine .... 1. *A. hastata*
1. High-climbing woody vines; lower surface of leaves white-woolly; calyx yellow, with a purplish orifice, over 15 mm long; roots without odor of turpentine .... 2. *A. tomentosa*

1. **Aristolochia hastata** Nutt. Gen. N. Am. Pl. 2:200. 1818. Fig. 42.
Aristolochia serpentaria L. var. *hastata* (Nutt.) Duchartre in DC. Prod. 15:434. 1864.
Aristolochia nashii Kearney, Bull. Torrey Club 21:485. 1894.

Perennial herbs from a short, aromatic rhizome; stems erect, slender, to 35 cm tall, usually slightly pubescent; leaves linear-lanceolate, acuminate at the apex, cordate and usually hastate at the base, entire, slightly pubescent to glabrous on both surfaces, to 10 cm long, to 2 cm broad, the lowermost leaves reduced to scales; petioles to 1.5 cm long, glabrous or slightly pubescent; flower solitary from near the base of the stem, on a slender pedicel up to 5 cm long, the pedicel bearing small scales; calyx tubular, strongly curved, up to 1.5 cm long, pubescent, expanded at the apex into 3 blunt, purple-brown, shallow lobes; capsules subglobose to some-what longer than broad, ridged, up to 1 cm in diameter; seeds obovoid, up to 5 mm long, with minute yellow-white papillae. May–July.

Swampy woods.

IL, KY, MO (the U.S. Fish and Wildlife Service considers this to be included in *A. serpentaria*, which they designate UPL in IL, KY, and MO).

Narrow-leaved snakeroot.

Although this species is usually considered to be a variety of *A. serpentaria*, its extremely narrow, often hastate leaves distinguish it from that species. This narrow-leaved plant invariably grows in swampy woods, which are occasionally inundated. Typical *A. serpentaria* is an upland woods species.

2. **Aristolochia tomentosa** Sims, Bot. Mag. 33:pl. 1369. 1811. Fig. 43.

Woody vines with twining, tomentose stems reportedly up to 30 m long; leaves alternate, simple, broadly ovate to suborbicular, obtuse to subacute at the apex, cordate at the base, entire, tomentose on both surfaces, up to 20 cm long, nearly as

Fig. 42. *Aristolochia hastata* (Narrow-leaved snakeroot). Habit (center, above). Leaves (far left). Seed (next to left). Capsule (above seed). Cluster of fruits (upper right of seed). Flowers (lower right).

broad; petioles up to 6 cm long, tomentose; flower solitary, axillary, on tomentose pedicels up to 4 cm long, without bracts; calyx tubular, strongly curved, tomentose, the tube yellow-green, up to 3 cm long, expanded at the tip into 3 acute, purple, spreading lobes; stigma 3-lobed, spreading; capsules cylindrical, tomentose, up to 8 cm long, up to 3 cm in diameter; seeds triangular, flat, notched at the apex. May–June.

Swampy woods.

IL, IN, KS, KY, MO (FAC).

Dutchman's pipevine.

This is a high-climbing woody vine with large, cordate, tomentose leaves. The common name alludes to the flower, which has the calyx in the configuration of a Dutchman's pipe.

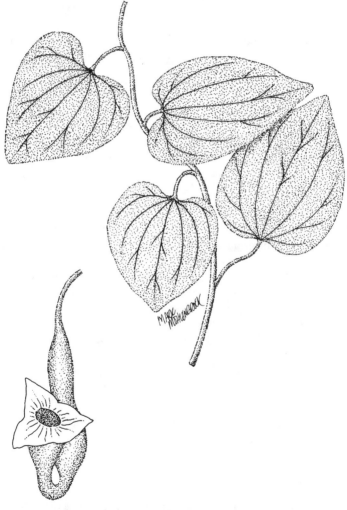

Fig. 43. *Aristolochia tomentosa* (Dutchman's pipevine). Leaves (right). Flower (lower left).

44. ASCLEPIADACEAE—MILKWEED FAMILY

Perennial herbs, vines, or less commonly shrubs, usually with latex; leaves opposite, less commonly alternate or whorled, without stipules; flowers perfect, actinomorphic, usually borne in umbels; calyx 5-parted; corolla 5-parted, the segments often reflexed; a 5-lobed corona often present between the stamens and the corolla; stamens 5, attached to the corolla; ovary superior, bicarpellate; styles 2; ovules numerous; fruit a follicle; seeds usually comose.

This family consists of 250 genera and about twenty-five hundred species, many of them in the subtropics. In addition to the genus described below, *Cynanchum* and *Matelea* also occur in the central Midwest.

1. Asclepias L.—Milkweed

Perennial herbs, usually with latex; leaves usually opposite, simple, entire; flowers perfect, actinomorphic, in terminal or axillary umbels; calyx 5-parted; corolla 5-parted, the segments reflexed; corona with 5 spreading or erect hoods, each with an incurved horn present; stamens 5; ovary superior; fruit a follicle; seeds usually comose.

Asclepias is a genus of 150 species, most of them in the New World. In addition to the two species described below, several others are present in the mid-central United States in primarily drier habitats.

1. Flowers pink or rosy; seeds comose .... 1. *A. incarnata*
1. Flowers white; seeds not comose .... 2. *A. perennis*

1. Asclepias incarnata L. Sp. Pl. 215. 1753. Fig. 44.

Perennial herbs with latex; stems erect, glabrous or pubescent in lines above, usually branched, to 1.3 m tall; leaves opposite, lanceolate to oblong-lanceolate, acuminate at the apex, tapering or somewhat rounded at the base, entire, glabrous or sparsely pubescent, to 15 cm long, to 10 cm wide, with prominent yellow or white veins; petioles 6–12 mm long; umbels numerous; calyx segments lanceolate, 2–3 mm long; corolla lobes pink or rosy, oblong, 3.5–4.5 mm long; hoods pink or purple, obtuse; horns longer than the hoods; follicles erect, 5–8 cm long, glabrous or usually sparsely pubescent; seeds comose. June–October.

Marshes, swamps, wet prairies, wet ditches, sometimes in shallow standing water.

IA, IL, IN, KY, MO, OH (OBL), KS, NE (FACW+).

Pink swamp milkweed.

This species is common throughout the central Midwest.

The veins are usually somewhat yellowish or whitish against the green background of the rest of the leaf. .

2. Asclepias perennis Walt. Fl. Carol. 107. 1788. Fig. 45.

Perennial herbs with latex; stems erect, usually branched, to 1 m tall, puberulent in lines above, glabrous below; leaves opposite, lanceolate, acuminate, tapering to the base, entire, glabrous or nearly so, to to 14 cm long, to 2.5 cm wide; petioles

Fig. 44. *Asclepias incarnata* (Pink swamp milkweed). Habit (right). Flower (left).

slender, to 8 mm long; umbels solitary or several, the peduncles up to 4 cm long; calyx lobes narrowly lanceolate, to 2 mm long; corolla lobes white, oblong, 3–4 mm long; hoods white, obtuse; horns longer than the hoods; follicle ovoid below, tapering to a long tip, glabrous, erect, 4–7 cm long; seeds not comose. May–September.

Swamps, swampy woods, sometimes in standing water.

IL, IN, MO (OBL), KY (not listed by the U.S. Fish and Wildlife Service for KY). White swamp milkweed.

The white flowers and absence of comose seeds readily distinguish this species.

45. ASTERACEAE—ASTER FAMILY

Annual or perennial herbs (in our area); latex sometimes present; leaves simple or compound, alternate, opposite, or whorled; flowers more than one arranged in one or more heads, all the flowers of a head sharing the same receptacle; flowers of two types: some flat, usually notched at the apex (rays or ligules), some tubular, usually 5-lobed, forming a disk; some heads with only ray flowers, or some with only disk flowers, or some with both ray and disk flowers; individual flowers perfect or unisexual or neutral, with 5 stamens and an inferior ovary, sometimes subtended by a small bract (chaff); styles 2-cleft; fruit an achene, often crowned with pappus; floral heads subtended by a group of bracts known collectively as the involucre.

This is one of the largest and most widespread families in the world, consisting of approximately eleven hundred genera and twenty thousand species. They occur in all variety of habitats. The ones that are described in this book may sometimes be found in habitats where there may be standing water for a part of the year.

Fig. 45. *Asclepias perennis* (White swamp milkweed). Habit, left. Hood with horn (upper center). Flower (upper right). Fruit (lower right).

1. Plants with latex; ray flowers present; disk flowers absent .... 25. *Prenanthes*
1. Plants without latex; ray flowers present or absent; disk flowers present.
  2. Some or all the leaves opposite or whorled.
    3. Some or all of the leaves whorled.
      4. Leaves petiolate; flowers pink-purple to purple .... 11. *Eupatoriadelphus*
      4. Leaves connate; flowers white .... 12. *Eupatorium*
    3. None of the leaves whorled.
      5. Plants submerged; leaves divided into filiform segments .... 19. *Megalodonta*
      5. Plants not submerged; leaves simple, lobed or divided but the segments not filiform.
        6. Vines; leaves simple, cordate, dentate .... 21. *Mikania*
        6. Plants erect, ascending, or prostrate; leaves divided or, if undivided, then not cordate.
          7. Plants prostrate to spreading or weakly ascending.
            8. Leaves linear-lanceolate to linear, obscurely serrate, sessile or nearly so; flowers white .... 10. *Eclipta*
            8. Leaves ovate-lanceolate to ovate, distinctly serrate, petiolate; flowers yellow ... . 1. *Acmella*
          7. Plants erect.
            9. Stems square.
              10. Leaves connate; flowers yellow .... 27. *Silphium*
              10. Leaves petiolate; flowers white .... 20. *Melanthera*
            9. Stems terete or nearly so.
              11. Lower leaves opposite, upper leaves alternate; plants over 2 m tall ....17. *Helianthus*
              11. All leaves opposite.
                12. Flowers yellow; pappus of 2 or 4 stout, barbed awns; stems often purple at the nodes ....4. *Bidens*
                12. Flowers white, pink, or rose; pappus of capillary bristles, or absent; stems usually without purple nodes.
                  13. Receptacle conical; flowers blue................................ 8. *Conoclinium*
                  13. Receptacle flat; flowers pink or white.
                    14. Flowers pink.................................... 14. *Fleischmannia*
                    14. Flowers white.
                      15. Bracts all the same length .... 2. *Ageratina*
                      15. Bracts of different lengths .... 12. *Eupatorium*
  2. None of the leaves opposite or whorled.
    16. Leaves prickly; flowers purple...7. *Cirsium*
    16. Leaves not prickly; flowers yellow, blue, rose-purple, or white.
      17. Stems winged.
        18. Leaves and stems harshly scabrous; plants more than 2 m tall ...... 31. *Verbesina*
        18. Leaves and stems glabrous or, if pubescent, not harshly scabrous; plants up to 1.5 m tall.
          19. Plants glabrous; flowers white; rays not notched .... 5. *Boltonia*
          19. Plants pubescent; flowers yellow; rays notched....................... 16. *Helenium*
      17. Stems unwinged.
        20. Plants with an unpleasant or medicinal odor.................................. 24. *Pluchea*
        20. Plants without an unpleasant or medicinal odor.
          21. Some of the leaves deeply divided.
            22. Ray flowers up to 1 cm long; plants up to 1 m tall.................23. *Packera*
            22. Ray flowers over 1 cm long; plants over 2 m tall .... 27. *Rudbeckia*
          21. None of the leaves deeply divided.
            23. At least the lower leaves sagittate-hastate .... 15. *Hasteola*

23. None of the leaves sagittate-hastate.
24. Basal leaved decidedly different from cauline leaves in shape.
25. Basal leaves cordate ....23. *Packera*
25. Basal leaves not cordate.
26. Basal leaves with conspicuous parallel veins; flowers white ...
.... 3.*Arnoglossum*
26. Basal leaves without conspicuous parallel veins; flowers yellow.
27. Inflorescence flat-topped; stems not angular ....
.... 22. *Oligoneuron*
27. Inflorescence paniculate or thyrsoid; stems often angular
.... 29. *Solidago*
24. All leaves similar in shape.
28. Ray and disk flowers present.
29. Flowers yellow.
30. Pappus of capillary bristles.
31. Leaves up to 5 mm wide; inflorescence flat-topped.......
.... 13. *Euthamia*
31. Leaves more than 5 mm wide; inflorescence paniculate
or thyrsoid... 29. *Solidago*
30. Pappus of disk flowers composed of 2 awns or with short
teeth.
32. Pappus of disk flowers composed of 2 awns; leaves up
to 1 cm wide ....17. *Helianthus*
32. Pappus of disk flowers with short teeth; leaves more
than 1 cm wide.....................................26. *Rudbeckia*
29. Flowers white, blue, or purple.
33. Pappus of capillary bristles.
34. Pappus in a double series ....9. *Doellingeria*
34. Pappus in a single series.............. 30. *Symphyotrichum*
33. Pappus composed of 2 awns and tiny bristles..................
.... 5. *Boltonia*
28. Only disk flowers present.
35. Flowers whitish .... 6. *Brachyactis*
35. Flowers purple.
36. Stems usually unbranched, up to 1.5 m tall ...... 18. *Liatris*
36. Stems branched, more than 2 m tall............. 32. *Vernonia*

1. **Acmella** Rich.—Spotflower

Annual or perennial herbs; leaves simple, opposite, toothed, petiolate; flowers in
heads from the axils of the leaves, bearing both ray and disk flowers, subtended by
2 series of bracts; ray flowers yellow or white, pistillate; disk flowers yellow, perfect;
receptacle convex or conical; achenes from ray flowers more or less trigonous, from
disk flowers compressed, with a pappus of 1–3 minute awns.

Because of its earlier date of publication, the name *Acmella* precedes that of
Spilanthes, the name used for this genus in the past. About forty species comprise
the genus.

1. **Acmella oppositifolia** (Lam.) R.K. Jansen, Syst. Bot. Monagr. 8:30. 1985. Fig. 46.
Anthemis oppositifolia Lam. Encycl. Meth. Bot. 1:576. 1783.
Anthemis repens Walt. Fl. Carol. 211. 1788.

Spilanthes repens (Walt.) Michx. Fl. Bor. Am. 2:131. 1803.
Spilanthes americana Hieron. Bot. Jahrb. Syst. 29:42. 1900.
Spilanthes americana (Hieron.) var. *repens* (Walt.) A. H. Moore, Proc. Am. Acad. 42: 547. 1907.

Perennial herbs; stems spreading to ascending, rooting at the lower nodes, slender, branched or unbranched, pubescent, to nearly 0.75 m long; leaves opposite, simple, lanceolate to ovate, acute to acuminate at the apex, tapering or rounded at the base, coarsely serrate to nearly entire, pubescent on both surfaces, to 8 cm long, to 4 cm wide, on petioles to 2 cm long; head solitary from the axils of the leaves, on peduncles up to 6 cm long, subtended by 2 series of bracts, the bracts oblong to lanceolate, acute or obtuse at the apex; ray flowers 8–12, yellow, 3-notched at the apex; disk flowers tubular, yellow, the disk 5–10 mm across; receptacle conical; achenes oblong, trigonous or compressed, 1.5–2.5 mm long, hispidulous, ciliate, the pappus of 1–2 minute awns. July–October.

Swampy woods.

MO (FACW+).

Spotflower.

This species is recognized by its spreading habit, its opposite, petiolate leaves, and its solitary yellow flower heads from the axils of the leaves.

Fig. 46. *Acmella oppositifolia* (Spotflower). Habit (center).
Achene (lower left). Disk flower (next to lower right). Ray flower (lower right).

2. **Ageratina** Spach

Perennial herbs with thickened roots; leaves opposite, simple, usually petiolate; ray flowers absent; disk flowers borne in numerous heads, tubular, perfect, subtended by unequal bracts in 1–2 series and not imbricate; style branches linear, obtuse, papillose; achenes angular, glandular.

Three species of this genus occur in the central Midwest, with the following sometimes occurring in habitats where there is standing water during the year.

1. **Ageratina aromatica** (L.) Spach, Hist. Nat. Veg. 10:286. 1841. Fig. 47.
Eupatorium aromaticum L. Sp. Pl. 2:839. 1753.

Perennial herb with thickened roots; stems erect, slender, branched or un-branched, somewhat pubescent to nearly glabrous, to 75 cm tall; leaves opposite, simple, rather thick, ovate, obtuse to acute at the apex, rounded or truncate at the base, crenate, usually somewhat pubescent on both surfaces, to 8 cm long, to 5 cm wide, the petioles up to 1.5 cm long; heads in corymbs arising from the reduced upper leaves, bearing up to 20 white flowers; involucre 4–5 mm tall, subtended by 1–2 irregular series of unequal bracts; corolla lobes pubescent on the outer face; achenes angular, 3–4 mm long. August–October.

Mostly in dry, sandy woods, but in Kentucky, I have seen it in swampy woods.

KY, OH (not listed by the U.S. Fish and Wildlife Service so that they consider it to be UPL in all areas).

Fragrant thoroughwort; small-leaved white snakeroot.

This species resembles the more widespread *A. altissima* (= *Eupatorium rugosum*), but differs primarily by its thicker, crenate, smaller leaves.

Because of the irregularly arranged unequal involucral bracts, this species has been segregated from the genus *Eupatorium*, which has imbricate involucral bracts in 3 distinct series.

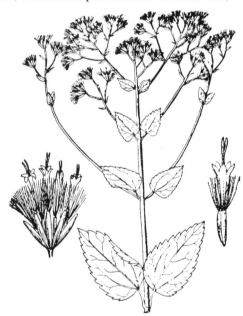

Fig. 47. *Ageratina aromatica* (Fragrant thoroughwort). Habit (center). Head of disk flowers (left). Disk flower (right).

3. **Arnoglossum** Raf.—Indian Plantain

Perennial herbs; leaves basal and/or alternate, usually glabrous; inflorescence corymbose, with several large heads, the heads 5-flowered, subtended by 5 bracts; ray flowers absent; disk flowers usually 5-lobed, white; stamens 5; ovary inferior; achenes subtended by numerous capillary bristles.

There are about forty-five species in the genus, all of them at one time placed in the genus *Cacalia*. Only the following occurs in wetlands of the central Midwest.

1. **Arnoglossum plantagineum** Raf. Fl. Ludov. 65. 1817. Fig. 48.
Cacalia tuberosa Nutt. Gen. N. Am. P1.2:138. 1818.
Mesadenia plantaginea Raf. New Fl. Bot. N. Am. 4:79. 1832.
Mesadenia tuberosa (Nutt.) Britt. Ill. Fl. N. U.S. 3:474. 1898.
Cacalia plantaginea (Raf.) Shinners, Field & Lab. 18:81. 1950.

Stout perennial from a tuberous-thickened base; stems glabrous, striate, ridged, to 20 dm tall; basal leaves lance-ovate, acute at the apex, tapering to the base, thick, firm, glabrous, entire, conspicuously parallel-veined, to 20 cm long, to 10 cm broad, the petioles sometimes as long as the blades; upper leaves smaller and sessile or subsessile; heads several in a flat-topped inflorescence, white, 5-flowered; disk to 7 mm across, the disk flowers 5-lobed; rays absent; achenes cylindric, glabrous, with numerous white capillary bristles. June–August.

Wet prairies, springy ground in marshes, bogs.

IA, IL, IN, MO (FAC), KY, OH (FACW), NE (NE).

Prairie Indian plantain; tuberous Indian plantain.

This species is readily recognized by its lower leaves that are lance-ovate with conspicuous thick parallel veins. It is often placed in the genus *Cacalia*.

4. Bidens L.—Beggar's-tick; Sticktight

Annual or biennial herbs; leaves opposite, simple or compound; flowers several to many in heads, the heads subtended by bracts in 2 series, with or without ray flowers but always with disk flowers; rays (when present) yellow, usually up to 10 per head, usually sterile; disk flowers perfect; achenes flattened to more or less 4-angled, usually with 2–4 barbed awns at apex.

This is a genus of nearly two hundred species found in most parts of the world. In the central Midwest, all but *B. bipinnata*, Spanish needles, are wetland plants.

1. Leaves simple and unlobed or with some of them at most 3- to 5-lobed, never completely pinnately divided.
  2. Rays 1.5–3.0 cm long .... 8. *B. laevis*
  2. Rays up to 1.5 cm long, or absent.
    3. Leaves sessile or connate at base; heads nodding at maturity .... 2. *B. cernua*
    3. Leaves petiolate or winged to base; heads erect.
      4. Achenes usually with 4 awns and tuberculate on mid-vein; lobes of disk flowers 5 ....
      .... 4. *B. connata*
      4. Achenes usually with 3 awns and smooth on mid-vein; lobes of disk flowers 4 ....
      .... 3. *B. comosa*
1. Most or all the leaves pinnately divided.
  5. Rays longer than the involucre; leaflets 3–7, deeply cleft or coarsely serrate.
    6. Achenes 2 1/2–4 times as long as wide, with ciliate margins; leaflets commonly 7, the ultimate segments up to 5 mm wide .... 5. *B. coronata*
    6. Achenes up to 2 1/2 times as long as wide, with scabrous, rarely ciliate margins; leaflets commonly 3 or 5, the ultimate segments more than 5 mm wide.
      7. Outer bracts of the involucre 8–12 .... 1. *B. aristosa*
      7. Outer bracts of the involucre more than 12 .... 9. *B. polylepis*

5. Rays shorter than the involucre, or absent; leaflets 3 or 5, usually uncleft.
  8. All leaves 3-parted...10. *B. tripartita*
  8. At least some of the leaves 5-parted.
    9. Outer bracts of the involucre 2–5, not ciliate .... 6. *B. discoidea*
    9. Outer bracts of the involucre 6 or more, most of them ciliate.
      10. Outer bracts of the involucre usually 6–10.......................................7. *B. frondosa*
      10. Outer bracts of the involucre more than 10.......................................11. *B. vulgata*

Fig. 48. *Arnoglossum plantagineum* (Prairie Indian plantain).
Upper part of plant (left). Basal leaf (right).

1. **Bidens aristosa** (Michx.) Britt. Bull. Torrey Club 20:281. 1893. Fig. 49.
Coreopsis aristosa Michx. Fl. Bor. Am. 2:140. 1803.
Coreopsis aristosa Michx. var. *mutica* Gray, Man. Bot. ed. 5, 260. 1867.
Bidens aristosa (Michx.) var. *fritcheyi* Fern. Rhodora 15:78. 1913.
Bidens aristosa (Michx.) var. *mutica* (Gray) Fern. Rhodora 15:78. 1913.

Annual or biennial herbs; stems much branched, up to 1.3 m tall, glabrous; leaves pinnately 3- to 5- (7-) divided, to 15 cm long, the segments lanceolate, serrate to pinnatifid, acuminate at the apex, tapering to the base, the petioles to 3 cm long; heads several, to 4 cm across, the outer bracts 8–12, linear, to 12 mm long, shorter than the inner bracts, glabrous or puberulent; rays usually 8, to 2.5 cm long; achenes flat, 5–7 mm long, strigose and ciliate, with 2 or rarely 4 awns at the top, or more rarely absent. August–October.

Marshes, swamps, wet ditches.

IA, IL, IN, KS, MO, NE (FACW), KY, OH (FACW–).

Swamp marigold; bur marigold; sticktight.

This is an extremely variable species that occurs in most types of wetlands. Typical plants have 8–12 outer involucral bracts per head and awns on the achenes that are upwardly barbed. Plants that have the awns downwardly barbed are common and may be known as var. *fritcheyi* Fern. Plants that have virtually no awns on the achenes may be known as var. *mutica* Gray. Very showy plants with 12 or more outer involucral bracts are segregated as a different species known as *B. polylepis* Blake.

2. **Bidens cernua** L. Sp. Pl. 832. 1753. Fig. 50.
Bidens minima Huds. Fl. Angl. 310. 1762.
Bidens cernua L. var. *minima* (Huds.) Pursh, Fl. Am. Sept. 2:566. 1814.
Bidens cernua L. var. *elliptica* Wieg. Bull Torrey Club 26:417–418. 1899.
Bidens cernua L. var. *integra* Wieg. Bull. Torrey Club 26:418–419. 1899.

Annual herbs; stems branched, up to 1 m tall, glabrous or sometimes hispid; leaves simple, lanceolate to oblong-lanceolate, acuminate at the apex, sessile and usually connate-perfoliate at the base, sharply serrate, glabrous, to 20 cm long, to 4 cm wide; heads relatively few, nodding at maturity, up to 2.5 cm across, the outer bracts 5–10, broadly linear, surpassing the disk; rays absent or up to 8 in number, up to 2.5 cm long; disk corollas 5-lobed; achenes 5–8 mm long, with usually 4 barbed awns at tip. August–October.

Wet ground, including bogs.

IA, IL, IN, KS, KY, MO, NE, OH (OBL).

Nodding bur marigold; sticktight.

When rays are present, this is an extremely beautiful plant. Frequently, however, the rays do not develop.

This species is also variable with respect to leaf shape and stem diameter. Plants with capillary stems, smaller disks, and 1–few flower heads have been found in Indiana and are called var. *minima*, although they may represent a different species. Typical var. *cernua* has sessile to connate leaves and coarse teeth. Var. *integra* is similar but with teeth less than 1 mm long; var. *elliptica* has coarsely toothed leaves that sometimes are short-petioled.

Fig. 49. *Bidens aristosa* (Swamp marigold). Upper part of plant (left). Achene (right).

Fig. 50. *Bidens cernua* (Nodding bur marigold). Upper part of plant (left).
Achene (lower right).

3. **Bidens comosa** (Gray) Wieg. Bull. Torrey Club 24:436. 1897. Fig. 51.
Bidens connata Muhl. var. *comosa* Gray, Man., ed. 5, 261. 1867.

Annual herbs; stems branched, to 1 m tall, glabrous; leaves simple, unlobed or deeply 3-lobed but not divided into separate leaflets, acute to acuminate at the apex, tapering to the sessile or short-petiolate base, glabrous, to 10 cm long, serrate, the petioles when present usually winged; heads several, to 2.5 cm across, the outer bracts 6–10, leafy, much longer than the disk, often unequal in length; rays absent; disk corollas 4-lobed, 4–5 mm long; achenes flat, cuneate, retrorsely ciliate, usually with 3 barbed awns. September–October.

Wet ground, occasionally in shallow standing water.

IA, IL, IN, KY, MO, OH (FACW), KS, NE (not listed for KS and NE by the U.S. Fish and Wildlife Service).

Swamp beggar's-tick.

This species is similar to *B. tripartita*, but most of the leaves are undivided. Those leaves that are 3-parted are not divided into separate leaflets but merely deeply lobed. *Bidens comosa* differs from *B. connata* by its 3-awned achenes and its 4-lobed disk flowers.

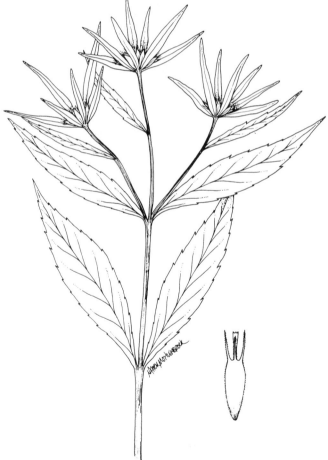

Fig. 51. *Bidens comosa* (Swamp beggar's-tick). Habit (center). Achene (lower right).

4. **Bidens connata** Muhl. in Willd. Sp. Pl. 3:1718. 1804. Fig. 52.

Bidens petiolata Nutt. Journ . Acad. Nat. Sci. Phila. 7:99–100. 1834.

Bidens tripartita L. var. *fallax* Warnst. Verh. Bot. Vev. Prov. Brandenburg 21:152. 1880.

Bidens connata Muhl. var. *petiolata* (Nutt.) Farw. Ann. Rep. Comm. Parks Detroit 11: 91. 1900.

Bidens connata Muhl. var. *fallax* (Warnst.) Sherff, Bot. Gaz. 76:154. 1923.

Annual herbs; stems usually branched, often purplish, up to 2.5 m tall, glabrous; leaves simple, lanceolate to oblong-lanceolate, acuminate at the apex, tapering to a petiolate base, sometimes lobed at base, or the uppermost leaves sometimes nearly sessile, sharply serrate, to 15 cm long, to 4 cm wide, glabrous; petioles up to 3 cm long, sometimes winged; heads several, up to 3 cm across, the outer bracts 4–9, somewhat longer than the inner bracts; rays absent, or 1–5 and very inconspicuous; disk flowers usually 5-lobed; achenes more or less angled, 3–7 mm long, tuberculate on the mid-vein, with usually 4 retrorsely barbed awns. September–October.

Wet ground.

IA, IL, IN, KY, MO, OH (FACW+), KS, NE (not listed by the U.S. Fish and Wildlife Service for KS and NE).

Swamp beggar's-tick; purple-stemmed beggar's-tick.

Plants of this species may have either sessile or short-petiolate leaves. Ray flowers are usually absent, but if they are present, they never exceed 5 mm in length.

The similar appearing *B. comosa* has 3-awned achenes and 4-lobed disk flowers.

Fig. 52. *Bidens connata* (Swamp beggar's-tick). Habit (center). Achene (lower right).

5. **Bidens coronata** (L.) Fisch. in Steud. Nom. ed. 2:202. 1840. Fig. 53.
Coreopsis coronata L. Sp. Pl. ed. 2, 1281. 1763.
Coreopsis trichosperma Michx. var. *tenuiloba* Gray, Syn. Fl. N. Am. 1:255. 1884.
Bidens coronata (L.) Benth. var. *tenuiloba* (Gray) Sherff, Bot. Gaz. 86:446. 1928.

Annual or biennial herb; stems branched, to 1.5 m tall, glabrous or nearly so; leaves pinnately (3-) 5- to 7- (9-) divided, to 15 cm long, the segments often linear to linear-lanceolate, the ultimate segments up to 5 mm wide, up to 20 mm wide, usually coarsely serrate, acute to acuminate at the apex, tapering to the base, the petioles up to 1.5 cm long; heads numerous, up to 4 cm across, the outer bracts 6–10, linear to linear-oblong, glabrous or with ciliate margins, rarely longer than the disk; rays 6–10, to 2.5 cm long; achenes flat, cuneate, 5–9 mm long, glabrous or puberulent, the pappus consisting of 2 short awns or scales. June–October.

Swamps, wet ground.

IA, IL, IN, KY, MO, OH (OBL), NE (not listed by the U.S. Fish and Wildlife Service for NE).

Tall swamp marigold.

This showy species usually has more leaflets and narrower leaflets than any of the other species of *Bidens* with compound leaves. Plants with achenes less than 6 mm long and larger leaflets less than 12 mm wide may be known as var. *tenuiloba* (Gray) Sherff. This latter is the more common variety in our area.

6. **Bidens discoidea** (Torr. & Gray) Britt. Bull. Torrey Club 20:281. 1893. Fig. 54.
Coreopsis discoidea Torr. & Gray, Fl. N. Am. 2:339. 1842.

Annual herbs; stems branched, to 2 m tall, glabrous; leaves trifoliolate, to 12 cm long, the segments lanceolate to oblong-lanceolate, acuminate, tapering to the base, dentate, glabrous, the petioles to 6 cm long; heads numerous, to 1.2 cm across, the outer bracts 3–5, linear-oblong, much longer than the disk, glabrous, rarely ciliate; rays absent; achenes flat, cuneate, 3–6 mm long, pubescent, the pappus consisting of 2 awns up to 2 mm long. August–October.

Swamps, often on submerged logs.

IA, IL, IN, MO (FACW–) KY, OH (FACW).

Swamp beggar's-tick; sticktight.

This species has the fewest outer bracts per head than any of our other species of *Bidens*.

7. **Bidens frondosa** L. Sp. Pl. 832. 1753. Fig. 55.
Bidens frondosa L. var. *anomala* Porter ex Fern. Rhodora 5:91. 1903.
Bidens frondosa L. f. *anomala* (Porter) Fern. Rhodora 40:352. 1938.

Annual herb; stems branched, often purplish, to 1.3 m tall, glabrous or nearly so; leaves pinnately 3- to 5-divided, to 15 cm long, the segments lanceolate to oblong-lanceolate, sharply serrate, acuminate at the apex, tapering to the base, glabrous or slightly pubescent on the lower surface, the petioles to 6 cm long; heads numerous, up to 1.5 cm across, the outer bracts 5–8 (–10), leafy, ciliate, longer than the disk; rays absent; achenes flat, cuneate, 5–10 mm long, usually short-hairy, the pappus consisting of 2 barbed awns. June–October.

Fig. 53. *Bidens coronata* (Tall swamp marigold). Habit (left). Leaf (right).
Achene (lower right).

Fig. 54. *Bidens discoidea* (Swamp beggar's-tick). Habit (left). Flowering head (right).

Marshes, swamps, wet prairies, disturbed moist soil.

IA, IL, IN, KS, KY, MO, NE, OH (FACW).

Common beggar's-tick; sticktight.

While most specimens have retrorsely barbed awns on the achene, a few may be found with antrorsely barbed awns. These latter have been called f. *anomala*.

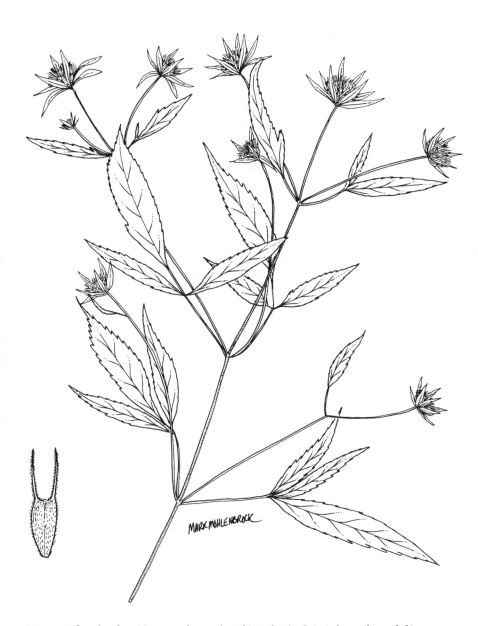

Fig. 55. *Bidens frondosa* (Common beggar's-tick). Habit (right). Achene (lower left).

8. **Bidens laevis** (L.) BSP. Prel. Cat. N. Y. 29. 1888. Fig. 56.
Helianthus laevis L. Sp. Pl. 906. 1753.

Annual herbs; stems branched, to 1 m tall, glabrous; leaves simple, lanceolate to oblong-lanceolate, acuminate at the apex, sessile or sometimes connate-perfoliate at the base, sharply serrate, glabrous, to 18 cm long, to 4.5 cm wide; heads relatively few, often nodding at maturity, up to 6 cm across, the outer bracts 8–10, obovate to oblong, obtuse to acute, glabrous; rays 1.5–3.0 cm long; achenes angled or flat, 6–12 mm long, with hispid margins, the pappus consisting of (2–) 4 retrorsely barbed awns. August–November.

Swamps, wet meadows.

IN, KY, MO, OH (OBL).

Showy bur marigold.

This species is similar to the more common *B. cernua* but is larger in all respects. It is more common on the coastal plain of the United States.

Fig. 56. *Bidens laevis* (Showy bur marigold). Habit (left).
Achene (lower right).

9. **Bidens polylepis** S. F. Blake, Proc. Acad. Sci. Wash. 35:78. 1922. Fig. 57.
Bidens polylepis S. F. Blake var. *retrorsa* Sherff, Bot. Gaz. 80:386. 1925.
Bidens aristosa (Michx.) Britt. var. *retrorsa* (Sherff) Wunderlin, Ann. Mo. Bot. Gard. 59: 472. 1973.

Annual or biennial herbs; stems much branched, up to 1.5 m tall, glabrous; leaves pinnately 3- to 5- (7-) divided, to 15 cm long, the segments lanceolate, serrate to pinnatifid, acuminate at the apex, tapering to the base, the petiole up to 3 cm long; heads several, to 4.5 cm across, the outer bracts 12 or more, ciliate, to 2 cm long; rays usually 8, to 3 cm long; achenes flat, 5–7 mm long, ciliate, usually with 2 barbed awns at the summit. August–October.

Marshes, swamps, wet ditches.

IA, IL, MO (FACW), KS, NE (FACW–).

Bur marigold; sticktight.

This is one of the showiest species of *Bidens* in the central Midwest because of its large yellow flowering heads. It differs from the very similar *B. aristosa* by its more numerous outer bracts of the flowering head.

Typical var. *polylepis* has antrorsely barbed awns on the achenes, while var. *retrorsa* has retrorsely barbed awns.

Fig. 57. *Bidens polylepis* (Bur marigold). Habit (left). Achene (lower right).

10. **Bidens tripartita** L. Sp. Pl. 832. 1753. Fig. 58.

Annual herbs; stems branched, to 1.5 m tall, glabrous; leaves all trifoliolate, acute to acuminate at the apex, tapering to the sessile or short-petiolate base, glabrous, to 12 cm long; heads several, to 3 cm across, the outer bracts 4–9, leafy, longer than the disk; rays absent; disk corollas 4-lobed, 3–4 mm long; achenes flat, cuneate, 6–9 mm long, retrorsely hairy, with 2 or 4 barbed awns at the summit. September–October.

Swamps, wet soil.

IA, IL, IN, KY, MO, OH (OBL). The U.S. Fish and Wildlife Service calls this plant *B. comosa*.

Swamp beggar's-tick.

The U.S. Fish and Wildlife Service considers *B. comosa* to be the same as *B. tripartita*. I prefer to maintain *B. comosa* as distinct because of its flatter and completely glabrous achenes, whereas *B. tripartita* has angled achenes that are pubescent.

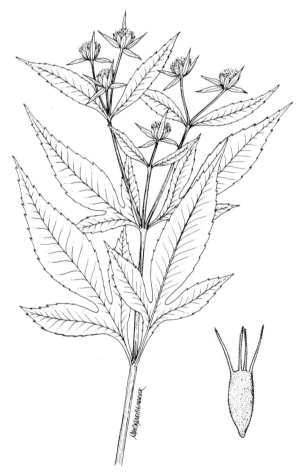

Fig. 58. *Bidens tripartita* (Swamp beggar's-tick). Habit (left). Achene (lower right).

11. **Bidens vulgata** Greene, Pittonia 4:72. 1899. Fig. 59.

Annual herbs; stems branched, to 3 m tall, glabrous to densely pubescent; leaves pinnately 3- to 5-divided, to 18 cm long, the segments lanceolate to oblonglanceolate, sharply serrate, acuminate at the apex, tapering to the base, glabrous or pubescent, the petioles to 6 cm long; heads numerous, up to 1.8 cm across, the outer bracts more than 10, linear to linear-oblong, pubescent at least near the tip; rays usually present but very short and inconspicuous; achenes flat, 7–12 mm long, hispid on the margins, the pappus consisting of 2 barbed awns. September–October.

Moist soil, often in disturbed areas.

IA, IL, IN, KS, KY, MO, NE, OH (the U.S. Fish and Wildlife Service does not list this species).

Tall beggar's-tick.

Although I have never seen this plant in standing water for any length of time, I am including it in this book because the other species of *Bidens*, which may be in standing water may be similar in appearance.

5. **Boltonia** L'Her.—False Aster

Perennial herbs; leaves alternate, simple, entire; flowers many in heads, the heads subtended by several nearly equal bracts; ray flowers white, pink, or bluish, pistillate; disk flowers usually yellow, perfect; achenes without pappus or with the pappus reduced to 2 very short awns or scales.

All five species that comprise this genus occur in the United States. The U.S. Fish and Wildlife Service recognizes two species in our area, although most botanists recognize three.

Fig. 59. *Bidens vulgata* (Tall beggar's-tick). Habit and leaves (center). Achene (lower left).
Ray flower (upper right).
Bract (lower right).

1. Leaves linear to oblanceolate, 5–20 mm wide; disk 6–10 mm wide.
    2. Leaves not decurrent .... 1. *B. asteroides*
    2. Leaves decurrent... 2. *B. decurrens*
1. Leaves linear, 1–5 mm wide; disk 3–6 mm wide .... 3. *B. diffusa*

1. **Boltonia asteroides** (L.) L'Her. Sert. Angl. 16. 1788. Fig. 60.
Matricaria asteroides L. Mant. 116. 1767.
Boltonia latisquama Gray, Am. Journ. Sci. II, 33:238. 1862.
Boltonia latisquama Gray var. *recognita* Fern. & Grisc. Rhodora 42:491–492. 1940.
Boltonia latisquama Gray var. *microcephala* Fern. & Grisc. Rhodora 42:493. 1940.
Boltonia asteroides (L.) L'Her. var. *recognita* (Fern. & Grisc.) Cronq. Bull Torrey Club 74. 149. 1947.

Perennial herbs, often with rhizomes; stems usually branched, to 2.5 m tall, glabrous; leaves alternate, simple, lanceolate to oblanceolate, acute at the apex, tapering to the sessile base, entire, glabrous, up to 15 cm long, up to 2 cm wide; heads several, to 1.5 cm across, usually arranged in corymbs; rays white or less commonly pink or bluish, 10–15 mm long; achenes winged, the pappus absent or consisting of 2 awns or scales up to 2 mm long. July–October.

Moist ground, marshes, prairies, sometimes in standing water.

IA, IL, IN, KS, KY, MO, NE, OH (FACW).

False aster.

Species of *Boltonia* resemble asters except that the achenes of asters have a pappus of capillary bristles.

Plants in the central Midwest fall into one of two varieties: var. *microcephala* has the bracts of the involucre up to 0.5 mm wide and rays 8–10 mm long; var. *recognita* has the bracts of the involucre 0.5–1.0 mm wide and rays 10–15 mm long. Typical var. *asteroides* apparently is confined to the coastal states.

2. **Boltonia decurrens** (Torr. & Gray) Wood, Am. Bot. Fl. 166. 1870. Fig. 61.
Boltonia glastifolia Michx. var. *decurrens* Torr. & Gray, Fl. N. Am. 21:188. 1842.
Boltonia asteroides (L.) L'Her. var. *decurrens* (Torr. & Gray) Engelm. Syn. Fl. N. Am. 1:166. 1884.
Boltonia latisquama Gray var. *decurrens* (Torr. & Gray) Fern. & Grisc. Rhodora 42:492. 1940.

Perennial herbs, usually with rhizomes; stems branched, to 2 m tall, glabrous; leaves alternate, simple, lanceolate to oblanceolate, acute at the apex, tapering to the decurrent base, entire, glabrous, up to 15 cm long, up to 2 cm broad; heads several, to 1.5 cm across; rays usually white, 10–15 mm long; achenes winged, usually with 2 short awns or scales at the summit. July–October.

Wet soil, occasionally in shallow water.

IL, MO (the U.S. Fish and Wildlife Service does not distinguish this species from *B. asteroides* in its designations).

Decurrent false aster.

This species apparently is confined to wet areas along the Illinois River and near the Mississippi River where the two rivers merge. It is readily distinguished by its decurrent leaf bases. It is considered Federally Threatened by the U.S. Fish and Wildlife Service.

Fig. 60. *Boltonia asteroides* (False aster). Habit (center). Achene (lower right).

Fig. 61. *Boltonia decurrens* (Decurrent false aster). Habit and leaves (center). Disk flower (lower left). Ray flower (lower right).

3. **Boltonia diffusa** Ell. Sketch Bot. S. Carol. 2:400. 1824. Fig. 62.
Boltonia diffusa Ell. var. *interior* Fern. & Grisc. Rhodora 42:490–491. 1940.

Perennial herbs; stems branched, glabrous, to 2 m tall; leaves alternate, simple, linear, acute at the apex, entire, glabrous, to 4 cm long, 1–5 mm broad, the lowest usually early deciduous; head solitary at the tips of long, subulate-bracted peduncles, up to 1.5 cm across; involucre 5–8 mm high, the bracts oblong, acute to obtuse, glabrous; rays 15–25, 5–8 mm long, white to pale purple; disk 3–6 mm wide; achenes glabrous, wing-margined, with awns absent or less than 1 mm long. July–September.

Moist or dry soil, occasionally in shallow depressions in wet woods.

IL, MO (FACW), KY (FAC).

Narrow-leaved false aster; doll's daisy.

Our plants belong to var. *interior*. Typical var. *diffusa* occurs along the coastal plain and is a more slender plant that is more branched than our plants.

6. **Brachyactis** Ledeb.

Annual herbs; leaves alternate, simple; heads few to many, subtended by well-developed herbaceous bracts; the ray flowers absent or reduced to vestiges, the disk flowers yellowish or whitish; achenes bearing pappus of capillary bristles.

When segregated from the genus *Aster, Brachyactis* consists of approximately five species. It differs from the genus *Aster* by the virtual absence of ray flowers.

1. **Brachyactis ciliata** (Ledeb.) Ledeb. Fl. Ross. 2:495. 1846. Fig. 63.
Erigeron ciliata Ledeb. Ic. Plant. pl. 24. 1829.
Aster brachyactis S. F. Blake, Contr. U.S. Natl. Herb. 25:564. 1925.

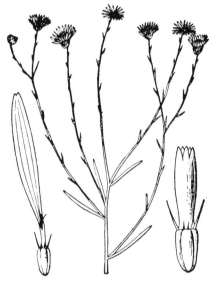

Fig. 62. *Boltonia diffusa* (Narrow-leaved false aster). Habit (center). Ray flower (lower left). Disk flower (lower right).

Annual herbs with a well-developed taproot; stems erect, sparsely to copiously branched, to 1 m tall, glabrous; leaves alternate, simple, linear to linear-lanceolate, to 10 cm long, up to 1 cm wide, acute at the apex, tapering to the base, entire, glabrous except sometimes with a ciliate margin, the lower leaves usually falling early; heads several to numerous in panicles or spikelike panicles; involucre 5–9 mm high, subtended by green bracts, the bracts in 2 series, linear, acute to acuminate at the apex, ciliate at the base, all more or less equal, or the outer somewhat longer than the inner; achenes 2–3 mm long, strigose, the pappus consisting of numerous capillary bristles. July–September.

Dry habitats but also edges of marshes.

Fig. 63. *Brachyactis ciliata* (Rayless aster). Ray flowers (lower left). Habit (center). Bract (upper right). Disk flower (lower right).

Fig. 64. *Cirsium muticum* (Swamp thistle). Flowering heads and leaf (center). Bract (lower left). Disk flower (lower right).

IA, IL, IN, MO (FAC), KS, NE, OH (NI).

Rayless aster.

This species is recognized by the absence or mere vestiges of ray flowers in the heads. Although *B. ciliata* is not generally considered to be a wetland species, I have observed it in shallow water at the edges of marshy areas in Iowa.

7. **Cirsium** Mill.—Thistle

Biennial or perennial, often coarse, herbs; leaves alternate, simple, toothed to pinnatifid, often prickly; heads several, subtended by numerous, often spine-tipped bracts; ray flowers absent; disk flowers tubular, usually perfect, often purplish; achenes often curved, nerveless, the pappus of plumose bristles.

This genus consists of approximately two hundred speices, most of them in the Northern Hemisphere. Only the following may sometimes occur in standing water in the central Midwest.

1. **Cirsium muticum** Michx. Fl. Bor. Am. 2:89. 1803. Fig. 64.

Biennial herbs with stout roots; stems branched, to 3 m tall, villous to arachnoid when young, becoming glabrate; leaves alternate, deeply pinnatifid, to 50 cm long, the lobes lanceolate to oblong, white-tomentose to glabrate on the lower surface, somewhat spiny-tipped; heads several, to 3.5 cm across, the bracts viscid and usually tomentose, not spiny; disk flowers purple or pinkish; achenes 4.5–5.5 mm long, the pappus consisting of capillary bristles. August–October.

Swamps, bogs, fens, wet meadows, wet woods.

IA, IL, IN, KY, MO, OH (OBL).

Swamp thistle; fen thistle.

This species is readily distinguished from all other thistles by its viscid flower heads. The tips of the leaf lobes are usually a little less spiny than those of most other thistles.

8. Conoclinium DC.

Perennial herbs from slender rhizomes; leaves opposite, simple, toothed; flowers in heads arranged in corymbs, the heads subtended by linear bracts and bearing only blue disk flowers; ovary inferior; receptacle conical; fruit an achene, the pappus of numerous capillary bristles.

Two species comprise this genus, which is often placed in the genus *Eupatorium*. *Conoclinium* differs from *Eupatorium* by its conical receptacle and blue flowers.

1. **Conoclinium coelestinum** (L.) DC. Prodr. 5:135. 1836. Fig. 65.
Eupatorium coelestinum L. Sp. Pl. 838. 1753.

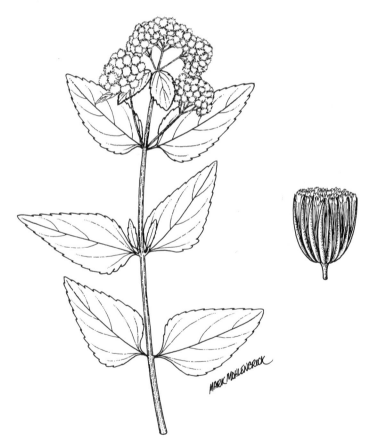

Fig. 65. *Conoclinium coelestinum* (Mist-flower). Habit (left). Flowering head (right).

Perennial herbs from slender rhizomes; stems erect, branched or unbranched, pubescent, to 1 m tall; leaves opposite, simple, deltoid to ovate, acute or subacute at the apex, truncate at the base, crenate, rugose, appressed-pubescent to nearly glabrous, to 10 cm long, to 3 cm wide, triple-nerved, on petioles to 3 cm long; heads in crowded corymbs, 3.5–5.0 mm high, subtended by an imbricate series of linear, acuminate bracts; receptacle conical; ray flowers absent; disk flowers blue; achenes with a pappus of numerous capillary bristles. July–October.

Swampy woods, wet meadows, damp fields.

IL, IN, MO (FAC+), KS, NE (FACW–), KY, OH (FAC).

Mist-flower.

This species is readily distinguished by its blue flower heads and rugose, deltoid, opposite leaves.

9. **Doellingeria** Nees—Flat-topped Aster

Perennial herbs; leaves alternate, simple, the lowermost often reduced to scales or sheaths; inflorescence corymbose, more or less flat-topped, consisting of numerous heads; heads hemispheric, subtended by several series of appressed bracts; ray flowers white, pistillate; disk flowers white, tubular, 5-lobed, perfect; stamens 5; ovary inferior; pappus double, the outer series consisting of scales or short bristles, the inner series consisting of long capillary bristles; fruit an achene.

Doellingeria differs from *Aster* by the flat-topped inflorescence and its double series of pappus. The genus consists of five species, all in eastern North America. Only the following may occur in shallow water in the central Midwest.

1. Stems, leaves, and bracts puberulent..1. *D. pubentior*
1. Stems, leaves, and bracts glabrous or nearly so ....2. *D. umbellata*

1. **Doellingeria pubentior** Cronq. Bull. Torrey Club 74:147. 1947. Fig. 66.
Aster umbellatus Mill. var. *pubens* Gray, Syn. Fl. N. Am. 1:197. 1884.
Doellingeria umbellata (Mill.) Nees var. *pubens* (Gray) Britt. Ill. Fl.N. U.S. 3:392. 1898.

Perennial herbs; stems branched, puberulent throughout, up to 1.5 m tall; leaves alternate, simple, lanceolate to oblong-lanceolate to elliptic, acuminate at the apex, tapering to the base, sessile or short-petiolate, entire, densely puberulent, at least on the lower surface, finely reticulate-veined, forming a submarginal nerve, to 15 cm long, to 3.5 cm wide; heads numerous, turbinate, 1.0–1.8 cm across, in a flat-topped corymb, each head 3.5–4.0 mm high, the bracts lanceolate, acute, puberulent, 1.0–1.2 mm wide; ray flowers 4–7 in number, 5–8 mm long, white; disk flowers 8–15 in number, white; achenes puberulent, the pappus double, the inner firm and broader than the outer ones. July–October.

Fens, wet meadows, wet woods.

IA, IL, MO, NE (the U.S. Fish and Wildlife Sevice does not distinguish this species from *D. umbellata*).

Northwestern flat-topped aster.

This species, often considered a variation of *D. umbellata*, differs by the general presence of pubescence on the stems, leaves, and involucral bracts and on the fewer flowers per head.

Fig. 66. *Doellengeria pubentior* (Northwestern flat-topped aster). Habit (center).
Leaf (lower left). Flower head (right).

2. **Doellingeria umbellata** (Mill.) Nees, Gen. & Sp. Ast. 178. 1832. Fig 67.
Aster umbellatus Mill. Gard. Dict. ed. 8, no. 22. 1768.

Perennial herbs; stems branched, at least above, glabrous or less commonly pubescent, up to 2 m tall; leaves alternate, simple, lanceolate to oblong-lanceolate to elliptic, acuminate at the apex, tapering to the base, sessile or short-petiolate, entire, glabrous above, usually pubescent on the veins beneath, finely reticulate-veined, to 15 cm long, to 3.5 cm wide; heads numerous, 1.2–2.0 cm across, in a flat-topped corymb; involucre 3–5 mm high, the bracts lanceolate, acute to obtuse, glabrous or ciliolate, less than 1 mm wide; rays 6–15, 5–8 mm long, white; achenes puberulent. July–October.

Bogs, fens, wet ground.

IA, IL, IN, KY, MO, OH (FACW), NE (OBL).

Flat-topped aster.

This handsome species is distinguished by its flat-topped inflorescence, white flowering heads, and entire leaves with a conspicuous reticulate vein pattern and a submarginal nerve. It differs from *D. pubentior* by its glabrous involucral bracts.

Fig. 67. *Doellengeria umbellata* (Flat-topped aster). Habit (center). Ray flower (next to lower right). Disk flower (lower right).

10. **Eclipta** L.

Annual herbs from fibrous roots; leaves simple, opposite; flowers in heads, with both ray and disk flowers, the ray flowers pistillate, white, the disk flowers perfect, short-tubular, 4- to 5-toothed; heads subtended by 1–2 series of bracts; receptacle flat; achenes angular, the pappus absent or of 2 short awns or a short erose crown.

Four species comprise this genus, all of them in the tropics.

1. **Eclipta prostrata** (L.) L. Mant. Pl. 2:286. 1771. Fig. 68.
Verbesina prostrata L. Sp. Pl. 2:902. 1753.
Verbesina alba L. Sp. Pl. 2:902. 1753.
Eclipta alba (L.) Hassk. Pl. Jav. Rar. 528. 1848.

Annual herbs from fibrous roots; stems prostrate to ascending, branched or unbranched, to 60 cm long, mauve, with appressed short white hairs; leaves opposite, simple, lanceolate to elliptic, acute at the apex, tapering to the base, entire or obscurely serrulate, usually with grayish pubescence, to 10 cm long, to 3 cm wide, sessile or nearly so; heads borne in the axils of the leaves, on peduncles to 4 cm long, bearing both ray and disk flowers, subtended by up to 12 bracts in 2 series, the bracts up to 4 mm long; ray flowers white, linear, pistillate, 1–2 mm long; disk flowers white, perfect, short-tubular, 4- to 5-lobed; receptacle flat; achenes numerous in a flattened cluster, angular, tuberculate, puberulent at the apex, 1.5–2.0 mm long, the pappus merely an erose crown. July–October.

Disturbed wet areas.

IA, IL, IN, KS, MO, NE (FACW), KY, OH (FAC).

Yerba de tajo.

This species is recognized by its prostrate mauve-colored stems with white appressed hairs, its small white flowering heads, and its flat cluster of crowded achenes.

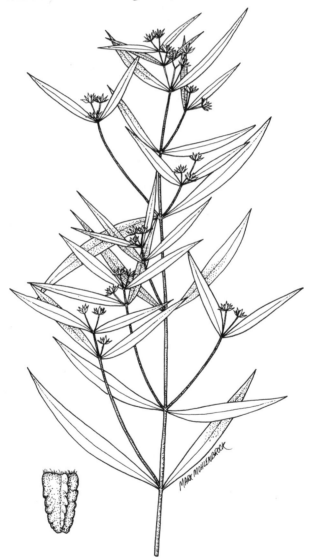

Fig. 68. *Eclipta prostrata* (Yerba de tajo). Habit. Achene (lower left).

11. **Eupatoriadelphus** R. M. King & H. Robins.—Joe-pye-weed

Stout perennial herbs; leaves whorled; inflorescence cymose to paniculate, consisting of numerous heads; heads discoid, the involucral bracts in several series; receptacle flat; disk flowers tubular, 5-lobed, pink or purple; stamens 5; ovary inferior; pappus of numerous capillary bristles; fruit an angular achene.

About six species comprise this genus. Most botanists in the past have included the species of *Eupatoriadelphus* in the genus *Eupatorium*.

1. Stems suffused with purple, glaucous; disk flowers 5–8 per head; heads in a rounded inflorescence .... 1. *E. fistulosus*
1. Stems purple-spotted, not glaucous; disk flowers 8–22 per hed; heads in a flat-topped inflorescence .... 2. *E. maculatus*

1. **Eupatoriadelphus fistulosus** (Barratt) R.M. King & H. Robins. Phytologia 19: 432. 1970. Fig. 69.

Eupatorium fistulosum Barratt, Eup. Vert. 1. 1841.

Perennial herbs; stems to 3 m tall, branched or unbranched, glaucous, usually suffused with purple but not purple-spotted, glabrous, hollow; leaves in whorls of 4–7, lanceolate, acute at the apex, tapering to the base, crenate, pubescent or glabrous on both surfaces, up to 25 cm long, up to 12 cm wide; petioles up to 1 cm long; heads numerous, in round-topped corymbs, the corymbs 20–30 cm across, the heads 6–9 mm high, subtended by imbricate bracts, the bracts obtuse to subacute glabrous or short-hairy; ray flowers absent; disk flowers pinkish purple, 5–8 per head; achenes cuneate, 3.0–4.5 mm long, the pappus consisting of numerous capillary bristles. July–September.

Swampy woods, wet meadows, mesic woods.

IL, IN, MO (OBL), KY, NE (FACW).

Hollow Joe-pye-weed; trumpetweed.

This species is distinguished by its hollow, glaucous stems, its leaves with crenate teeth, and some of the leaves often more than 5 per whorl.

Because of the hollow stems, early pioneers would make holes in the stems and use the stems as musical instruments, hence the common name of trumpetweed.

2. **Eupatoriadelphus maculatus** L. Amoen. Acad. 4:288. 1755. Fig. 70.

Eupatorium maculatum L. Amoen. 4:288. 1755.

Eupatorium bruneri Gray, Syn. Fl. 1 (2):96. 1884.

Perennial herbs; stems to 2 m tall, usually purple-spotted, occasionally entirely purple, more or less pubescent; leaves in whorls of 4 or 5, lanceolate to ovate, thin or rather thick, acute to acuminate at the apex, tapering to the base, coarsely dentate, glabrous to sparsely hairy, less commonly densely hairy, to 20 cm long, to 8 cm wide; petioles up to 1 cm long; heads numerous, in flat-topped cymes; involucre 6–9 mm high, the bracts imbricate, obtuse, glabrous or short-hairy, often purplish; ray flowers absent; disk flowers purple to rose-purple, 8–22 per head; achenes cuneate, 3–5 mm long, the pappus consisting of capillary bristles. June–October.

Marshes, fens.

IA, IL, IN, MO (OBL), KY, OH (FACW). Not listed for KS and NE by the U.S. Fish and Wildlife Service.

Fig. 69. *Eupatoriadelphus fistulosus* (Hollow Joe-pye-weed). Habit. Disk flower (upper left).

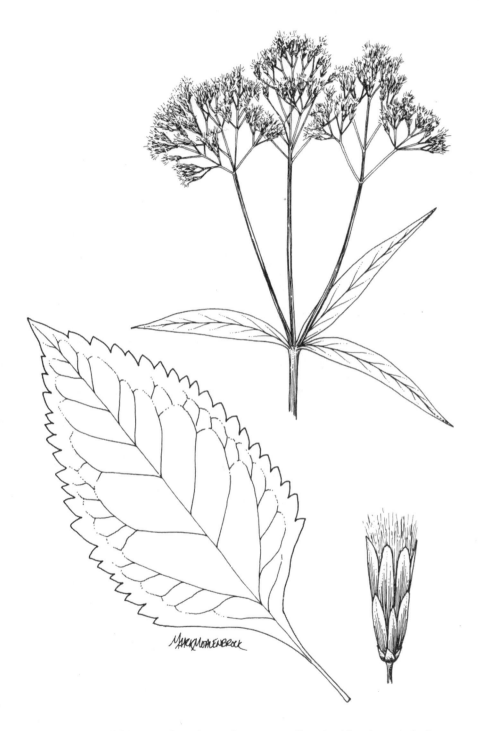

Fig. 70. *Eupatoriadelphus maculatus* (Spotted Joe-pye-weed). Habit (above). Lower leaf (lower left). Disk flower (lower right).

Spotted Joe-pye-weed.

This species is sometimes known as *Eupatorium maculatum*. Plants from Iowa and Nebraska tend to have narrower, thicker, hairier leaves and are sometimes segregated as *Eupatorium bruneri* Gray.

Eupatoriadelphus maculatus differs from the other species of the genus by having 8–22 flowers per head, a flat-topped inflorescence, and a stem that is not glaucous.

12. **Eupatorium** L.—Thoroughwort

Perennial herbs; leaves simple, opposite or whorled; heads numerous, subtended by numerous bracts in several series; receptacle flat; ray flowers absent; disk flowers tubular, perfect, white, pink, blue or purple; achenes angular, crowned by a pappus of capillary bristles.

This mostly New World genus consists of nearly one thousand species. Only the following sometimes occur in shallow water in the central Midwest.

1. Leaves connate-perfoliate; plants densely white-villous.................................. 2. *E. perforatum*
1. Leaves petiolate; plants not densely white-villous.
    2. Some of the leaves whorled; leaves entire, linear to linear-lanceolate.... 1. *E. hyssopifolium*
    2. Leaves opposite, toothed, ovate... 3. *E. semiserratum*

1. **Eupatorium hyssopifolium** L. var. **laciniatum** Gray, Syn. Fl. N. Am. 1:98. 1884. Fig. 71.
Eupatorium torreyanum Short & R. Peter, Trans. Journ. Med. Assoc. Sci. 8:575. 1836.

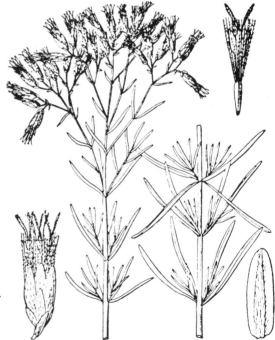

Fig. 71. *Eupatorium hyssopifolium* (Hyssop-leaved thoroughwort). Habit (next to left). Leaves (next to right). Flowering head (lower left). Disk flower (upper right). Bract (lower right).

Perennial herbs; stems erect, robust, to 1.5 m tall, strigose to puberulent; leaves simple, opposite or in whorls of 3 or 4, the uppermost leaves alternate, linear to linear-lanceolate, acute to acuminate at the apex, tapering to the sessile or subpetiolate base, sharply serrate to shallowly lobed, glabrous above, usually pubescent on the veins beneath, glandular-punctate, to 10 cm long, to 1.5 cm wide, often with fascicles of reduced leaves in the axils inflorescence a much branched corymb composed of numerous heads; heads 4–6 mm high, subtended by 2 series of bracts, the bracts obtuse to subacute, pubescent, with scarious margins; each head with 5 white disk flowers; achenes 2–3 mm long, the pappus of numerous capillary bristles. August–September.

Wet barrens, dry woods, fields.

KY (not listed for any region by the U.S. Fish and Wildlife Service, indicating that it is considered UPL in all areas).

Hyssop-leaved thoroughwort.

Although this plant is most often found in dry habitats, it does occur in wet barrens in Kentucky. The leaves range from opposite to whorls of 3 or 4, with the uppermost leaves often alternate.

Typical var. *hyssopifolium*, which does not occur in wet barrens in our area, has entire or fewer teeth and shorter stature.

2. **Eupatorium perfoliatum** L. Sp. Pl. 838. 1753. Fig. 72.
Eupatorium cuneatum DC. Prodr. 5:149. 1836.

Stout perennial herbs; stems erect, branched, to 1.5 m tall, densely white-villous or rarely short-hirtellous; leaves simple, opposite, less commonly 3 in a whorl, connate-perfoliate, rarely merely sessile, rugose, lanceolate, acuminate at the apex, serrate to crenate, densely white-villous or rarely short-hirtellous, to 20 cm long, to 4.5 cm wide; inflorescence a flat-topped corymb of numerous crowded heads; heads 10- to 40-flowered, white to gray-white, 3–5 mm high, subtended by 2–3 series of lanceolate, pubescent bracts with scarious tips; pappus of numerous capillary bristles; fruit an achene. July–October.

Wet woods, wet prairies, marshes, fens.

IA, IL, IN, KS, KY, MO, NE, OH (FACW+).

Perfoliate boneset; perfoliate thoroughwort.

This species is easily recognized by its rugose, connate-perfoliate leaves and dense white-villous pubescence. Whorled leaves are occasionally encountered. Plants from southern Indiana, southern Illinois, and southern Missouri, with short-hirtellous pubescence and non-connate leaves have been called var. *cuneatum*. They may represent a distinct species.

3. **Eupatorium semiserratum** DC. Prodr. 5:177. 1836. Fig. 73.
Euopatorium cuneifolium var. *semiserratum* (DC.) Fern. & Grisc. Rhodora 37:179. 1935

Perennial herbs from short rhizomes; stems erect, branched, to 1.2 m tall, densely pubescent; leaves simple, opposite, elliptic to lanceolate, acute at the apex, tapering to the base, sharply serrate, strongly 3-nerved from the base, to 8 cm long,

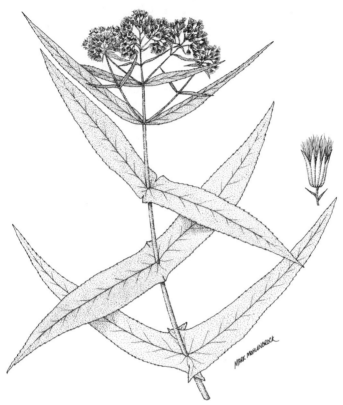

Fig. 72. *Eupatorium perfoliatum* (Perfoliate boneset). Habit.
Flowering head (right).

Fig. 73. *Eupatorium semiserratum*
(Sharp-toothed thoroughwort).
Habit. Disk flower (center right).
Leaf (lower right).

to 3 cm wide, densely pubescent on both surfaces, twisted at base, short-petiolate; heads in round-topped corymbs, subtended by 2 series of pubescent, scarious, obtuse bracts, the outer series of bracts much shorter than the inner series; heads 3–4 mm high; receptacle flat; ray flowers absent; disk flowers 5, white; achenes 2–3 mm long, the pappus of numerous capillary bristles. August–September.

Swampy woods.

KY, MO (FACW).

Sharp-toothed thoroughwort.

This species differs from other species of *Eupatorium* by its tapered leaf bases, non-whorled leaves that are strongly triple-nerved, and its relatively short involucres.

13. **Euthamia** Nutt.

Perennial herbs from rhizomes; leaves alternate, simple, narrow, entire, punctate, sessile or nearly so; heads numerous in small clusters forming a flat-topped corymb, each head subtended by several series of chartaceous bracts and bearing ray and disk flowers; ray flowers yellow, pistillate, fertile; disk flowers yellow, perfect, fertile; achenes pubescent with a pappus of numerous capillary bristles.

This genus, at one time merged with *Solidago*, consists of about eight North American species. It differs from *Solidago* by its flat-topped inflorescencese and punctate leaves.

1. Leaves 3- or 5-veined; heads with 20 or more flowers ....1. *E. graminifolia*
1. Leaves 1-veined, or obscurely 3-veined; heads with up to 20 flowers ....2. *E. leptocephala*

1. **Euthamia graminifolia** (L.) Nutt. Gen. 2:162. 1818. Fig. 74.
Chrysocoma graminifolia L. Sp. Pl. 841. 1753.
Solidago graminifolia (L.) Salisb. Prodr. 109. 1796.
Euthamia nuttallii Greene, Pittonia 5:75. 1902.
Solidago graminifolia (L.) Salisb. var. *nuttallii* (Greene) Fern. Rhodora 10:92. 1908.
Euthamia graminifolia (L.) Nutt. var. *nuttallii* (Greene) Sieren, Rhodora 83:564. 1981.

Perennial herbs from rhizomes; stems erect, branched above, to 1.5 m tall, glabrous to pubescent; leaves alternate, simple, numerous, linear-lanceolate, acute to acuminate at the apex, tapering to a sessile or subsessile base, entire, punctate, 3- or 5-veined, glabrous or hirtellous on both surfaces, to 12 cm long, to 1 cm wide; heads numerous in small clusters, arranged in a flat-topped corymb, each head 3–5 mm high, subtended by several series of bracts, with up to 30 very short yellow ray flowers and up to 10 yellow disk flowers; achenes hispidulous, 2–3 mm long, with a pappus of capillary bristles. July–October.

Wet soil, fields, peaty marshlands, fens, interdunal areas around Lake Michigan.

IA, IL, IN, MO (FACW–), KY, OH (FACW).

Grass-leaved goldenrod.

Euthamia graminifolia is the only *Euthamia* in our area with 3- or 5-veined leaves. Plants that are glabrous or nearly so are var. *graminifolia*; plants with hirtellous leaves may be called var. *nuttallii*.

Fig. 74. *Euthamia graminifolia* (Grass-leaved goldenrod). Habit (left). Flowering head (right).

2. **Euthamia leptophylla** (Torr. & Gray) Greene, Mem. Torrey Club 5:321. 1894. Fig. 75.

Solidago leptophylla Torr. & Gray, Fl. N. Am. 2:226. 1841.

Perennial herbs from rhizomes; stems erect, branched above, to 1 m tall, glabrous; leaves alternate, simple, linear-lanceolate, acute to acuminate at the apex, tapering to the sessile or subsessile base, entire, punctate, 1-veined or obscurely 3-veined, glabrous or nearly so, to 8 cm long, to 6 mm wide; heads numerous in small clusters, arranged in a flat-topped corymb, each head 4.5–6.0 mm high, subtended by 4 series of stramineous bracts, the outer bracts broader than the inner bracts, with up to 20 short yellow rays flowers and 4–6 yellow disk flowers; achenes hispidulous, 2–3 mm long, with a pappus of numerous capillary bristles. August–October.

Wet meadows, swampy woods.

IL, MO (FACW), KY (NI). Mississippi Valley grass-leaved goldenrod.

This species has fewer leaf veins, fewer flowers per head, and taller involucres than *E. graminifolia*.

14. Fleischmannia Sch.-Bip.

Perennial herbs; leaves opposite, simple, petiolate; heads several in corymbs, subtended by short and long bracts; receptacle flat; ray flowers absent; disk flowers tubular, perfect, pink; achenes with a pappus of numerous capillary bristles.

This genus is often merged with *Eupatorium* but differs by having flowers that are not white.

Approximately one hundred species comprise this genus that is often merged into *Eupatorium*.

Fig. 75. *Euthamia leptophylla* (Mississippi Valley grass-leaved goldenrod). Habit (left). Ray flower (bottom center). Disk flower (lower right).

1. **Fleischmannia incarnata** (Walt.) R.M. King & H. Rob. Phytologia 19:203. 1970. Fig. 76.

Eupatorium incarnatum Walt. Fl. Carol. 200. 1788.

Perennial herbs; stems ascending to erect, to 1.5 m tall, much branched, puberulent throughout; leaves simple, opposite, ovate to deltoid, acute to acuminate at the apex, cordate or truncate at the base, crenate-serrate, to 8 cm long, to 5.5 cm wide, puberulent on both surfaces, on petioles up to 4 cm long; heads in small terminal clusters, each head 3–5 mm high, subtended by several linear, acute, bicostate, pubescent bracts in 1–2 series, bearing about 20 pink disk flowers; ray flowers absent; achenes angular, 2–3 mm long, several-ribbed, glabrous or nearly so, the pappus of numerous capillary bristles. August–October.

Swampy woods, wet ditches, mesic woods.

IL, IN, KY, MO, OH (FAC).

Pink thoroughwort.

This species is often placed in *Eupatorium* but differs by its 1–2 series of involucral bracts and its pink flowers.

One of the habitats for this species is swampy woods, although it also occurs in mesic woods.

15. Hasteola Raf.—Sweet Indian Plantain

Perennial herbs with thickened roots; leaves alternate, often hastate; heads several, forming a flat-topped inflorescence, each head subtended by 2 series of bracts and consisting only of many white disk flowers; receptacle flat; achenes cylindric, the pappus of numerous capillary bristles. There are about fifty species in this genus.

1. **Hasteola suaveolens** (L.) Pojark. Bot. Mater. Gerb. Bot. Inst. Kom. Acad. Nauk SSSR 20:381. 1960. Fig. 77.
Cacalia suaveolens L. Sp. Pl. 2:835. 1753.
Synosma suaveolens (L.) Raf. in Loud. Gard. Mag. 8:247. 1832.

Perennial herbs from thickened roots; stems erect, sometimes branched, glaucous, grooved, up to 1.5 m tall; leaves alternate, simple, deltoid, usually hastate,

Fig. 76. *Fleischmannia incarnata* (Pink thoroughwort). Habit. Disk flower (upper left).

Fig. 77. *Hasteola suaveolens* (Sweet-scented Indian plantain). Habit and leaf. Disk flower (lower right).

acute to acuminate at the apex, more or less truncate at the base, coarsely toothed, glabrous, to 25 cm long, to 15 cm wide, the lower on winged petioles, the uppermost sessile or nearly so; heads several to numerous in large, flat-topped corymbs, each head 6–10 mm high, subtended by 2 series of linear, acute bracts, and bearing only disk flowers; disk flowers white, 20–30 in number; achenes ellipsoid, glabrous, with a pappus of numerous capillary bristles. July–September.

Swampy woods, along streams. IA, IL, IN, MO (OBL), KY, OH (FAC–).

Sweet-scented Indian plantain. This species is readily recognized by its hastate, toothed leaves, its glaucous stems, and its large flat-topped corymbs of white flower heads.

16. **Helenium** L.—Sneezeweed

Annual or perennial herbs; leaves alternate, simple, often decurrent, punctate; heads on long peduncles, subtended by reflexed or spreading, linear or subulate bracts in 1–2 series; receptacle convex; ray flowers pistillate and fertile, or sometimes neutral, with several yellow, usually 3-lobed, rays; disk flowers perfect, tubular, 4- to 5-lobed; stamens 5; ovary inferior; pappus of 5–8 aristate scales; achenes ribbed.

Helenium is a genus of about forty species native to North and Central America. Three species occur in the central Midwest, but only the following may be found sometimes in shallow water.

1. Disk yellow; leaves up to 5 cm wide ....1. *H. autumnale*
1. Disk purple; leaves up to 2 cm wide .... 2. *H. flexuosum*

1. **Helenium autumnale** L. Sp. Pl. 886. 1753. Fig. 78.

Helenium canaliculatum Lam. Journ. Hist. Nat. 2:213. 1792.
Helenium parviflorum Nutt. Trans. Am. Phil. Soc. n.s. 7:384. 1841.
Helenium autumnale L. var. *canaliculatum* (Lam.) Torr. & Gray, Fl. N. Am. 2:384. 1842.
Helenium autumnale L. var. *parviflorum* (Nutt.) Fern.Rhodora 45:492. 1943.

Perennial herbs; stems erect, branched or unbranched, to 1.5 m tall, usually pubescent, winged by the decurrent leaf bases; leaves alternate, simple, linear to elliptic to lanceolate, acute top acuminate at the apex, tapering to the decurrent base, firm to membranaceous, entire to coarsely toothed, to 15 cm long, to 5 cm broad, glabrous or pubescent; heads 1- many on long peduncles; rays yellow, reflexed, to 2.5 cm long, to 12 mm wide; disk globose, yellow, 8–25 mm across; achenes pubescent, ribbed, with aristate scales at the summit. July–October.

Swamps, wet meadows, damp thickets.

IA, IL, IN, KY, MO, OH (FACW+), KS, NE (FACW–).

Yellow sneezeweed.

This species is readily recognized by its winged stems due to the decurrent leaf bases and by its yellow, globose disk. Considerable variation occurs in the leaves. Those plants with leaves firm and up to seven times longer than wide, and with rays more than 13 mm long, are var. *autumnale*. Plants with membranaceous leaves at least seven times longer than wide and with rays less than 13 mm long may be known as var. *parviflorum*.

Var. *canaliculatum* has firm leaves more than 7 times longer than broad.

Fig. 78. *Helenium autumnale* (Yellow sneezeweed). Habit. Disk flower (right).

2. Helenium flexuosum Raf. New Fl. 4:81. 1838. Fig.79.
Helenium nudiflorum Nutt. Trans. Am. Phil. Soc. n.s. 7:384–385. 1841.

Perennial herbs; stems erect, branched, to 1 m tall, pubescent, winged by the decurrent leaf bases; leaves alternate, simple, linear to lanceolate, acute to sub-acute at the apex, tapering to the decurrent base, firm, entire or more often coarsely

Fig. 79. *Helenium flexuosum* (Purple-headed sneezeweed). Habit.
Disk flower (lower right).

toothed, to 10 cm long, to 2 cm wide, pubescent on both surfaces, scabrous above; heads few–several on pubescent peduncles, forming a corymb, each head subtended by 2–3 series of linear, pubescent bracts, the bracts often reflexed; heads bearing both ray and disk flowers; ray flowers 10–15 in number, yellow with a purple base, drooping, 3-lobed at the apex; disk globose, purple, up to 1 cm in diameter, many-flowered; achenes oblanceolate, hispidulous, 2.0–3.5 mm long, bearing a pappus of ovate, aristate scales. June–October.

Wet meadows, along streams.

IL, IN, MO (FAC+), KS (FACW), KY, OH (FAC–).

Purple-headed sneezeweed.

The purple globose disk in the center of each flower head distinguishes this species from *H. autumnale*, which has a yellow disk.

17. Helianthus L.—Sunflower

Stout perennials or sometimes annuals; leaves simple, undivided, opposite or the uppermost sometimes alternate; inflorescence of 1–many heads; heads usually more than 2 cm across, subtended by several series of narrow, green bracts, bearing both ray flowers and disk flowers; receptacle flat; ray flowers yellow, neutral; disk flowers yellow or purple, perfect, with 5-lobed corollas; stamens 5; ovary inferior; pappus of 2–4 scales or short awns, soon deciduous; fruit a compressed or 4-angled achene.

Helianthus consists of about fifty North American species. Only the following may occasionally be found in shallow water in the central Midwest.

1. Leaves linear, slightly scabrous, 1-nerved; disk purple .... 1. *H. angustifolius*
1. Leaves lanceolate to ovate, harshly scabrous, 3-nerved; disk yellow.
  2. Leaves ovate to lance-ovate; rhizomes tuber-bearing .... 4. *H. tuberosus*
  2. Leaves lanceolate; rhizomes not tuber-bearing.
    3. Stems glabrous or nearly so, glaucous .... 3. *H. grosseserratus*
    3. Stems densely pubescent, not glaucous .... 2. *H. giganteus*

1. Helianthus angustifolius L. Sp. Pl. 906. 1753. Fig. 80.

Stout perennials from slender rootstocks; stems erect, usually branched, scabrous, sometimes hirsute, to 2 m tall; leaves alternate above, opposite below, firm, linear, acute to acuminate at the apex, to 15 cm long, to 15 mm wide, entire, 1-nerved, usually slightly scabrous above, sessile; heads 1-few, to 6 cm across, subtended by several linear-lanceolate, acute to acuminate, pubescent bracts; rays 12–20, yellow, to 2.5 cm long; disk purple; achenes truncate, 3.5–4.0 mm long, glabrous, bearing 2 short awns at the summit. August–October.

Swampy woods.

IL, IN, MO (FACW–), KY, OH (FACW).

Narrow-leaved sunflower; swamp sunflower.

This readily recognizable sunflower is distinguished by its linear leaves and purple disks of the flowering heads.

Fig. 80. *Helianthus angustifolius* (Narrow-leaved sunflower). Habit (left). Leaf (right).

2. Helianthus giganteus L. Sp. Pl. 905. 1753. Fig. 81.

Stout perennial herbs from short rhizomes; stems erect, sometimes branched, to 3 m tall, scabrous and often hirtellous; upper leaves alternate, lower leaves opposite, lanceolate, acuminate at the apex, tapering to the sessile or short-petiolate base, 3- nerved, harshly scabrous, entire or sparingly toothed, to 15 cm long, to 3 cm broad; heads 1-several, subtended by several spreading or recurved, linear bracts; rays 10–20, yellow, to 2 cm long; disk yellow, up to 2 cm across; achenes truncate, with 2 subulate awns at summit. July–October.

Swampy woods, fens, wet soil.

IA, IL, IN, KY, OH (FACW).

Tall sunflower; late sunflower.

This is often the most common sunflower in wetlands in the central Midwest. It is similar to *H. grosseserratus*, which has glabrous or nearly so stems.

Fig. 81. *Helianthus giganteus* (Tall sunflower). Habit (center). Leaves (far left). Disk flower (next to far left). Bract (third from left). Achene (next to right). Ray flower (lower right).

3. Helianthus grosseserratus Martens, Sel. Sem. Hort. Loven. 1839. Fig. 82.

Robust perennials from thickened rhizomes and fleshy roots; stems erect, branched or unbranched, to 4.5 m tall, glabrous or nearly so, but usually somewhat pubescent in the inflorescence, glaucous; leaves simple, the lower often opposite, the upper often alternate, lanceolate, acute to acuminate at the apex, tapering to the base, to 20 cm long, to 8 cm wide, usually coarsely toothed, scabrous and strigose on the upper surface, paler and densely pubescent on the lower surface, the petioles up to 4 cm long; heads few to several, up to 6 cm across, subtended by several series of linear-subulate, loosely spreading, hirsute bracts, bearing both ray and disk flowers; ray flowers up to 20, yellow, 2.5–4.5 cm long; disk flowers yellowish; achenes glabrous or nearly so, the pappus consisting of 2 lanceolate awns. July–October.

Prairies, wet meadows, wet ditches.

IA, IL, IN, MO (FACW–), KS, NE (FAC), KY, OH (FACW).

Sawtooth sunflower.

This robust, attractive sunflower resembles *H. giganteus* but differs primarily by its glabrous and glaucous stems. Despite the specific epithet *grosseserratus*, some plants have leaves with few teeth.

4. Helianthus tuberosus L. Sp. Pl. 2:905. 1753. Fig. 83.
Helianthus tuberosus L. var. *subcanescens* Gray, Syn. Fl. N. Am. 1:280. 1884.
Helianthus subcanescens (Gray) E. Wats. Pap. Mich. Acad. Sci. 9:430. 1929.

Perennial herbs from rhizomes and swollen tubers; stems stout, erect, usually branched, to 3 m tall, pubescent, scabrous; leaves simple, the lower opposite, the upper alternate, or all opposite, lance-ovate to ovate, acute to acuminate at the apex, tapering to the base, to 25 cm long, to 12 cm wide, serrate, harshly scabrous above, pusbescent beneath, strongly triple-nerved with a prominent white midvein, the petioles winged, up to 8 cm long; heads few to several , up to 8 cm across, subtended by several series of lanceolate, acuminate, pubescent bracts, bearing both ray and disk flowers; ray flowers up to 20, yellow, 2.5–4.0 cm long; disk flowers yellowish, forming a disk up to 15 mm across; achenes pubescent, the pappus consisting of minute teeth. August–October.

Fig. 82. *Helianthus grosseserratus* (Sawtooth sunflower). Habit (right). Leaf (left).

Fig. 83. *Helianthus tuberosus* (Jerusalem artichoke). Habit. Underground tuber (lower right).

A variety of habitats including swampy woods.
IA, IL, IN, KS, KY, MO, NE, OH (FAC).
Jerusalem artichoke.
The fleshy tuber is edible and may be prepared as a vegetable.
Plants with all leaves opposite and softly pubescent on the lower surface may be known as var. *subcanescens*.

18. **Liatris** Schreb.—Blazing-star

Perennial herbs with underground corms; leaves simple, alternate, narrow, numerous, usually punctate; heads borne in racemes or spikes, subtended by several series of imbricate bracts, bearing purple disk flowers and no ray flowers; receptacle flat; achene ribbed, with a pappus of 1–2 series of barbellate or plumose bristles.

Approximately thirty-five species, all in North America, comprise the genus. Most species are found in dry habitats.

1. **Liatris spicata** (L.) Willd. Sp. Pl. 3:1636. 1804. Fig. 84.
Serrulata spicata L. Sp. Pl. 819. 1753.

Perennial herbs with underground corms; stems erect, stiff, unbranched, to 2 m tall, glabrous or nearly so; leaves simple, alternate, numerous, entire, linear to linear-lanceolate, acute to subacute at the apex, tapering to the sessile base, usually punctate, to 40 cm long, to 2 cm wide, glabrous or nearly so; flowers in axillary heads, forming a dense spike up to 40 cm long, each head 8–10 mm high, sessile, oblong-cylindric, subtended by several series of bracts and bearing up to 12 (–14) flowers; bracts oblong, broadly rounded at the apex, scarious-margined, often purplish; ray flowers absent; disk flowers purple; achenes 4–6 mm long, hispidulous, with a pappus of numerous barbellate bristles. July–September.

Wet meadows, marshes, savannas, mesic prairies, fens.
IL, IN, MO (FAC), KY, OH (FAC+).
Spicate blazing-star; marsh blazing-star.
This is the only *Liatris* that has spicate inflorescences, sessile heads, obtuse bracts, and barbellate pappus. Unlike other species in the genus, *L. spicata* occurs mostly in damp or wet habitats.

19. **Megalodonta** Greene—Water Marigold

Aquatic perennial herbs; leaves opposite or whorled, the submersed ones finely divided into filiform segments, the emersed ones, if present, serrate to laciniate; heads subtended by bracts in 2 series; ray flowers neutral, yellow; disk flowers perfert, 5-lobed, yellow; achenes terete, the pappus consisting of 3–6 barbed awns.

In addition to the species described below, one other species occurs in Washington state. Some botanists merge this genus with *Bidens*.

1. **Megalodonta beckii** (Torr.) Greene, Pittonia 4:271. 1901. Fig. 85.
Bidens beckii Torr. in Spreng. Neue Entdeck. 2:135. 1821.

Aquatic perennial herbs; stems scarcely branched, glabrous, up to 2 m long; submersed leaves much divided into filiform segments, glabrous; emersed leaves

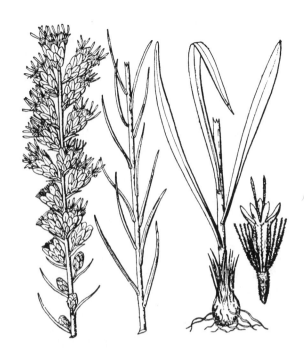

Fig. 84. *Liatris spicata* (Spicate blazing-star). Inflorescence (far left). Leafy stem (next to left). Lower leaves and corm (next to right). Disk flower (far right).

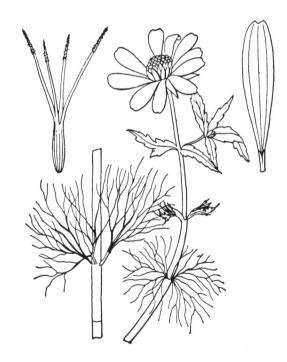

Fig. 85. *Megalodonta beckii* (Water marigold). Habit (center). Leaves (lower left). Fruit (upper left). Ray flower (upper right).

simple, lanceolate to oblong, serrate, acute to acuminate at the apex, glabrous, up to 3 cm long; heads solitary or few, subtended by ovate to oblong, more or less obtuse, glabrous bracts; rays 12–18, 1.0–1.5 cm long, yellow; achenes terete, 10–14 mm long, the pappus consisting of 3–6 retrorsely barbed awns. June–August.

Ponds, deep, clean lakes.

IA, IL, IN, MO, OH (OBL).

Water marigold.

This is the only truly aquatic member of the aster family in the central Midwest. The emergent leaves are very different than the submersed ones.

20. **Melanthera** Rohr.

Perennial herbs; leaves simple, opposite, petiolate; flowers borne in heads, each head subtended by a series of bracts and bearing only disk flowers; disk flowers white, perfect, tubular; receptacle flat, chaffy; achenes without a pappus.

Only the following species comprises the genus.

1. **Melanthera nivea** (L.) Small, Fl. S.E. U.S. 1251. 1933. Fig. 86.
Bidens nivea L. Sp. Pl. 2:833. 1753.

Perennial herbs; stems erect, branched, square, to 1.5 m tall, glabrous; leaves simple, opposite, or the uppermost alternate, lance-ovate to ovate, or the uppermost lanceolate, acute to acuminate at the apex, tapering to the base, sharply serrate, to 12 cm long, to 6 cm wide, glabrous, on petioles up to 2 cm long, or the uppermost sessile; flowers crowded into 1–few globose heads on long peduncles, the heads up to 2 cm in diameter, subtended by a series of bracts, the peduncles up to 10 cm long, glabrous; ray flowers absent; disk flowers white, tubular, perfect, 5–20 in number; achenes without a pappus. June–October.

Swampy woods.

IL, KY (NI).

Snow squarestem.

This is a very handsome species that is also relatively rare in the central Midwest. It is readily recognized by its square stems, opposite, sharply serrate leaves, and globose white flowering heads.

Fig. 86. *Melanthera nivea* (Snow squarestem). Habit. Fruit (lower left).

21. **Mikania** Willd.—Climbing Hempweed

Vines (in the United States), without tendrils; leaves opposite, simple, toothed; inflorescence cymose-paniculate, with numerous heads, each head 4-flowered and subtended by four narrow bracts; ray flowers absent; disk flowers pink or sometimes white, tubular, 5-lobed; stamens 5; ovary inferior; pappus of numerous capillary bristles; achenes 5-angled.

This primarily tropical American genus consists of more than two hundred species. Only the following occurs in the central Midwest.

1. **Mikania scandens** (L.) Willd. Sp. Pl. 3:1743. 1804. Fig. 87.
Eupatorium scandens L. Sp. Pl. 836. 1753.

High-climbing, twining vines; stems glabrous or nearly so; leaves opposite, simple, ovate, acute to acuminate at the apex, cordate at the base, shallowly dentate, glabrous or nearly so, to 10 cm long, to 5 cm wide, on petioles shorter than the blades; heads several, 4–5 mm high, pink, subtended by four acuminate bracts; ray flowers absent; disk flowers present; achenes 4–5 mm long, 5-angled, somewhat resinous. July–October.

Swamps, wet woods, damp thickets.

IA, IL, IN, MO (OBL), KY, OH (FACW+).

Climbing hempweed.

This is the only vine in the wetlands of the central Midwest that has opposite, cordate leaves that are shallowly dentate.

22. **Oligoneuron** Small—Flat-top Sunflower

Perennial herbs; leaves simple, alternate and basal, the basal leaves considerably larger than the upper leaves and long-petiolate, the upper leaves sessile or on short petioles; inflorescence a flat-topped corymb, with up to 30 flowering heads; heads subtended by several series of thick, obtuse bracts; rays yellow, up to 20 in number, pistillate; disk yellow, perfect; stamens 5; ovary inferior; pappus of numerous capillary bristles; fruit an achene.

Oligoneuron consists of about six species, some of which have previously been placed in the genus *Solidago* and others in the genus *Aster*. *Oligoneuron* differs from both *Solidago* and *Aster* by its flat-topped inflorescence. *Euthamia*, which also has a flat-topped inflorescence, has all leaves alike.

The following species may sometimes occur in shallow water.

1. Inflorescence glabrous; leaves flat...1. *O. ohiense*
1. Inflorescence puberulent; leaves folded .... 2. *O. riddellii*

1. **Oligoneuron ohiense** (Riddell) Small, G. N. Jones, Trans. Ill. Acad. Sci. 35:62. 1942. Fig. 88.
Solidago ohiensis Riddell, Syn. Fl. West. States 57. 1835.

Perennial herbs from a branched rootstock; stems unbranched up to the inflorescence, to 1 m tall, glabrous; leaves basal and alternate, lanceolate to oblong-lanceolate, obtuse at the apex, tapering to the base, entire or low-serrulate, glabrous, the

Fig. 87. *Mikania scandens* (Climbing hempweed). Habit. Flowering head (lower left).

Fig. 88. *Oligoneuron ohiense* (Ohio goldenrod).
Inflorescence and leaf.

largest lowest ones more than 20 cm long, the lowest on long petioles; inflorescence glabrous; heads numerous, in flat-topped corymbs; involucre 4–5 mm high, the bracts glabrous, obtuse; rays 6–9; disk flowers about 20; pappus of numerous capillary bristles; achenes 3–4 mm long, glabrous. August–October.

Fens, low sand flats, other moist areas.

IL, IN, OH (OBL).

Ohio goldenrod.

This species is sometimes known as *Solidago ohiensis*. It differs from the similar-appearing *O. riddellii* by its flat rather than folded leaves, its slightly shorter involucre, and its flat-topped inflorescence.

2. **Oligoneuron riddellii** (Frank) Rydb. Fl. Pr. & Plains 799. 1932. Fig. 89. *Solidago riddellii* Frank in Riddell, Syn. Fl. West. States 57. 1835.

Perennial herb from a thickened rootstock and usually creeping rhizomes; stems unbranched up to the inflorescence, to 1 m tall, glabrous or becoming slightly puberulent above; leaves basal and alternate, lanceolate, thick, conduplicate, acute to acuminate at the apex, tapering to the base, entire, glabrous except

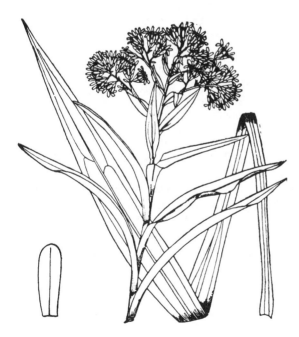

Fig. 89. *Oligoneuron riddellii* (Riddell's goldenrod).
Inflorescence and leaf. Bract (lower left).

for the scabrous margins, to 20 cm long, to 2 cm wide, the lowest on long petioles, the uppermost sometimes clasping; inflorescence puberulent; heads up to 30, in somewhat rounded corymbs; involucre 5–6 mm high, the bracts glabrous, obtuse; rays 7–9; disk flowers 20–25; pappus of numerous capillary bristles; achenes 4–5 mm long, glabrous or nearly so. August–November.

Prairies, fens, other moist ground.

IA, IL, IN, MO, OH (OBL).

Riddell's goldenrod.

This plant is sometimes known as *Solidago riddellii*. It differs from the similar *O. ohiense* by its folded leaves and puberulent branches of the inflorescence.

23. **Packera** A. Love & D. Love

Annual or perennial herbs; leaves simple or compound, alternate, the basal different in shape from the upper; inflorescence usually a corymbose cyme, with several heads; heads subtended by 1–2 series of bracts, each head with ray and disk flowers, the flowers yellow; achenes with a pappus of numerous capillary bristles.

Packera for many years was included within the genus *Senecio*. As a separate genus, is contains approximately eighty species.

1. Basal leaves cordate; stems not hollow .... 1. *P. aurea*
1. None of the leaves cordate; lower leaves pinnatifid; stems hollow .... 2. *P. glabella*

1. **Packera aurea** (L.) A. Love & D. Love, Bot. Nos. 128:520. 1975. Fig. 90.
Senecio aureus L. Sp. Pl. 870. 1753.
Senecio gracilis Pursh, Fl. Am. Sept. 529. 1814.
Senecio aureus L. var. *gracilis* (Pursh) Hook. Fl. Bor. Am. 333. 1834.
Senecio semicordatus Mack. & Bush, Ann. Rep. Mo. Bot. Gard. 16:107. 1905.
Senecio aureus L. var. *semicordatus* (Mack. & Bush) Greenm. Ann. Mo. Bot. Gard. 3:129. 1916.
Senecio aureus L. var. *intercursus* Fern. Rhodora 45:499. 1943.
Senecio aureus L. var. *aquilonius* Fern. Rhodora 45:500. 1943.

Perennial herbs from creeping rhizomes and often with stoloniferous offshoots; stems erect, slender, branched above, glabrous or sometimes floccose-tomentose, to 1 m tall; leaves basal and alternate, the basal ones ovate, obtuse to subacute at the apex, cordate or rarely rounded at the base, often purplish, crenate or rarely sharply serrate, to 10 cm long, 2–10 cm wide, glabrous or sometimes floccose-to-mentose at first, on petioles as long as or longer than the blade, the cauline leaves lanceolate, acute to acuminate at the apex, tapering to the sessile base; inflorescence of several heads arranged in a corymb, each head 5–10 mm high, subtended by 1–2 series of nearly equal, linear, glabrous bracts, and bearing 8–12 ray flowers and numerous disk flowers, the ray flowers yellow, 6–12 mm long, the disk flowers yellow; achenes glabrous, 2–4 mm long, with a pappus of numerous capillary bristles. May–July.

Swampy woods, wet meadows.

IA, IL, IN, KY, MO, OH (FACW), as *Senecio aureus*.

Golden ragwort; heart-leaved ragwort.

Fig. 90. *Packera aurea* (Golden ragwort). Habit. Ray flower (right).

This species is easily recognized by its heart-shaped basal leaves and its golden yellow flowering heads.

Considerable variation occurs in *P. aurea*. Typical var. *aurea* has strongly cordate, purple basal leaves and achenes 3.5–4.0 mm long. Var. *gracilis* is similar but lacks the purple offshoots and has achenes 2.0–3.5 mm long; var. *intercursus* also lacks the purple offshoots and has achenes 2.0–3.5 mm long, but has wider basal leaves over 3 cm wide. Var. *subcordatus* has basal leaves only slightly cordate or even rounded at the base. Var. *aquilonius* has leaves that are sharply serrate.

2. **Packera glabella** (Poir.) C. Jeffrey, Kew Bull. 47:101. 1992. Fig. 91.
Senecio glabellus Poir. in Lam. Encycl. 7:102. 1806.

Annual herbs from fibrous roots; stems erect, branched, hollow, succulent, glabrous except at the leaf axils, to 1 m tall; leaves alternate, the lower ones deeply pinnatifid, up to 20 cm long, up to 6 cm wide, the lobes rounded, toothed, the terminal lobe much larger than the others, glabrous the upper leaves toothed; heads numerous in corymbs, each head 4–6 mm high, subtended by several linear, acute bracts, the branches of the corymb with cobwebby hairs at first, becoming glabrous; ray flowers 6–12 in number, yellow, showy; disk up to 10 mm across, yellow; achenes ellipsoid, hispidulous, 2.5–4.0 mm long, the pappus of numerous capillary bristles. April–June.

Fig. 91. *Packera glabella* (Butterweed). Habit. Flowering head (lower left).

Swampy woods, wet ditches, low fields.

IA, IL, IN, KY, MO, OH (OBL), KS, NE (FACW), as *Senecio glabellus*.

Butterweed.

This showy species often completely dominates low fields. It is readily distinguished by its succulent hollow stems, deeply pinnatifid basal leaves with the terminal lobe the largest, and its bright yellow flowering heads.

24. **Pluchea** Cass.—Stinkweed

Aromatic annual or perennial herbs (in our area); leaves simple, alternate, toothed; inflorescence a corymbose cyme, with several heads; heads discoid, subtended by several series of appressed bracts; receptacle flat; ray flowers absent; outer disk flowers perfect, inner disk flowers perfect or sterile, 5-lobed, white, pink, purple, or yellow; stamens 5; ovary inferior; pappus a single series of capillary bristles; fruit an angular achene.

Pluchea consists of about fifty species in tropical and temperate climates. Only the following may occur in shallow water in the central Midwest.

1. Leaves petiolate; disk 3–6 mm across, purplish; annual .... 1. *P. camphorata*
1. Leaves sessile, clasping; disk 6–12 mm across, yellowish white; perennial .... 2. *P. foetida*

1. **Pluchea camphorata** (L.) DC. Prodr. 5:451. 1836. Fig. 92.

Erigeron camphoratum L. Sp. Pl., edd. 2, 1212. 1763.

Pluchea petiolata Cass. Dict. Sci. Nat. 42:2. 1826.

Fig. 92. *Pluchea camphorata* (Stinkweed). Habit. Flowering head (lower right).

Foul-smelling annual or perennial herbs; stems branched, viscid-pubescent, grooved, to 1.5 m tall; leaves alternate, simple, lanceolate to ovate-lanceolate, acute to acuminate at the apex, tapering to the petiolate base, up to 4 cm long, up to 3 cm wide, serrate, on petioles up to 1.5 cm long; heads several in a rounded corymb, 3–4 mm high, subtended by narrow glabrous or granular bracts; ray flowers absent; disk flowers pink or purple; pappus of numerous capillary bristles; achenes 1.5–2.5 mm long, pubescent. July–October.

Swamps, sloughs, marshes.

IL, IN, KY, MO, OH (FACW), KS (FACW–).

Stinkweed; camphorweed; marsh fleabane.

This species has a strong foul-smelling or medicinal-smelling odor. The flowering heads range from pink to purple. It differs from *P. foetida*, which has yellowish white flowers and sessile, clasping leaves.

2. Pluchea foetida (L.) DC. Prodr. 5:452. 1816. Fig. 93.
Baccharis foetida L. Sp. Pl. 861. 1753.

Perennial herbs with a strong odor, with fibrous roots; stems erect, usually branched above, to 1 m tall, glandular-pubescent and often viscid; leaves simple, alternate, lanceolate to oblong to lance-ovate, acute or obtuse at the apex, rounded at the sessile and often clasping base, sharply denticulate, strongly reticulate-veined, puberulent, to 9 cm long, to 3.5 cm wide; inflorescence composed of a few sessile heads arranged in a corymb, each head 5–8 mm high, subtended by several series of appressed, lanceolate, acute, viscid-pubescent bracts, bearing only disk flowers; disk 6–12 mm across, yellowish white; achenes 1–2 mm long, hispidulous, 4- to 6-angled, with a single whorl of capillary bristles. July–September.

Swampy woods, wet ditches.

MO (OBL).

Sticky marsh fleabane.

This primarily southeastern species barely reaches our area in southern Missouri. Its strongly foul odor, yellowish white flowering heads, and sessile, clasping leaves are distinctive.

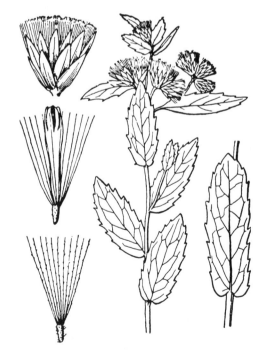

Fig. 93. *Pluchea foetida* (Sticky marsh fleabane). Habit (center). Flowering head (upper left). Disk flower (center left). Fruit (lower left). Leaf (lower right).

25. **Prenanthes** L.—White Lettuce

Perennial herbs from thickened tuberous roots and with latex; leaves alternate, simple, variously lobed or toothed; flowering heads in spikelike or open panicles or in the axils of the leaves, each head subtended by one row of bracts and a few bractlets and bearing only ray flowers; receptacle flat; achenes strongly ribbed.

This genus of about twenty-five species occurs in North America, Europe, Asia, and Africa.

Only the following species may be found in wetland habitats.

1. **Prenanthes racemosa** Michx, Fl. Bor. Am. 2:83. 1803. Fig. 94.
Prenanthes racemosa Michx. var. *multiflora* (Cronq.) Dorn, Vasc. Pl. Wyoming 295. 1988.

Perennial herbs from thickened rootstocks, with latex; stems erect, branched, glabrous except for pubescence on the inflorescence branches, glaucous, striate, to 1.7 m tall; leaves simple, alternate, firm, glabrous, glaucous, the lowest oval to oblong to obovate, denticulate, to 20 cm long, to 10 cm wide, on winged petioles to 10 cm long, the middle and upper leaves smaller, sessile, clasping; heads numerous in narrow leafy thyrses, pendulous at least at maturity, each head 10–15 mm high, subtended by 8–10 hirsute bracts and several smaller bracts, bearing up to 15 ray flowers; rays 12–16 per head, pink to pale purple, 3-toothed at the apex; achenes 4.5–6.5 mm long, glabrous, 8- to 12-ribbed, with a pappus of yellowish white capillary bristles. July–August.

Fig. 94. *Prenanthes racemosa* (Glaucous white lettuce). Inflorescence and leaf (left). Fruit (bottom center). Bract (next to lower right). Ray flower (lower right).

Wet meadows, prairies, along streams.

IA, IL, IN, MO (FACW), OH (FACW−), NE (not listed).

Glaucous white lettuce; rattlesnake root.

This species is distinguished by its glaucous and glabrous stems and leaves, its pink or purplish flowering heads, and the presence of latex.

Plants from Iowa and Nebraska, with somewhat larger flowering heads, may be called var. *multiflora*.

26. Rudbeckia L.—Coneflower

Perennial herbs; leaves alternate, simple, sometimes deeply lobed; flowering heads on long peduncles, showy, subtended by usually 2 rows of spreading bracts and bearing both ray flowers and disk flowers; ray flowers yellow, neutral; disk flowers tubular, perfect, fertile; receptacle conical or columnar; achenes 4-angled, glabrous, the pappus consisting of a short crown or absent.

About thirty North American species comprise the genus.

The following may occur in wetland habitats in the central Midwest.

1. Disk brown or purple; leaves not lobed ....1. *R. fulgida*
1. Disk yellow; leaves lobed ....2. *R. laciniata*

1. **Rudbeckia fulgida** Ait. Hort. Kew. 3:251. 1789. Fig. 95.

Rudbeckia speciosa Wenderoth, Ind. Sem. Hort. Marb. 1828.

Rudbeckia deamii Blake, Rhodora 19:113. 1917.

Rudbeckia fulgida Ait. var. *deamii* (Blake) Perdue, Rhodora 59:297. 1957.

Rudbeckia umbrosa C. L. Boynt. & Beadle, Biltmore Bot. Stud. 1:16. 1901.

Rudbeckia palustris Eggert in C. L. Boynt. & Beadle, Biltmore Bot. Stud. 1:16. 1901.

Rudbeckia sullivantii C. L. Boynt. & Beadle, Biltmore Bot. Stud. 1:16. 1901.

Rudbeckia speciosa Wenderoth var. *sullivantii* (C. L. Boynt. & Beadle) B. L. Robins. Rhodora 10:68. 1908.

Rudbeckia fulgida Ait. var. *sullivantii* (C. L. Boynt. & Beadle) Cronq. Rhodora 47:400. 1945.

Rudbeckia fulgida Ait. var. *umbrosa* (C. L. Boynt. & Beadle) Cronq. Rhodora 47:400. 1945.

Rudbeckia fulgida Ait. var. *speciosa* (Wenderoth) Perdue, Rhodora 59:297. 1957.

Perennial stoloniferous herbs, often with basal offshoots; stems erect, branched, variously pubescent, to 1 m tall; leaves simple, basal and alternate, usually pubescent, mostly scabrous, the basal ones lanceolate to ovate, acute at the apex, toothed to entire, to 8 cm long, tapering to a long petiole or less commonly cordate, the lower leaves similar but with a short, winged petiole, the middle and upper leaves lanceolate to oblanceolate, to 6 cm long, sessile; flowering heads 1–few per stem, on long, usually scabrous peduncles, the involucre 1.0–2.4 mm high, subtended by deltoid to ovate, subacute to obtuse, glabrous or strigose, reflexed, green bracts to 7 mm long; rays 1–4 cm long, sometimes 3-lobed at the apex, golden yellow; disk ovoid, 1–2 cm across, purple-brown to black; achenes 2–3 mm long, strongly angled, glabrous, the pappus a low crown. August–October.

Fig. 95. *Rudbeckia fulgida* (Coneflower). Habit. Disk flower (right).

Wet prairies, swamps, woods, fens, dry habitats.

IL, IN, MO (OBL), KY, OH (FAC).

Coneflower; showy black-eyed Susan.

This species has the general appearance of R. *hirta*, the black-eyed Susan, but differs by its spreading-hirsute to nearly glabrous stems and its nearly glabrous or strigose bracts. *Rudbeckia hirta* has rough-hispid stems and bracts with a hispid apex.

There is much variation in this species, and some of the varieties may actually qualify as distinct species. Typical var. *fulgida* has rays 1.0–2.5 cm long and basal leaves elliptic, entire or denticulate, 2–5 cm wide. Variety *tenax* has rays 1.0–2.5 cm long and basal leaves elliptic-lanceolate, entire, or denticulate, up to 2 cm wide. Variety *umbrosa* has rays 1.0–2.5 cm long and basal leaves ovate, 3.5–5.0 cm wide, sharply toothed. Variety *palustris* has rays 1.0–2.5 cm long and glabrous bracts. Varieties *speciosa*, *sullivantii*, and *deamii* have rays 2.5–4.0 cm long. In addition, var. *speciosa* has sparsely hirsute stems, glabrous or sparsely pubescent bracts, and cauline leaves lanceolate; var. *sullivantii* has sparsely hirsute stems, glabrous or sparsely pubescent bracts, and cauline leaves ovate; var. *deamii* has densely retrorse-hirsute stems and densely pubescent bracts.

2. **Rudbeckia laciniata** L. Sp. Pl. 906. 1753. Fig. 96.

Robust perennial herbs from woody rootstocks; stems stout, erect, branched, glabrous or nearly so, glaucous, to 3.5 m tall; leaves simple, basal and alternate, deeply 3- to 7-cleft, the lobes entire or toothed, or some of them undivided, to 45 cm long to 25 cm wide, glabrous or sometimes pubescent, particularly on the lower surface, the basal leaves on long petioles; heads several to numerous on long peduncles, 4.0–8.5 cm across, subtended by several unequal foliaceous bracts, bearing both ray flowers and disk flowers; rays 6–15 in number, 3–6 cm long, bright yellow, drooping; disk conical to cylindrical, to 3 cm long, at maturity greenish yellow, to 2.5 cm across; achenes flattened, truncate at the apex, 5–6 mm long, glabrous or nearly so, the pappus consisting of a short crown. July–September.

Swampy woods, floodplain forests, fens.

IA, IL, IN, MO (FACW+), KS, NE (FAC), KY, OH (FACW).

Goldenglow.

This robust perennial is distinguished by its large 3- to 7-cleft leaves, its glabrous and often glaucous stems, and its large yellow flowering heads with drooping rays. Plants with double the number of rays, producing a large, beautiful flowering head, are cultivated.

27. **Silphium** L.—Rosinweed

Robust perennial herbs; leaves simple, mostly opposite; flowering heads large, on long peduncles, subtended by firm bracts in several series, bearing both ray flowers and disk flowers; ray flowers yellow, pistillate, fertile; disk flowers sterile; receptacle flat; achenes compressed, winged, glabrous, the pappus consisting of 2 awns or absent.

Fifteen species, all in the United States, comprise this genus. Although the flowering heads resemble those of *Helianthus* (sunflower), its bracts are broad based rather than narrow.

Only the following species occurs in wetland habitats.

Fig. 96. *Rudbeckia laciniata* (Goldenglow). Habit (right). Leaf (lower left).

1. **Silphium perfoliatum** L. Sp. Pl., ed. 2, 1301. 1761. Fig. 97.

Robust perennials from thickened rootstocks; stems stout, erect, glabrous or nearly so, square, up to 2.5 m tall; leaves simple, opposite, ovate, acute at the apex, rounded at the base, connate-perfoliate forming a cup, scabrous and usually pubescent on both surfaces, coarsely dentate, to 30 cm long, to 20 cm wide; heads numerous in corymbs, up to 6.5 cm across, 12–25 mm high, subtended by several ovate, ciliate bracts, bearing both ray flowers and disk flowers; rays up to 30 (–35) in number, up to 4 cm long, yellow; disk yellow; achenes obovate, glabrous, winged, 5–8 mm wide, the pappus usually reduced to 2 small teeth. July–September.
Floodplains, fens.
IA, IL, IN, MO (FACW–), KS, NE (NI), KY, OH (FACU).
Cup-plant.
The connate-perfoliate leaves and the square stem are distinctive.

28. **Solidago** L.—Goldenrod

Perennial herbs; leaves alternate, simple; flowers in numerous heads, subtended by numerous imbricate bracts in several series; ray flowers usually yellow, pistillate, up to 1 cm long; disk flowers tubular, perfect, yellow; achenes terete or angled, the pappus consisting of capillary bristles.

There are approximately one hundred species in the genus, depending upon whether the genus is broken up into smaller genera or not. Most species are in the Northern Hemisphere. The majority of our species do not occur in wetlands.

1. Leaves essentially uniform in size, 3-veined above the base.................................1. *S. gigantea*
1. Basal leaves much larger than the middle and upper leaves, none of them 3-veined above the base, the lowest sometimes deciduous at flowering time.
   2. Stems below the inflorescence glabrous or nearly so; leaves not rugose.
      3. Branches of the panicle spreading or recurved; stems strongly angled ....*S. patula*
      3. Branches of the panicle erect or ascending; stems not strongly angled...........................
      ....4. *S. uliginosa*
   2. Stems below the inflorescence pubescent; leaves rugose .... 3. *S. rugosa*

1. **Solidago gigantea** Ait. Hort. Kew. 3:211. 1789. Fig. 98.
Solidago serotina Ait. Hort. Kew. 3:211. 1789.
Solidago gigantea Ait. var. *leiophylla* Fern. Rhodora 41:457. 1939.

Perennial from long, creeping rhizomes; stems erect, to 2 m tall, glabrous except in the puberulent inflorescence, often glaucous; basal leaves absent; cauline leaves lanceolate to elliptic, acute at the apex, tapering to the sessile base, sharply serrate, glabrous on both surfaces or sometimes pilose on the veins beneath, 3-veined above the base, to 15 cm long, to 3.5 (–4.0) cm wide; inflorescence a dense terminal panicle, the lower branchlets recurved with the heads secund, the branchlets of the panicle puberulent; flowering heads 2.5–4.0 mm high; bracts in 3–5 series, linear, obtuse to acute, glabrous, more or leass green-tipped; ray flowers 10–17, 1–3 mm long, yellow; disk flowers 3–8, 3–4 mm long, yellow; achenes glabrous or nearly so, 1.3–1.6 mm long; pappus of numerous capillary bristles. August–November.

Swamps, marshes, fens, wet meadows.

IA, IL, IN, KY, MO, OH (FACW), KS, NE (FAC).

Late goldenrod.

This is the only species of *Solidago* that has triple-nerved leaves and a glabrous stem and that may occur in shallow water. The stems are often glaucous. Typical var. *gigantea* has leaves with pubescence on the veins on the lower surface. Leaves of var. *leiophylla* are completely glabrous.

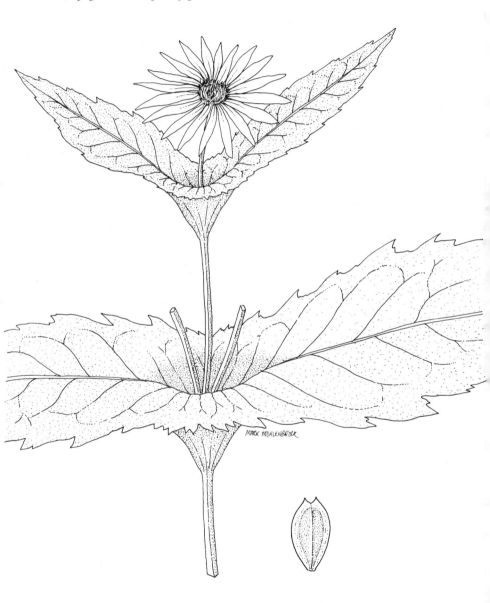

Fig. 97. *Silphium perfoliatum* (Cup-plant). Habit. Achene (lower right).

Fig. 98. *Solidago gigantea* (Late goldenrod). Habit (center). Inflorescence (lower left). Flowering head (upper right).

2. **Solidago patula** Muhl. in Willd. Sp. Pl. 3:2059. 1804. Fig. 99.

Perennial herb from a short rootstock; stems unbranched until the inflorescence, to 2.5 m tall, glabrous or puberulent at least above, strongly angled; leaves rather thick, scabrous on the upper surface, glabrous on the lower surface, the lowest leaves elliptic to ovate, acute at the apex, tapering to an elongated winged petiole, sharply serrate, to 30 cm long, to 10 cm wide, the upper leaves progressively smaller and less sharply serrate, eventually sessile; heads numerous, secund, on wide-spreading branches; involucre 3–5 mm high, the bracts obtuse to acute; ray flowers 5–12; disk flowers 8–20 (–23); achenes 3–4 mm long, sparsely hairy, the pappus consisting of capillary bristles. August–October.

Swamps, marshes, fens, wet meadows.

IA, IL, IN, KY, MO, OH (OBL).

Swamp goldenrod; spreading goldenrod.

The distinguishing features of this goldenrod are the extremely large, scabrous basal leaves and the strongly angled stems.

Fig. 99. *Solidago patula*
(Swamp goldenrod).
Habit (center).
Lower leaf (lower left).
Flowering head
(upper right).

3. **Solidago rugosa** Mill. Gard. Dict. ed. 8, 25. 1768. Fig.100.

Solidago aspera Ait. Hort. Kew. 3:212. 1789.

Solidago villosa Pursh, Fl. Am. Sept. 2:537. 1814.

Solidago rugosa Mill. var. *villosa* (Pursh) Fern. Rhodora 10:91. 1908.

Solidago rugosa Mill. var. *aspera* (Ait.) Fern. Rhodora 17:7. 1915.

Perennial from a thickened caudex and long, creeping rhizomes; stems erect, to 2 m tall, hispid to villous; lowest leaves early deciduous; cauline leaves lanceolate to elliptic, acute to acuminate at the apex, tapering to the base, serrate, rugose, villous or scabrous or sometimes glabrous on the upper surface; inflorescence a panicle with recurved branches composed of numerous secund heads; involucre 3.0–5.5 mm high; bracts linear-oblong, obtuse to acute, glabrous; ray flowers 6–10, 4.5–7.0 mm long, yellow; disk flowers 4–7, 2.5–3.5 mm high, yellow; achenes flat, short-pubescent. August–November.

Wet or dry soil, occasionally in shallow depressions in wet woods.

IL, IN, MO (FAC+), KY, OH (FAC).

Rough-leaved goldenrod.

This species is distinguished by its rugose-veiny leaves and its uniformly pubescent stems. The stems may be villous in the typical variety, or hispid in var. *aspera*.

4. **Solidago uliginosa** Nutt. Journ. Phil. Acad. 7:101. 1834. Fig. 101.

Perennial herb from an elongated rootstock; stems unbranched up to the inflorescence, to 1.5 m tall, glabrous except in the inflorescence; leaves oblong-lanceolate to lanceolate, acute to acuminate at the apex, tapering to the base, entire to low-serrulate, glabrous except for the scabrous margins, to 35 cm long, to 6 cm wide, the lowest much larger than the upper and on long petioles, the upper sessile; heads numerous, usually in a terminal thyrse longer than broad, not spreading; involucre 3–5 mm high, the bracts acute to obtuse, usually glabrous; rays 1–8; disk flowers 6–20; achenes 3–4 mm long, short-hairy. August–November.

Bogs, fens.

IA, IN, IL, OH (OBL).

Swamp goldenrod; bog goldenrod.

This species is distinguished by its elongated inflorescence that does not have spreading branches. The flowers are never secund. Plants with more than 9 disk flowers per head and with a more compact inflorescence are sometimes segregated as *S. purshii* Porter. This last element occurs in the Chicago area.

29. **Symphyotrichum** Nutt.—Aster

Perennial herbs; leaves alternate, simple; heads few to many, each with rays around a disk, subtended by an involucre of bracts in 2 or more rows; rays flowers pistillate, white or blue; disk flowers perfect, usually yellow; achenes bearing pappus of capillary bristles.

Symphyotrichum is primarily a genus of North American plants numbering about 175 species. The species in this genus have traditionally been placed in *Aster*. In addition to the ones described below, many others live in wetlands and in dry habitats in the central Midwest.

Fig. 100. *Solidago rugosa* (Rough-leaved goldenrod). Habit (left). Flowering head (right).

Fig. 101. *Solidago uliginosa* (Swamp goldenrod). Habit (left).
Leaf (right). Flowering head (upper center).

1. Stem leaves clasping or frequently auriculate at base.
  2. Involucral bracts glandular; upper parts of stem often viscid at maturity ....
  .... 6. *S. novae-angliae*
  2. Involucral bracts eglandular; upper parts of stem never viscid.
    3. Plants very slender, the stems at most 2.5 mm in diameter; involucral bracts acute; rays white or pale lavender .... 1. *S. boreale*
    3. Plants stout, the stems more than 2.5 mm in diameter; involucral bracts long-attenuate, if acute, then squarrose; rays blue or white.
      4. Stems coarsely hispid; rhizomes short; flowers blue .... 9. *S. puniceum*
      4. Stems glabrous or pubescent, not hispid; rhizomes elongated; flowers blue or white.
        5. Stems glabrous; flowers usually white; bracts long-attenuate .... 3. *S. firmum*
        5. Stems pubescent; flowers blue; bracts acute or obtuse .... 8. *prenanthoides*
1. Leaves neither clasping nor auriculate at base.
  6. Leaves conspicuously reticulate-veined below; rays usually blue .... 7. *S. praealtum*
  6. Leaves not conspicuously reticulate-veined below; rays usually white or pale blue.
    7. Involucre up to 4 mm high .... 10. *S. racemosum*
    7. Involucre 4–7 mm high.
      8. Leaves pubescent on the veins on the lower surface .... 5. *S. lateriflorum*
      8. Leaves glabrous on the lower surface.
        9. Heads 1-several, solitary at the tips of stiff, ascending peduncles ....
        .... 2. *S. dumosum*
        9. Heads few to many, on short, flexible peduncles .... 1. *S. boreale*

1. **Symphyotrichum boreale** (Torr. & Gray) A. Love & D. Love, Taxon 31:358–359. 1982. Fig. 102.

Aster junceus Ait. Hort. Kew. 3:204. 1789, misapplied.

Aster laxiflorus var. *borealis* Torr. & Gray. Fl. N. Am. 2:138. 1841.

Aster borealis (Torr. & Gray) Prov. Fl. Can. 1:308. 1862.

Aster junciformis Rydb. Bull. Torrey Club 37:142. 1910.

Perennial from long, thin rhizomes; stems erect, slender, to 80 cm tall, glabrous in the lower half, pubescent in the upper half, distinctly grooved, often reddish, at least at the base; basal leaves broadly elliptic, acute at the tip, tapering to the petiolate base, early deciduous; cauline leaves narrowly lanceolate to elliptic, acute to acuminate at the tip, tapering to the sessile or sometimes clasping base, entire but scabrous along the margins, glabrous except for the pubescent midvein, to 13 cm long, to 5.5 (–6.0) mm wide; inflorescence an open corymb, with 5–20 heads, rarely with solitary heads; involucre hemispherical, 5–7 mm high, up to 10 mm wide; bracts in 3–5 series, linear, acute, glabrous, often with a purplish tip and purplish margin; ray flowers (20–) 30–50, 7–15 mm long, white to lavender; disk flowers about 30, yellow-brown; achenes flat, glabrous or strigose, yellow, several-nerved, about 2 mm long; pappus of numerous white capillary bristles. August–October

Calcareous fens, bogs.

IA, IL, IN, NE, OH (OBL), as *Aster borealis*.

Rush aster.

This species, which enters the northern part of our range, for years was known as *Aster junciformis*, but that binomial is predated by *A. borealis*. The main range of this species is north of the central midwest. The flowers are white to pale lavender. The slender stems are less than 1 m tall.

Fig. 102. *Symphyotrichum boreale* (Rush aster). Habit (left). Disk flower (lower right).

2. **Symphyotrichum dumosum** (L.) Nesom, Phytologia 77:280. 1994. Fig. 103.
Aster dumosus L. Sp. Pl. 873. 1753.

Perennial from slender rhizomes; stems erect, much branched, stiff, glabrous, to 1 m tall; leaves alternate, simple, stiff, linear to linear-lanceolate, more or less entire, acute at the apex, cuneate at the base, glabrous but with a scabrous margin, up to 5 cm long, to 8 mm wide, those of the flowering branchlets numerous, small, and bractlike; involucre 4–7 mm high; bracts linear, acute to subulate, with a green tip; rays 15–30, usually white but occasionally pale blue, 4–5 mm long; pappus white; achenes pubescent. August–October.

Dry, sandy soil, but occasionally in depressions in wet woods.

IL, IN, MO (FAC+), KY (FAC), as *Aster dumosus*.

Rice button aster; bushy aster.

Fig. 103. *Symphyotrichum dumosum* (Rice button aster). Habit and leaf (center). Ray flower (lower left). Bract (next to lower right). Disk flower (lower right).

The numerous small, bractlike leaves on the flowering branches resemble those of *S. racemosum*, but the involucres of *S. racemosum* are smaller than those of *S. dumosum*, not reaching 4 mm high.

This species is most often found in dry, usually sandy habitats, but it does occur occasionally in shallow depressions in wet woods where water stands for several weeks.

3. **Symphyotrichum firmum** (Nees) Nesom, Phytologia 77:282. 1994. Fig. 104.
Aster firmus Nees, Syn. Asterac. 25. 1818.
Aster puniceus L. var. *firmus* (Nees) Torr. & Gray, Fl. N. Am. 2 (1):141. 1841.
Aster puniceus L. var. *lucidulus* Gray, Syn. Fl. N. Am. 1:195. 1886.
Aster lucidulus (Gray) Wieg. Rhodora 26:4. 1924.
Aster puniceus L. ssp. *firmus* (Nees) A. G. Jones, Phytologia 55:384. 1984.

Perennial from long, creeping rhizomes, often forming colonies; stems erect, to 2.5 m tall, glabrous except sometimes for a line of pubescence below the leaves; lowest leaves early deciduous; cauline leaves lanceolate to ovate-lanceolate, acute at the tip, sessile and clasping at the base, entire or nearly so, glabrous on the upper surface, glabrous or pubescent on the midnerve beneath, 4–7 cm long, 1.0–2.5

Fig. 104. *Symphyotrichum firmum* (White swamp aster). Habit (center).
Achene (lower left).

(–3.5) cm wide; inflorescence an open panicle, with 60–100 heads; involucre 6–12 mm high, 6–10 mm wide; bracts imbricate, linear, long-attenuate, loosely arranged, glabrous except for the ciliate margins, with elongated green tips; ray flowers 30–60, 10–18 mm long, white, rarely blue; disk flowers about 40, yellow; achenes flat, glabrous or nearly so, about 1.5 mm long; pappus of numerous white capillary bristles. August–October.

Calcareous fens, wet prairies, moist soil.

IL, IN, MO (FACW+), as *Aster lucidulus*.

White swamp aster.

This species is similar in appearance to *S. puniceum* but is readily distinguished by its nearly glabrous stems and usually white flowers.

4. **Symphyotrichum lanceolatum** (Willd.) G. L. Nesom, Phytologia 77:284. 1994. Fig. 105.

Aster paniculatus Lam. Encycl. 1:306. 1783, misapplied.
Aster lanceolatus Willd. Sp. Pl. ed. 4, 3:2050. 1803.
Aster simplex Willd. Enum. Pl. 2:887. 1809.
Aster interior Wieg. Rhodora 35:25. 1933.
Aster lanceolatus Willd. var. *simplex* (Willd.) A. G. Jones, Phytologia 63:132. 1987.
Aster lanceolatus Willd. var. *interior* (Wieg.) Semple & Chmiel. Can. Journ. Bot. 66: 1058. 1987.

Perennial from long-creeping rhizomes, often forming colonies, stems erect, to 2.4 m tall, glabrous or with a few lines of pubescence; lowest leaves early deciduous; cauline leaves simple, alternate, linear-lanceolate to oblong-lanceolate, acute at the apex, sessile and sometimes slightly clasping at the base, glabrous on both surfaces, 6–20 cm long, 3–35 mm wide; inflorescence an open panicle, with 60–100 heads; involucre 3–6 mm

Fig. 105. *Symphyotrichum lanceolatum* (Small white aster). Habit (center). Disk flower (lower left).

high, 4–8 mm broad; bracts imbricate, linear to linear-oblong, long-attenuate, loosely arranged, glabrous exact for the ciliate margins; ray flowers, 20–40, white, 6–12 mm long; disk flowers about 40, yellow; achenes flat, pubescent, about 1.5 mm long; pappus of numerous white capillary bristles. August–October.

Swampy woods, floodplain forests, wet meadows.

IA, IL, IN, KS, KY, MO, NE, OH (FACW).

Small white aster.

This species in the past has been called *Aster simplex*. It has several variations in our area. Typical var. *lanceolatum* has involucres 4–6 mm high and leaves 3–12 mm wide. Var. *simplex* has involucres 4–6 mm high and leaves 10–35 mm wide. Var. *interior* have involucres 3–4 mm high.

Symphyotrichum lanceolatus differs from the similar appearing *S. lateriflorus* by its glabrous leaves.

Wet woods.

IA, IL, IN, KS, KY, MO, NE, OH (FACW), as *Aster simplex*.

5. **Symphyotrichum lateriflorum** (L.) A. Love & D. Love, Taxon 31:359. 1982. Fig. 106.

Solidago lateriflora L. Sp. Pl. 879. 1753.

Aster lateriflorus (L.) Britt. Trans. N. Y. Acad. Sci. 9:10. 1889.

Perennial; stems branched, glabrous, puberulent, or rarely hirsute, to 1.5 m tall; leaves glabrous above, pubescent on the midvein beneath, serrate, the lowest ovate, acute at the apex, rounded at the petiolate base, to 8 cm long, to 4 cm wide, the middle and upper leaves linear-lanceolate to broadly lanceolate, acute to acuminate at the apex, tapering to the sessile base, to 15 cm long, to 3 cm wide; heads several to numerous, 6–10 mm across; involucre 4.0–5.5 mm high, the bracts obtuse to acute, glabrous, appressed, with a green tip, sometimes tinged with purple; ray flowers 9–14, 4.0–6.5 mm long, white or less commonly pale purple; disk flowers purple; achenes puberulent. July–October.

Floodplain woods, fens; also drier habitats.

IA, IL, IN, KY, MO, OH (FACW–), KS, NE (FACW+), as *Aster lateriflorus*.

Calico aster; side-flowering aster.

This relatively common species is distinguished by the leaves, which are glabrous except for the midvein on the lower surface.

6. **Symphyotrichum novae-angliae** (L.) Nesom, Phytologia 77:287. 1994. Fig. 107.

Aster novae-angliae L. Sp. Pl. 875. 1753.

Perennial from a short, thickened rhizome and a hardened caudex; stems erect, up to 1.8 (–2.0) m tall, glandular-hispid to hirsute; lowest leaves early deciduous; cauline leaves lanceolate to oblong, acute at the apex, clasping at the base, entire, strigose and scabrous above, softly strigose beneath, to 12 cm long, to 2 cm wide; inflorescence a panicle, with 30–65 heads; involucre 6–10 mm high, 6–10 mm wide; bracts in 2–3 series, glandular-hairy, often spreading at the apex, the tips green or purplish; ray flowers 45–80 (–100), 10–20 mm long, reddish purple, rarely white; disk flowers about 50, reddish purple; achenes flat, sericeous, with obscure nerves, 1.8–2.0 mm long; pappus of numerous white capillary bristles. July–October.

Fig. 106. *Symphyotrichum lateriflorum* (Calico aster). Habit (left). Achene (lower right).

Fig. 107. *Symphyotrichum novae-angliae* (New England aster). Habit (center). Disk flower (lower left).

Fens, bogs, wet prairies, wet meadows, pastures.

IA, IL, IN, KS, MO, NE (FACW), KY, OH (FACW–), as *Aster novae-angliae*. New England aster.

This is one of the most attractive wetland species when in flower and is sometimes planted as an ornamental. Flower color encompasses various shades of blue or purple. This species is similar to *S. puniceum* and *S. firmum* by its clasping leaves but differs in having glandular, often viscid, pubescence in the inflorescence.

7. **Symphyotrichum praealtum** (Poir.) Nesom, Phytologia 77:289. 1994. Fig. 108.

Aster salicifolius Lam. Encycl. 1:306. 1783, misapplied.

Aster praealtus Poir. Encycl. Suppl. 1:493. 1810.

Aster subasper Lindl. Comp. Bot. Mag. 1:97. 1835.

Aster salicifolius Lam. var. *subasper* (Lindl.) Gray, Syn. Fl. 1:188. 1884.

Aster nebraskensis Britt. Ill. Fl. N. U.S. 3:375. 1898.

Aster praealtus Poir. var. *angustior* Wieg. Rhodora 35:24. 1933.

Aster praealtus Poir. var. *subasper* (Lindl.) Wieg. Rhodora 35:24. 1933.

Aster praealtus Poir. var. *nebraskensis* (Britt.) Wieg. Rhodora 35:25. 1933.

Perennial from long, creeping rhizomes; stems erect, to 1.5 m tall, densely pubescent, at least on the upper half, but usually becoming glabrate; cauline leaves linear-lanceolate to elliptic, acute at the apex, rounded at the sessile or very rarely subclasping base, entire or sparingly dentate, glabrous but sometimes scabrous above, glabrous or puberulent beneath, conspicuously reticulate-nerved beneath, to 12 cm long, to 18 mm wide; inflorescence 5–8 mm high, 6–9 mm wide; bracts in several series, linear, acute, glabrous or sparsely pubescent, greenish or reddish at the tip; ray flowers 6–15, 6–15 mm long, blue or less commonly white; disk flowers about 35, yellow; achenes flat, pubescent, few-nerved, 1–2 mm long; pappus of numerous white capillary bristles. August–October.

Fig. 108. *Symphyotrichum praealtum* (Willow-leaved aster). Habit (center). Ray flower (upper left). Disk flower (upper right). Bract (lower right).

Wet meadows, calcareous fens.

IA, IL, IN, KS, KY, MO, NE, OH (FACW), as *Aster praealtus*.

Willow-leaved aster; net-leaved aster.

This species is recognized by its conspicuous reticulate venation of the lower surface of the leaves. Some of the leaves are often more than four times longer than broad, hence the common name of willow-leaved aster.

The leaves vary considerably in their texture, shape, and pubescence. Typical var. *praealtum* has lanceolate to elliptic leaves that are rather thick and firm and generally smooth on the upper surface. Plants with lanceolate to elliptic leaves that are rather thick and firm but scabrous on the upper surface may be called var. *subasper*. Plants with thin leaves that are linear to linear-lanceolate have been called var. *angustior*. Some plants from northern Iowa and Nebraska with shaggy pubescence on the lower surface of the leaves may be known as var. *nebraskensis*, or as a separate species—*S. nebraskense*.

8. **Symphyotrichum prenanthoides** (Muhl.) G. L. Nesom, Phytologia 77:290. 1994. Fig. 109.
Aster prenanthoides Muhl. in Willd. Sp. Pl. 3:2046.

Perennial herbs with long creeping rhizomes; stems erect and often zigzag, to 1 m tall, pubescent above, pubescent in lines below; leaves alternate, simple, lanceolate to ovate, acuminate at the apex, tapering to the base and often clasping the stem, serrate or entire, glabrous or scabrous on the upper surface, glabrous or pubescent on the veins beneath, to 20 cm long, to 5 cm wide, the uppermost leaves smaller and sessile; heads several to numerous in terminal leafy bracted clusters; involucre 5–7 mm long, glabrous, the bracts more or less loose and squarrose, acute to obtuse; rays 20–35 in number, to 15 mm long, blue or pale blue; disk yellow; achenes strigose, 2–3 mm long, with a yellowish pappus. August–October.
Wet meadows, fens, wet woods.
IA, IL, IN, KY, OH (FAC).
Zigzag aster.
This species is recognized by its zigzag stems with sparse pubescence and its auriculate-clasping leaves.

9. **Symphyotrichum puniceum** (L.) A. Love & D. Love, Taxon 31:359. 1982. Fig. 110.
Aster puniceus L. Sp. Pl. 875. 1753.

Perennial from short, stout rhizomes and a thickened caudex; stems erect, to 2.5 m tall, hispid, often becoming glabrate except on the upper part of the stem, usually purplish; lowest leaves early deciduous; cauline leaves narrowly lanceolate to ovate, acute at the tip, auriculate and clasping at the base, scabrous on the upper surface, glabrous or pilose on the lower surface, entire to irregularly serrate, up to 15 cm long, up to 4 cm wide; inflorescence an open corymb, with 30–50 heads; involucre 5–12 mm high, 7–14 mm wide; bracts in 2–3 series, linear, acute to acuminate, the outer ones tending to become more or less leafy, glabrous or nearly so; ray flowers 30–60, 7–15 mm long, dark blue to purple; disk flowers about 50, yellow; achenes flat, pubescent, becoming glabrate, several-nerved, 1.4–1.7 mm long; pappus of numerous white capillary bristles. August–October.
Calcareous fens, moist soil.
IA, IL, IN, KY, MO, OH (OBL), as *Aster puniceus*.

Fig. 109. *Symphyotrichum prenanthoides* (Zigzag aster). Habit (center).
Achene (lower right).

Fig. 110. *Symphyotrichum puniceum* (Swamp blue aster). Habit (center).
Disk flower (lower left).

Swamp blue aster; bristly aster; purple-stemmed aster.

Symphyotrichum puniceum is distinguished by its clasping leaves and harshly pubescent stems. The stems or at least the nodes are usually dark purple.

10. **Symphyotrichum racemosum** (Ell.) Nesom, Phytologia 77:290. 1994. Fig. 111.
Aster vimineus Lam. Encycl. 1:306. 1783, misapplied.
Aster racemosus Ell. Sketch Bot. S. Carol. 2:348–349. 1824.

Perennial; stems erect, slender, often purple-tinged, to 1.5 m tall, glabrous or with lines of pubescence; leaves linear to lanceolate, acute at the apex, cuneate at the base, entire or sparingly serrate, glabrous; inflorescence spicate to racemose, on pedicels up to 10 mm long, with numerous small, bractlike leaves on the branches; heads spicate to racemose; involucre 3.0–3.6 mm high; bracts narrowly linear, acute, with a central green midrib; rays 15–30, white, 3–6 mm long, less than 1 mm wide; disk corollas 2.5–3.5 mm long; achenes puberulent. August–September.

Moist ground, sometimes in standing water in shallow depressions of wet woods. IL, IN, MO (FACW), KY, OH (FAC), as *Aster dumosus*.

Small white aster.

The numerous, small bractlike leaves on the flowering branches resemble those of *S. dumosum*, but *S. racemosum* has involucres only 3.0–3.6 mm high.

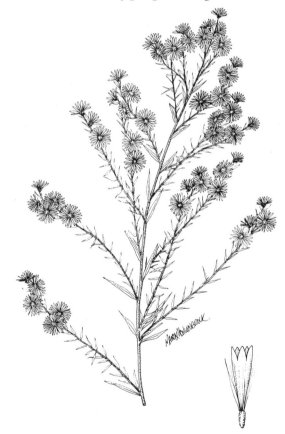

Fig. 111. *Symphyotrichum racemosum* (Small white aster). Habit (left). Disk flower (lower right).

30. **Verbesina** L.—Crownbeard

Perennials from thickened roots or taproots; leaves simple, alternate or opposite, toothed; flowering heads several, subtended by 2 or more sets of herbaceous bracts, bearing both ray flowers and disk flowers; ray flowers neutral or pistillate, white or yellow; disk globose to convex, yellowish, the disk flowers perfect and fertile; achenes flattened, winged, the pappus consisting of 2 awns.

Verbesina consists of approximately sixty species, all in the New World. Only the following species occurs in wetland habitats.

I. **Verbesina alternifolia** (L.) Britt. in Kearney, Bull. Torrey Club 20:485. 1893. Fig. 112.

Coreopsis alternifolia L. Sp. Pl. 909. 1753.

Actinomeris alternifolia (L.) DC. Prodr. 5:575. 1836.

Perennial herbs with thickened fibrous roots; stems erect, branched or unbranched, glabrous or pubescent, usually winged, to 3 m tall; leaves simple, alter-

Fig. 112. *Verbesina alternifolia* (Yellow wingstem). Habit (left). Fruiting head (near right). Achene (far right).

nate, lanceolate to ovate, acute to acuminate at the apex, tapering to the petiolate base, scabrous, hirsute, usually sharply serrate, to 25 cm long, to 8 cm wide, the petioles winged and decurrent on the stem; heads several to numerous, up to 4 cm across, subtended by 2 or more series of herbaceous, glabrous, deflexed bracts, bearing both ray flowers and disk flowers; ray flowers 2–10 in number, 1.0–2.5 cm long, deflexed, yellow; disk globose, 1.0–1.5 cm across, yellowish; achenes flattened, obovate, sparsely pubescent, usually winged, the pappus consisting of a pair of awns. August–October.

Bottomland woods, swampy woods.

IA, IL, IN, MO (FACW), KS, KY, NE, OH (FAC).

Yellow wingstem; yellow crownbeard.

The winged stem and deflexed yellow rays distinguish this species. In the past, this species has been segregated into the genus *Actinomeris*.

31. **Vernonia** Schreb.—Ironweed

Robust perennials; leaves simple, alternate, sessile or nearly so, toothed; heads numerous in corymbiform cymes, each head subtended by numerous appressed bracts and bearing only disk flowers; disk flowers tubular, purple; receptacle flat or convex; achenes ribbed, pubescent, the pappus consisting of a double row, the inner bristlelike, the outer of short scales or short bristles.

About 1,000 species, found throughout much of the world, comprise this genus.

1. Bracts with a prolonged, filiform tip.
   2. Leaves pitted; flowers more than 55 per head .... 1. *V. arkansana*
   2. Leaves not pitted; flowers with up to 55 per head.................................. 4. *V. noveboracensis*
1. Bracts obtuse to acute to acuminate at the tip.
   3. Lower surface of leaves glabrous or nearly so, pitted.....................................2. *V. fasciculata*
   3. Lower surface of leaves pubescent, the pits obscured .... 3. *V. missurica*

1. **Vernonia arkansana** DC. Prodr. 7:264. 1838. Fig. 113.
Vernonia crinita Raf. New Fl. N.Am. 4:77. 1836, misapplied

Robust perennial herbs with rhizomes; stems stout, erect, branched, glabrous or nearly so, glaucous, to 3 m tall; leaves simple, alternate, linear to linear-lanceolate, acute to acuminate at the apex, tapering to the nearly sessile base, to 20 cm long, to 2.5 cm wide, denticulate or nearly entire, glabrous or nearly so, pitted on the lower surface; heads several to numerous in a cyme, on thick peduncles, each head 1.0–1.8 cm across, 9–15 mm high, subtended by linear filiform-tipped ciliate bracts, bearing more than 55 disk flowers; disk flowers purple; achenes cylindrical, 5–6 mm long, ribbed, glabrous or hispidulous, the pappus consisting of numerous purplish capillary bristles. July–October.

Swampy woods, rocky streambeds.

IL, IN, MO, KS (FAC), KY (OH).

Ozark ironweed.

This and *V. noveboracensis* are the two species of *Vernonia* with filiform-tipped bracts. The leaves of *V. arkansana* are pitted on the lower surface.

The earlier binomial *V. crinita* has been misapplied to this species.

Fig. 113. *Vernonia arkansana* (Ozark ironweed). Habit (next to left). Achene (lower left). Bracts (upper right). Flowering head (lower right).

2. **Vernonia fasciculata** Michx. Fl. Bor. Am. 2:94. 1803. Fig. 114.
Vernonia corymbosa Schwein. Narr. Exp. Long. 2:394. 1824.
Vernonia fasciculata Michx. var. *corymbosa* (Schwein.) Daniels, Univ. Mo. Stud. Sci. Ser. 1:403. 1907.

Robust perennial herbs with rhizomes; stems stout, erect, branched, glabrous or sparsely pubescent at least above, to 2 m tall; leaves simple, alternate, linear-lanceolate to oblong-lanceolate, acute to acuminate at the apex, tapering to the sessile or short-petiolate base to 20 cm long, to 3 cm wide, denticulate or sharply serrate, glabrous or sometimes puberulent on the lower surface, pitted on the lower surface; heads several to numerous in a cyme, on glabrous peduncles, each head 8–15 mm across, 5–9 mm high, subtended by ovate, obtuse to subacute, ciliate bracts, bearing up to 30 disk flowers; disk flowers purple; achenes cylindrical, 3.0–3.5 mm long, ribbed, glabrous or puberulent, the pappus consisting of numerous purplish capillary bristles. July–September.

Marshes, wet prairies, alluvial terraces.

IA, IL, IN, MO (FACW), KS, NE (FAC), KY, OH (FAC+).

Smooth ironweed.

This species differs from *V. arkansana* and *V. noveboracensis* by its obtuse to subacute bracts, and from *V. missurica* by its glabrous stems and leaves.

In Iowa and Nebraska, some plants have lance-ovate leaves and heads 8–9 mm high. These may be called var. *corymbosa*.

Fig. 114. *Vernonia fasciculata* (Smooth ironweed). Habit (center). Flowering head (near right). Disk flower (far right).

3. Vernonia missurica Raf. Herb. Raf. 28. 1833. Fig. 115.

Robust perennial herbs with rhizomes; stems stout, erect, branched, pubescent throughout, to 2 m tall; leaves simple, alternate, lanceolate, acuminate at the apex, tapering to the sessile or short-petiolate base, to 20 cm long, to 5 cm wide, serrate or nearly entire, pubescent and scabrous above, densely tomentose beneath with crooked hairs, the pitted surface obscured by the pubescence; heads several to numerous in a cyme, on long pubescent peduncles, each head 6–12 mm across, 4–6 mm high, subtended by ovate, obtuse to acute, pubescent, ciliate bracts, bearing 30–55 disk flowers; disk flowers purple; achenes 4.0–4.5 mm long, ribbed, hispidulous, the pappus consisting of tawny or brownish capillary bristles. July–September.

Wet prairies, wet ditches, pastures.

IA, IL, IN, MO (FAC+), KS, NE (FACW), KY, OH (FACU+).

Missouri ironweed.

This ironweed is distinguished by its densely tomentose lower leaf surfaces and its obtuse to acute bracts. It is similar to and often intergrades with *V. gigantea* (*V. altissima*) but differs by its crooked hairs on the lower leaf surface and its more numerous flowers per head.

4. Vernonia noveboracensis (L.) Willd. Sp. Pl. 3:1632. 1804. Fig. 116.

Robust perennial herbs with rhizomes; stems stout, erect, branched, glabrous or sparsely pubescent, to 2 m tall; leaves simple, alternate, linear-lanceolate to oblong-lanceolate, acute to acuminate at the apex, tapering to the sessile or short-petiolate base, to 30 cm long, to 6 cm wide, denticulate or nearly entire, the lower surface densely covered by crooked hairs, not pitted on the lower surface; heads several to

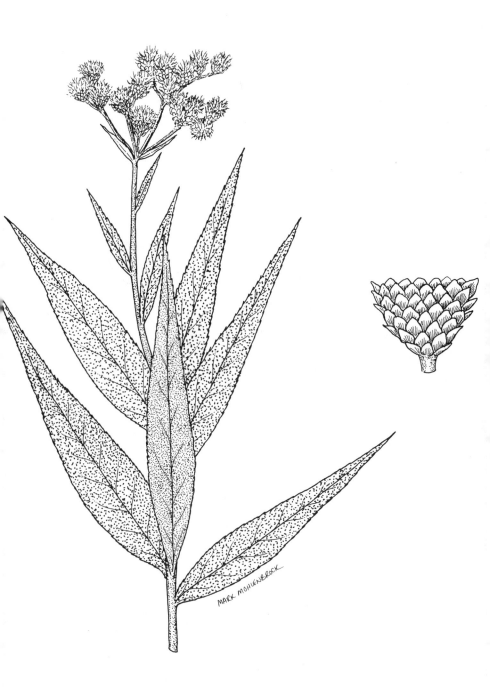

Fig. 115. *Vernonia missurica* (Missouri ironweed). Habit (left). Bracts (right).

numerous in a cyme, on pubescent peduncles, each head 8–16 mm across, 7–16 mm high, subtended by ovate to oblong bracts tapering to a filiform tip, bearing 30–55 disk flowers; disk flowers purple; achenes 4.0–4.5 mm long, ribbed, glabrous or hispidulous, the pappus consisting of purple or brownish purple capillary bristles. August–October.

Swampy woods, marshes.

KY, OH (FACW+).

New York ironweed.

Because of its filiform-tipped bracts, this species is closely allied to *V. arkansana* but differs by its non-pitted leaves and fewer flowers.

Fig. 116. *Vernonia noveboracensis* (New York ironweed). Habit (left). Flowering head (lower right).

46. BALSAMINACEAE—JEWELWEED FAMILY

Semisucculent herbs with thick, sticky, clear sap; leaves alternate, simple, toothed, without stipules; flowers zygomorphic, perfect, with a spurred calyx; stamens 5; ovary superior.

There are two genera in this family. The other is found in Asia.

1. Impatiens L.—Touch-me-not; Jewelweed

Semisucculent herb with thick, sticky, clear sap; leaves alternate or sometimes opposite; flowers perfect, zygomorphic; sepals 3, the upper 2 small, the lower saccate and spurred at back; petals 3, free from each other; stamens 5; ovary superior; fruit an explosive capsule.

This primarily tropical genus has approximately 450 species. Several of them are grown as ornamentals in this country. Only the following two species occur in the eastern United States, where they are found rarely in shallow water.

1. Flowers orange ....1. *I. capensis*
1. Flowers pale yellow .... 2. *I. pallida*

1. **Impatiens capensis** Meerb. Afbeel. Gew. pl. 10. 1775. Fig. 117.
Impatiens biflora Walt. Fl. Carol. 219. 1788.

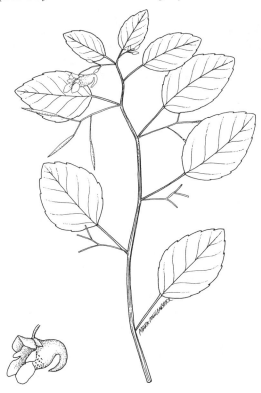

Fig. 117. *Impatiens capensis* (Spotted touch-me-not). Habit (center). Flower (lower left).

Semisucculent perennial to 2 m tall, sometimes glaucous, glabrous; leaves alternate or sometimes opposite, to 10 cm long, elliptic, crenate, the petioles often longer than the blades; flowers orange, with brown spots, up to 3 cm long, the saccate sepal longer than broad, the spur strongly curved forward, 6 mm long or longer; capsules up to 2 cm long, glabrous. June–October.

Shaded woods, marshes, fens, rarely in standing water.

IA, IL, IN, KS, KY, MO, NE, OH (FACW).

Spotted touch-me-not; spotted jewelweed; orange touch-me-not.

The gelatinous sap seems to alleviate temporarily the itch of poison ivy and the sting of stinging nettle for some people. I am unable to distinguish the two species of *Impatiens* in the vegetative condition.

2. **Impatiens pallida** Nutt. Gen.1:146. 1818. Fig. 118.

Semisucculent perennial up to 2 m tall, often glaucous, glabrous; leaves alternate or sometime opposite, to 10 cm long, elliptic, crenate, the petioles often longer than the blades; flowers pale yellow, up to 3 cm long, the saccate sepal broader than long, the spur spreading at right angles, up to 5 mm long; capsules up to 2 cm long, glabrous. June–October.

Shaded woods, rarely in standing water.

IA, IL, IN, KS, KY, MO, NE, OH (FACW).

Pale touch-me-not; pale jewelweed; yellow touch-me-not.

The lemon-yellow flower is about the only accurate way to tell this species from the one above.

Fig. 118. *Impatiens pallida* (Pale touch-me-not). Habit (center). Flower (lower left).

47. BETULACEAE—BIRCH FAMILY

Trees or shrubs; leaves alternate, simple, stipulate; flowers unisexual, the staminate usually in elongated, pendulous catkins, the pistillate pendulous or erect, short, sometimes woody; calyx in the staminate flowers 2- to 4-parted, absent in the pistillate flowers; petals absent; stamens 2 or 4; ovary superior; ovule 1; fruit a nutlet subtended by leathery or woody bracts.

This family is made up of two genera and about eighty-five species found mostly in the Northern Hemisphere. Some botanists combine the Betulaceae with the Corylaceae, but there seem to be several significant differences between the two families to justify their separation.

1. Bracts woody, persistent; leaves orbicular to obovate .... 1. *Alnus*
1. Bracts leathery, deciduous; leaves ovate to oblong .... 2. *Betula*

1. **Alnus** Mill.—Alder

Shrubs or trees, the bark usually remaining tight; winter buds with few scales; leaves alternate, simple, toothed, stipulate; flowers unisexual, borne in catkins, the staminate catkins slender and pendulous, the pistillate catkins shorter, thicker, and more or less erect; staminate flowers usually borne 3 or 6 together, subtended by a bract, each flower composed of a 4-lobed calyx and 4 stamens; pistillate flowers usually 2–3 together, subtended by a bract, each flower composed only of an inferior ovary and 2 styles; fruit composed of woody bracts subtending small, compressed, winged or wingless nuts.

Alnus is a genus of about fifteen species native to the Northern Hemisphere of both the Old and New World and in the Andes of South America. The following two shrubs may sometimes be found in shallow standing water.

1. Leaves ovate to elliptic, usually whitened beneath .... 1. *A. rugosa*
1. Leaves obovate, green beneath .... 2. *A. serrulata*

1. **Alnus rugosa** (DuRoi) Spreng. var. **americana** (Regel) Fern. Rhodora 47:350. 1945. Fig. 119.
Alnus incana β *americana* Regel, Mem. Soc. Imp. Nat. Moscou Nouv. 13:155. 1861.
Alnus americana (Regel) Czerep. Notul. Syst., Inst. Bot. Komarov. Acad. Sci. URSS 17: 103. 1955.

Shrubs to 5 m tall; bark smooth, brown to dark gray, bearing conspicuous white lenticels; twigs glabrous or nearly so; leaves simple, alternate, ovate to elliptic, obtuse to subacute at the apex, more or less rounded at the base, singly or doubly serrulate, glabrous or nearly so, sometimes pubescent on the veins below, green on the upper surface, whitened on the lower surface, up to 12 cm long, up to 10 cm wide, with petioles up to 2 cm long; catkins appearing before the leaves, the staminate up to 7 cm long, the pistillate ovoid to oblongoid, up to 1 cm long; fruiting "cone" up to 1.5 cm long, sessile or borne on a short stalk; nut orbicular, narrowly marginate. March–May.
Wet woods, bogs, rarely in standing water.
IA, IL, IN, (OBL), OH (FACW).

Fig. 119. *Alnus rugosa*
(Speckled alder).

a. Branch and cones.
b. Staminate catkins.
c. Staminate flower.

d. Bract from cone.
e. Seed.

Speckled alder; tag alder.

This species is distinguished from *A. serrulata* by its doubly toothed leaves that are whitened on the lower surface.

2. **Alnus serrulata** (Ait.) Willd. Sp. Pl. 4:336. 1805. Fig. 120.

Betula serrulata Ait. Hort. Kew. 3:338. 1789.

Shrubs or several-stemmed small trees to 7 m tall; bark smooth, brown to dark gray, bearing pale lenticels; young branchlets usually pubescent, at maturity becoming glabrous; leaves simple, alternate, obovate, obtuse at the apex, tapering to the base, singly serrate, glabrous on the upper surface, usually pubescent, at least on the veins below, green on both sides, up to 10 cm long, up to 8 cm broad, with petioles up to 2 cm long; catkins appearing before the leaves, the staminate up to 10 cm long, the pistillate ovoid, up to 1 cm long; fruiting "cone" up to 1.5 cm long, sessile or on a short stalk; nut ovoid, narrowly marginate. February–May.

Along and in streams.

IL, IN, KS, KY, MO, OH (OBL).

Brookside alder.

This species is distinguished from other members of the genus by its obovate leaves that are green on the lower surface.

2. **Betula** L.—Birch

Trees or shrubs, often aromatic, the bark sometimes peeling into papery shreds; winter buds with many scales; leaves alternate, simple, toothed, stipulate; flowers unisexual, the staminate in elongated, pendulous catkins, the pistillate in short, erect spikes; staminate flowers 3 together, subtended by a bract, with a (2-) 4-toothed calyx, no petals, and 2 stamens; pistillate flowers 2–3 together, subtended by a 3-lobed bract, with no sepals and no petals, and with a sessile ovary with 2 styles; fruit a small, winged nut.

Betula is a genus of thirty-five often attractive woody plants found in north temperate and arctic regions of both the Old and the New World. Many of the arctic species are dwarfed shrubs. Several species are grown as ornamentals. Three species may sometimes be found in standing water in the central Midwest.

1. Leaves with 8 or more pairs of lateral veins; bark brownish to pinkish, peeling into shaggy pieces .... 1. *B. nigra*
1. Leaves with up to 7 (–8) pairs of lateral veins; bark peeling into white layers, or bark dark and not peeling.
    2. Bark white, peeling into thin layers; trees ....2. *B. papyrifera*
    2. Bark dark, not peeling into thin layers; shrubs .... 3. *B. pumila*

1. **Betula nigra** L. Sp. Pl. 982. 1753. Fig. 121.

Trees to 25 m tall, with a trunk diameter up to 0.7 m, the crown irregularly rounded; bark curling, shredding, brownish pink to reddish brown; twigs slender, reddish brown, with several short hairs, the buds up to 0.5 mm long, pointed, hairy; leaves simple, alternate, rhombic to ovate, coarsely doubly toothed, paler and densely hairy on the lower surface, up to 6 cm long, up to 4 cm wide, acute at the apex,

Fig. 120. *Alnus serrulata*
(Brookside alder).

a. Twig with mature fruiting cones.
b. Young branch with staminate
and pistillate catkins.

c. Staminate flower.
d. Pistillate catkin and bract.
e. Seed.

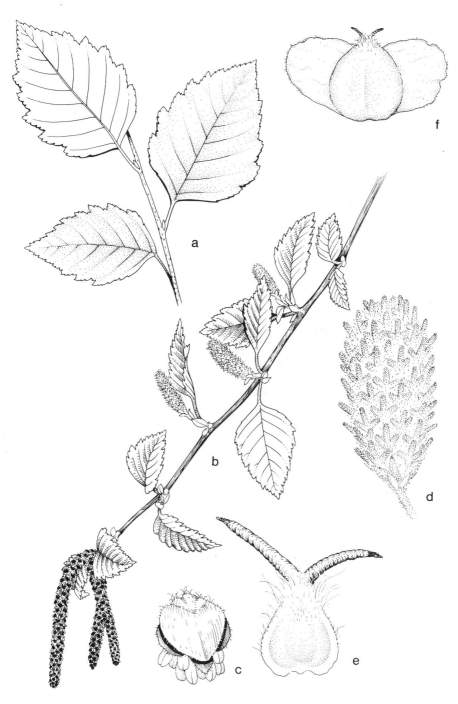

Fig. 121. *Betula nigra*
(River birch).

a. Leaves.
b. Flowering branch.
c. Staminate flower with calyx.
d. Pistillate catkin, mature.
e. Pistillate flower.
f. Winged seed.

truncate or tapering to the base, the petioles woolly; staminate and pistillate flowers on same tree, inconspicuous, the staminate in slender drooping clusters, the pistillate in short, conelike, woolly clusters; nutlets pubescent, each with a 3-lobed wing, crowded together in a cylindrical cone up to 3 cm long and 1.5 cm thick. April–May.

Along rivers and streams; bottomland woods.

IA, IL, IN, KS, KY, MO, OH (FACW).

River birch; red birch.

The shaggy, peeling reddish brown bark readily distinguishes this tree from any other in the central Midwest.

2. **Betula papyrifera** Marsh. Arbust. Am. 19. 1785. Fig. 122.
Betula papyracea Ait. Hort. Kew. 3:337. 1789.
Betula alba L. var. *papyrifera* (Marsh.) Spach, Ann. Sci. Nat. II, 15:187. 1841.

Trees to 20 m tall, with a trunk diameter up to 0.75 m, with slender, pendulous branches; bark cream or white or sometimes orange-tinted, separating into thin, papery layers; twigs at first orange, then changing to deep orange-brown and finally white, lenticellate, the winter buds obovoid, acute, brown, resinous, up to 5 mm long; leaves firm, ovate, acute to acuminate at the apex, more or less rounded at the base, coarsely doubly serrate, dark green on the upper surface, pale yellow-green on the lower surface, the lower surface glabrous or with tufts of axillary hairs, usually dotted with black glands, up to 7 cm long, up to 4.5 cm wide; petioles stout, yellow, glabrous or pubescent, glandular, up to 16 mm long; staminate catkins usually several in a cluster, pendulous, up to 8 cm long, up to 7 mm thick, red-brown; pistillate catkins arching or becoming somewhat pendulous, pedunculate, oblongoid-cylindrical, up to 2.5 cm long, becoming 3–4 cm long in fruit, the bracts usually 3-lobed; nut ellipsoid, 1.5–2.0 mm long, much narrower than the wing. April–May.

Wooded ravines, rarely in fens and rarely in standing water.

IA, IL, IN (FACU+), OH (FACU).

Paper birch; canoe birch.

Although not usually considered to be a wetland plant, paper birch occasionally occurs in degraded fens. The white bark is the most distinguishing characteristic.

3. **Betula pumila** L. Mant. 124. 1767. Fig. 123.
Betula pumila L. var. *glabra* Regel, Bull. Soc. Nat. Moscou 38, pl. 2. 1861.
Betula pumila L. var. *glandulifera* Regel, Bull. Soc. Nat. Moscou 38:410. 1861.
Betula glandulifera (Regel) Butler, Bull. Torrey Club 36:425. 1909.

Shrubs to 3 m tall; twigs slender, brown, pubescent or rarely glabrous, glandular or eglandular, the winter buds ovoid, more or less acute, up to 5 mm long; leaves obovate to orbicular, obtuse to subacute at the apex, rounded or tapering to the base, coarsely toothed, dull green and usually glabrous on the upper surface, paler and pubescent or glabrous on the lower surface, usually up to 3 cm long, up to 2.5 cm wide; petioles pubescent or glabrous, up to 6 mm long; staminate catkins few together, pendulous, up to 4 cm long, up to 5 mm thick, brown; pistillate catkins up to 1.5 cm long, becoming up to 3 cm long and 1 cm broad in fruit, with bracts pubescent, 3-lobed, with the middle lobe much elongated; nut oblongoid, up to 4 mm long, usually a little broader than the wings. May.

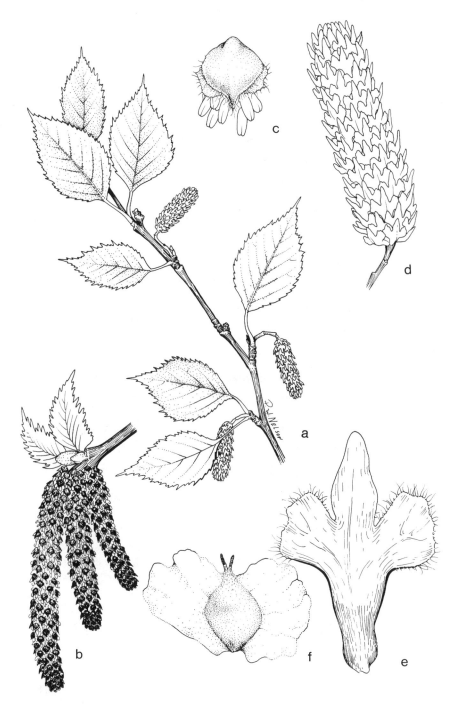

Fig. 122. *Betula papyrifera* (Paper birch).

a. Mature branch.
b. Young branch with staminate catkins.
c. Staminate flower with calyx.

d. Pistillate catkin.
e. Pistillate bract.
f. Winged seed.

Bogs.

IA, IL, IN, OH (OBL).

Bog birch; dwarf birch.

This is the only birch in the central Midwest that is a shrub and not a tree. Some variation exists on vegetative characteristics. Typical var. *pumila* has glabrous, eglandular twigs with leaves that are pubescent and eglandular on the lower surface. Var. *glandulifera* has glandular, glabrous twigs and leaves that are glandular on the lower surface. Var. *glabra* has glabrous and eglandular twigs and leaves.

48. BIGNONIACEAE—BIGNONIA FAMILY

Woody vines (in our area); leaves opposite, compound, without stipules; flowers large, showy; sepals 5, united below; petals 5, united below into a tube; stamens 2 or 4; ovary superior, 2-locular; fruit a 2-valved capsule, with flat winged seeds.

Approximately one hundred genera and eight hundred species, mostly tropical, comprise this family.

Two genera have species in wetland habitats in the central Midwest.

1. Trees; leaves simple, whorled, large, cordate; flowers white ....3. *Catalpa*
1. Woody vines; leaves compound, opposite, not cordate; flowers red or yellow.
   2. Leaflets 2; tendrils present ....1. *Bignonia*
   2. Leaflets 5 or more; tendrils absent... 2. *Campsis*

1. **Bignonia** L.—Crossvine

Woody vine with tendrils; leaves opposite, compound, with 2 leaflets; inflorescence in axillary cymes; flowers large, showy; calyx 5-toothed; corolla tubular, with 5 lobes; stamens 4; ovary superior; fruit a capsule with several flat, winged seeds.

Only the following species comprises the genus.

1. **Bignonia capreolata** L. Sp. Pl. 624. 1753. Fig. 124.

Woody vine to 20 m long, glabrous; leaves opposite, terminated by a branched tendril with adhesive disks, petiolate; leaflets 2, oblong to ovate, acute to acuminate at the apex, subcordate or rounded at the base, entire, to 15 cm long, to 7 cm wide, sometimes with small leaflets in the axils resembling stipules; flowers 1–5 in cymes, large, showy; calyx campanulate, 5–8 mm long, with 5 minute teeth; corolla short-tubular, to 4.5 cm long, expanding into 5 subacute lobes, red-orange and puberulent on the outside, yellow on the inside; stamens 4, 2 longer than the others, included within the corolla tube; capsules somewhat woody, flat, to 20 cm long, to 1.7 cm wide; seeds several, flat, broadly winged, to 4 cm long. April–June.

Swampy woods, mesic woods.

IL, IN, MO (FACW), KY, OH (FAC).

Crossvine.

This is the only woody vine in the central Midwest with opposite leaves, each of which is bifoliolate.

A cross-section of the woody stem reveals a cross-shaped structure, accounting for the common name.

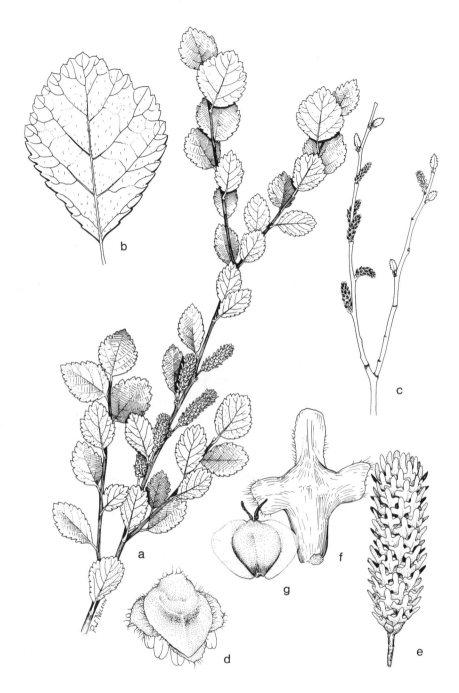

Fig. 123. *Betula pumila* (Bog birch).

a. Habit with pistillate catkins.
b. Leaf.
c. Twig with staminate catkins.
d. Staminate flower with calyx.

e. Pistillate catkin.
f. Pistillate bract.
g. Winged seed.

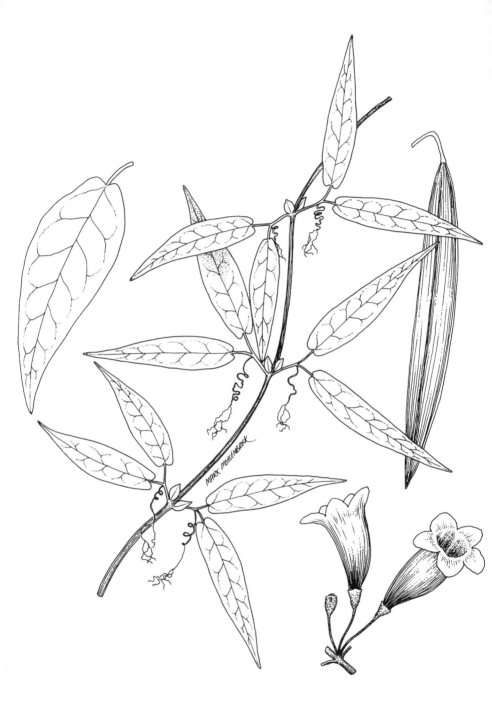

Fig. 124. *Bignonia capreolata* (Crossvine). Habit (center). Flowers (lower right). Leaf (upper left). Fruit (upper right).

2. Campsis Lour.—Trumpetvine

Woody vine without tendrils, but with aerial roots; leaves opposite, pinnately compound, with several leaflets; inflorescence a few-flowered corymb; flowers large, showy; calyx 5-parted; corolla tubular, with 5 lobes; stamens 4; ovary superior; fruit a capsule with several flat, winged seeds.

The following species and one in Asia comprise the genus.

1. **Campsis radicans** (L.) Seem. Journ. Bot. 5:362. 1867. Fig. 125.

Bignonia radicans L. Sp. Pl. 624. 1753.

Tecoma radicans (L.) DC. Prodr. 9:223, 1845.

Woody vine to 15 m long, glabrous; leaves opposite, petiolate; leaflets 5–13, lanceolate to ovate, acute to acuminate at the apex, somewhat rounded at the base, strongly reticulate-veined, sharply serrate, to 8 cm long, to 5 cm wide; flowers large, showy; calyx campanulate, 5-lobed at the apex, tubular, 10–15 mm long; corolla tubular, with 5 obtuse shallow lobes at the apex, to 8 cm long, red-orange on the outside, yellow on the inside; stamens 4, 2 longer than the others, included within the corolla tube; capsules woody, ellipsoid, to 15 cm long; seeds numerous, broadly winged, the wings erose at tip. June–September.

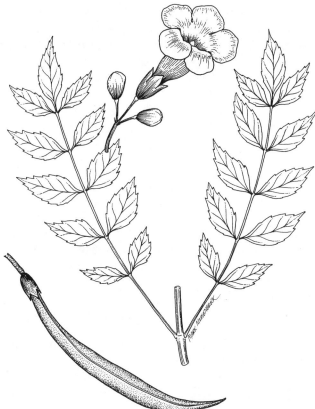

Fig. 125.
Campsis radicans
(Trumpetvine).
Leaves with flower
(center). Fruit (lower
left).

Swampy woods, mesic woods, old fields, fence rows.

IA, IL, IN, KS, KY, MO, NE, OH (FAC).

Trumpetvine; trumpet creeper.

This is the only vine in the central Midwest with opposite leaves and 5–13 leaflets. The showy tubular flowers are attractive to hummingbirds.

Trumpetvines occur in a variety of wet and dry habitats, particularly swampy woods.

3. **Catalpa** Scop.—Catalpa

Trees; leaves simple, whorled or opposite; flowers in large terminal panicles of corymbs, zygomorphic, perfect; calyx 2-lipped; corolla campanulate, 2-lipped, 5-lobed; stamens 2, with 3 staminodia; ovary superior, bilocular; capsule elongated, with flat, winged seeds.

Catalpa contains approximately six species in North America and eastern Asia.

1. **Catalpa speciosa** Warder ex Engelm. Bot. Gaz. 5:1. 1880. Fig. 126.

Medium to tall trees to 25 m tall, with a trunk diameter up to 2 m and with a broad, widely spreading crown; bark light brown, dark brown, or blackish, usually with rather deep furrows; twigs stout, glabrous, brown, with conspicuous lenticels, the buds round, brown to black, glabrous, to 2 mm long; leaves simple, whorled, ovate, acuminate at the apex, cordate at the base, up to 30 cm long, up to 20 cm wide, entire, dark green and glabrous or sparsely pubescent on the upper surface, softly pubescent on the lower surface, the petioles stout, up to 15 cm long; flowers large, showy, several in an elongated cluster, appearing after the leaves have expanded, the inflorescence up to 15 cm long; each flower zygomorphic, up to 4.5 cm long, the petals white and lined with purple; capsules elongated, up to 45 cm long, up to 1.5 cm wide, brown, splitting into 2 parts to reveal several winged, pubescent seeds about 2 cm long. May–June.

Swampy woods.

IA, IL, IN, MO (FACU), KS, NE (FAC–) KY, OH (FAC).

Northern catalpa.

This tree is recognized by its large cordate, whorled leaves that lack an unpleasant odor and by its clusters of large white flowers.

49. BORAGINACEAE—BORAGE FAMILY

Herbs; leaves alternate, simple, often entire; flowers perfect, actinomorphic, often in scorpioid cymes; sepals 5, free or united; petals 5, attached to each other; stamens 5, attached to the corolla tube; ovary superior, usually deeply 4-lobed; fruit usually a cluster of four 1-seeded nutlets.

Although the flowers of this family usually have deeply 4-parted ovaries as in the mint family, the leaves are always alternate.

This family consists of approximately one hundred genera and about two thousand species found worldwide. Most of the species are found in the uplands of the central Midwest.

1. Leaves ovate; flowers crowded in coiled spikes... 1. *Heliotropium*
1. Leaves lanceolate to oblong to elliptic; flowers loosely arranged in racemes........ 2. *Myosotis*

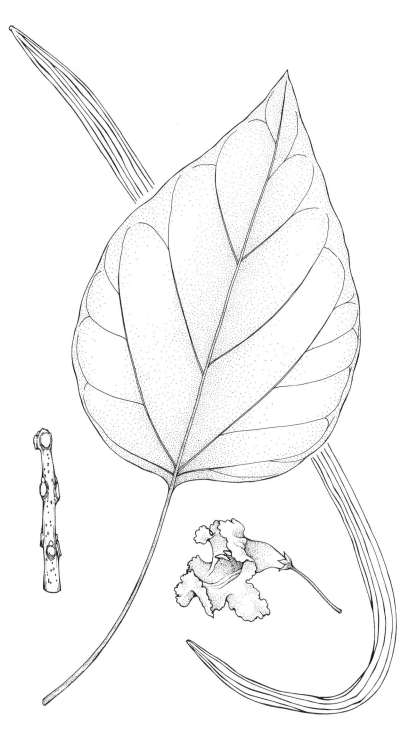

Fig. 126. *Catalpa speciosa* (Northern catalpa). Leaf and fruit (center). Twig (left). Flower (lower right).

1. **Heliotropium** L.—Heliotrope

Herbs (in our area) or shrubs; leaves alternate, simple, entire, petiolate; flowers in coiled spikes, crowded or scattered; sepals 5, united below; corolla salverform or funnelform, 5-lobed; stamens 5; ovary superior; fruit 2- or 4-lobed, separating into 1-seeded nutlets.

More than two hundred species, mostly tropical and subtropical, comprise this genus. One non-native species occurs in wetlands in the central Midwest.

1. **Heliotropium indicum** L. Sp. Pl. 1:130. 753. Fig. 127.

Annual herbs with a taproot; stems erect, branched, hirsute, to 75 cm tall; leaves simple, alternate, ovate to oval, obtuse to acute at the apex, tapering or rounded at the base, strigose on both surfaces, to 15 cm long, to 8 cm wide, on slightly winged, hirsute petioles up to 4 cm long; inflorescence a coiled spike up to 10 cm long, bearing numerous flowers; sepals 5, lanceolate, acute, pubescent, united at base, 1–2 mm long; corolla 5-lobed, tubular below, blue, 3–5 mm long, strigose; stamens 5; ovary superior, 2-lobed; fruit deeply 2-lobed, glabrous, 3–4 mm long, separating at maturity into 4, 1-seeded nutlets. July–October.

Disturbed wet areas.

IL, IN, KS, MO (FACW), KY, OH (FAC+).

Indian heliotrope.

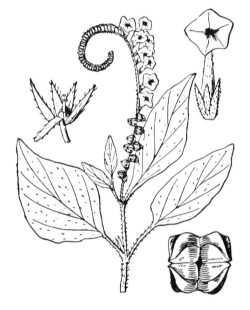

Fig. 127. *Heliotropium indicum* (Indian heliotrope). Habit (center). Calyx and pistil (upper left). Flower (upper right). Fruit (lower right).

The coiled inflorescence of blue flowers and the hirsute stems are distinctive for this species native of Asia.

2. **Myosotis** L.—Forget-me-not; scorpion grass.

Herbs; leaves alternate, simple, entire; flowers perfect, actinomorphic; sepals 5, attached at least at base; petals 5, the free parts about equaling the tube; stamens 5, not exserted; ovary superior, 4-lobed; nutlets compressed.

Myosotis consists of about fifty species in temperate and boreal regions. Two of them may occur in standing water in the central Midwest.

1. Plants not stoloniferous, the stems terete; branches of inflorescence subtended by small bracts; corolla 3–6 mm across .... 1. *M. laxa*
1. Plants stoloniferous, the stems angular; branches of inflorescence not bracteate; corolla 6–9 mm across .... 2. *M. scorpioides*

1. **Myosotis laxa** Lehm. Asperif. 83. 1818. Fig. 128.

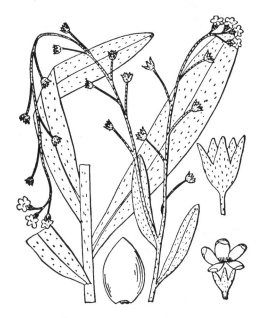

Perennial herbs without stolons; stems terete, to 40 cm long, with appressed hairs; leaves simple, alternate, lanceolate to oblong to elliptic, entire, to 7 cm long, 5–18 mm wide, with appressed pubescence; at least the lowest branches of the inflorescence subtended by bracts; calyx lobes equaling the calyx tube; corolla blue, usually with a yellow center, 3–6 mm across; style shorter than the nutlets. May–September.

Shallow water.

IN, OH (OBL).

Small forget-me-not.

Although similar in appearance to *M. scorpioides*, this species has bracts at the base of the lowest flowers and has the tubes of the calyx about as long as the lobes. This species also lacks stolons.

Fig. 128. *Myosotis laxa* (Small forget-me-not). Leaves (left). Upper part of plant (right of center). Seed (bottom center). Calyx (upper right). Flower (lower right).

2. **Myosotis scorpioides** L. Sp. Pl. 131. 1753. Fig. 129.

Perennial herbs with stolons; stems angular, to 70 cm tall, with appressed pubescence; leaves simple, alternate, linear-oblong to lanceolate, scabrous but with appressed pubescence, entire, to 8 cm long, 7–20 mm wide, entire; inflorescence not subtended by bracts; calyx lobes shorter than the calyx tube; corolla blue, usually with a yellow center, 6–9 mm across; style longer than the nutlets. May–October.

Shallow water.

IL, IN, KY, MO, OH (OBL)

Forget-me-not; aquatic scorpion-grass.

This species often is found in clear spring water where it is frequently associated with water cress (*Nasturtium officinale*). This European native differs from *M. laxa* by its stolons, larger flowers, and calyx lobes shorter than the calyx tube.

50. BRASSICACEAE—MUSTARD FAMILY

Herbs, often with stellate pubescence; leaves alternate, simple or pinnatifid or palmately lobed or pinnately divided, without stipules; inflorescence mostly in bractless racemes, the flowers bisexual, usually actinomorphic; sepals 4, free, in 2 whorls; petals 4, free; stamens mostly 6, free, the outer whorl of two usually shorter than the inner whorl of four; pistil one, the ovary superior, 2-locular and 2-carpel-

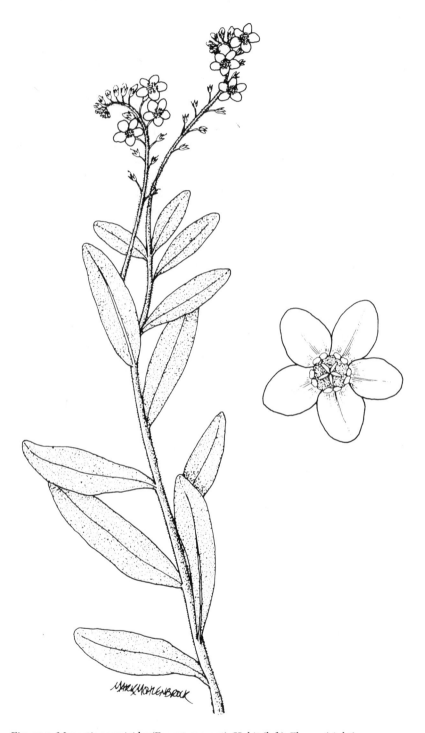

Fig. 129. *Myosotis scorpioides* (Forget-me-not). Habit (left). Flower (right).

late, the 2 locules separated by a membranous replum, the placentation parietal, with few to several ovules; fruit dehiscent, valvate, either an elongated silique or a short silicle; seeds several.

This family is composed of about 350 genera and 3,400 species, mostly in the temperate regions of the Northern Hemisphere. Most botanists in the past have called this family the Cruciferae. Four genera in the central Midwest may have species that sometimes live in standing water.

The Brassicaceae is one of the most easily recognized families because of the four free petals, six stamens of two different lengths, and a fruit separated into two compartments by a membranous partition known as a replum. In addition, the leaves are almost invariably alternate.

1. Flowers white or pinkish purple.
  2. Cauline leaves simple .... 1. *Cardamine*
  2. All leaves pinnately divided or divided into capillary divisions.
    3. Leaves divided into capillary divisions.. 3. *Neobeckia*
    3. Leaves divided but not into capillary divisions.
      4. Petals 3–6 mm long; pedicels spreading or recurving .... 2. *Nasturtium*
      4. Petals up to 4 mm long; pedicels ascending.. 1. *Cardamine*
1. Flowers yellow .... 4. *Rorippa*

1. **Cardamine** L.—Bitter Cress

Perennial (less commonly annual) herbs from bulbous rootstocks or fibrous roots; leaves cauline and basal, simple or pinnately divided; inflorescence mostly racemose, with white or purple flowers; sepals 4, free; petals 4, free; stamens (4–) 6; pistil one, the ovary superior, the style small; siliques linear, flat, dehiscing elastically from the base, the valves essentially nerveless, the seeds compressed, wingless, arranged in one row in each cell.

There are nearly two hundred species in this genus, which are found worldwide. Those that may live in shallow water in the central Midwest may be separated by the following key:

1. Petals 5 mm long or longer; cauline leaves simple.
  2. Stems glabrous; sepals green; petals white.
    3. Stems erect; leaves sessile or nearly so; petals 7–15 mm long .... 4. *C. rhomboidea*
    3. Stems decumbent; leaves petiolate; petals 5–10 mm long .... 5. *C. rotundifolia*
  2. Stems hirsute; sepals purplish; petals usually pinkish purple..................... 1. *C. douglassii*
1. Petals up to 4 mm long; cauline leaves pinnate to pinnatifid (rarely simple).
  4. Terminal leaflet about the same width as the lateral leaflets; bases of leaflets not decurrent along the rachis .... 2. *C. parviflora*
  4. Terminal leaflet broader than the lateral leaflets; bases of leaflets decurrent along the rachis .... 3. *C. pensylvanica*

1. **Cardamine douglassii** (Torr.) Britt. Trans. N. Y. Acad. Sci. 9:8. 1889. Fig. 130.
Arabis rhomboidea Pers. var. *purpurea* Torr. Am. Journ. Sci. 4:66. 1822.
Arabis douglassii Torr. in Torr. & Gray, Fl. N. Am. 1:83. 1838, *pro syn.*
Cardamine rhomboidea (Pers.) DC. var. *purpurea* (Torr.) Torr. Fl. N. Y. 1:56. 1843.
Cardamine bulbosa (Schreb.) BSP. var. *purpurea* (Torr.) BSP. Prel. Cat. N. Am. 4. 1888.

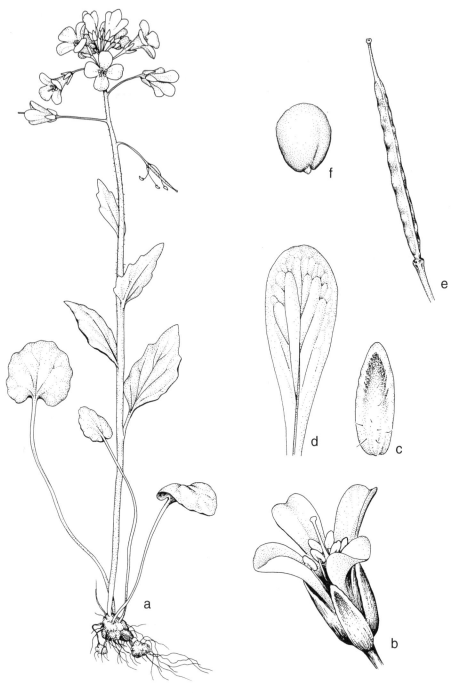

Fig. 130. *Cardamine douglassii*
(Purple spring cress).

a. Habit.
b. Flower.
c. Sepal.

d. Petal.
e. Fruit.
f. Seed.

Perennials from a swollen bulbous base and from tuber-bearing rootstocks; stems erect, usually unbranched, to 30 cm long, hirsute, at least in the upper half; basal leaves ovate to orbicular, cordate at the base, glabrous or sparsely pubescent, the blade to 2.5 cm long, angulate-lobed, the petiole longer than the blade; cauline leaves ovate, obtuse to subacute at the apex, rounded to subcordate at the base, entire to dentate, to 3 cm long, glabrous or sparsely pubescent, all but the lowest 1–2 sessile; racemes terminal, the flowers several, to 2 cm broad, on spreading-ascending pedicels to 2.5 cm long; sepals 4, elliptic to lanceolate, 3–6 mm long, purple; petals 4, rose-purple or pink, to 2 cm long; siliques nearly straight, ascending, linear to narrowly lanceoloid, to 2.5 cm long, the style persistent as beak 3–5 mm long, tipped by a prominent stigma, the seeds several, oval. March–May.

Wet woods, around springs, seldom in standing water.

IA, IL, IN, MO (FACW), KY, OH (FACW+).

Purple spring cress.

This species, generally less common than *C. rhomboidea*, differs by its hirsute stems and usually pinkish purple petals and purple sepals.

2. **Cardamine parviflora** L. var. **arenicola** (Britt.) O. E. Schulz, Bot. Jahrb. 32:485. 1903. Fig. 131.
Cardamine virginica Michx. Fl. Bor. Am. 2:29. 1803, non L. (1753).
Cardamine arenicola Britt. Bull. Torrey Club 19:220. 1892.

Annual or biennial herbs from fibrous roots; stems erect, usually branched, to 35 cm tall, glabrous; basal leaves several, to 10 cm long, on glabrous petioles, pinnately divided into 5–13 leaflets, the leaflets all nearly uniform in size and shape, linear to linear-oblong or occasionally obovate, entire or few-toothed, glabrous; cauline leaves pinnately divided into 5–13 usually uniform leaflets, linear to oblanceolate, entire, glabrous; racemes terminal, several-flowered, the flowers 4–5 mm broad, on ascending pedicels; sepals 4, elliptic, glabrous, 1–2 mm long, green; petals 4, 2.5–3.5 mm long; stamens 6; siliques straight, erect, linear, glabrous, to 3 cm long, the short slender style persistent as a beak up to 1 mm long, the seeds numerous, 1.0–1.5 mm long, wingless. April–August.

Dry or moist soil, rarely in standing water except sometimes in shallow depressions where water may stand for several months.

IA, IL, IN, MO (FAC), KS (FACW), KY, OH (FACU).

Small-flowered bitter cress.

North American plants are referred to as var. *arenicola*, differing from typical var. *parviflora* by their larger petals, which are 2.5–3.5 mm long.

Sometimes, at least in southern Illinois, this species may occur in shallow water in depressions on ledges and in swampy woods.

3. **Cardamine pensylvanica** Muhl. ex Willd. Sp. Pl. 3:486. 1800. Fig. 132.

Annual or biennial herbs from fibrous roots; stems erect, usually branched, to 75 cm tall, glabrous; basal leaves several, to 15 cm long, on glabrous petioles, pinnately divided into 3–17 leaflets, the terminal leaflet obovate to suborbicular, larger and broader than the lateral leaflets, dentate or undulate, rarely entire, glabrous; cau-

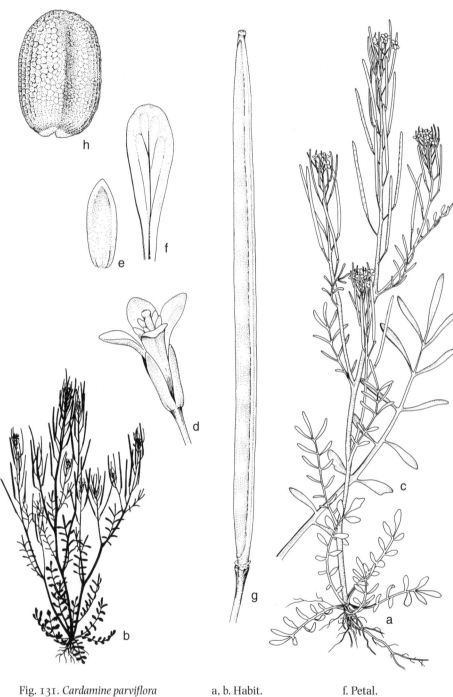

Fig. 131. *Cardamine parviflora*
(Small-flowered bitter cress).

a, b. Habit.
c. Leaf.
d. Flower.
e. Sepal.

f. Petal.
g. Fruit.
h. Seed.

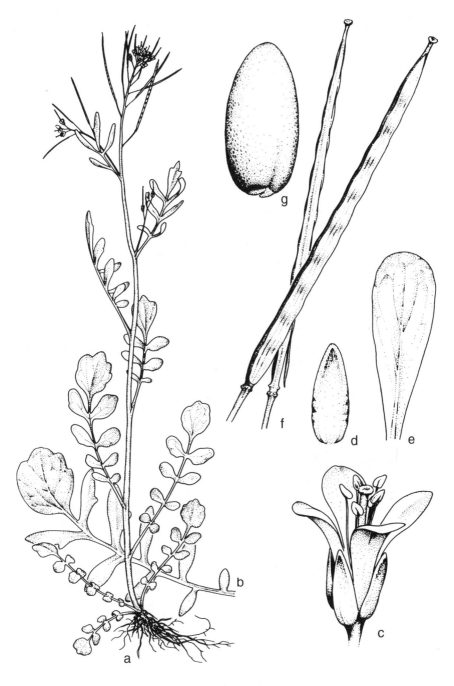

Fig. 132. *Cardamine pensylvanica*
(Bitter cress).

a. Habit.
b. Leaf.
c. Flower.
d. Sepal.

e. Petal.
f. Fruits.
g. Seed.

line leaves pinnately divided into 9–17 leaflets, the leaflets linear to oblanceolate, entire or dentate or undulate, glabrous; racemes terminal, several-flowered, the flowers 4–5 mm broad, on ascending pedicels; sepals 4, elliptic, glabrous, 1–2 mm long, green; petals 4, white, 2–4 mm long; stamens 6; siliques straight, erect, linear, glabrous, to 3.5 cm long, the slender style persistent as a beak up to 2 mm long, the seeds numerous, 1.0–1.5 mm long, wingless. March–October.

Swamps, wet woods, base of wet cliffs, damp fields.

IA, IL, IN, MO (FACW+), KS, NE (FACW), KY, OH (OBL).

Bitter cress.

This species resembles *C. parviflora* var. *arenicola* but differs by its terminal leaflet much larger than the lateral leaflets.

4. **Cardamine rhomboidea** DC. Syst. Veg. 2:246. 1821. Fig. 133.
Arabis bulbosa Schreb. ex. Muhl. Trans. Am Phil. Soc. 3:174. 1793, misapplied.
Cardamine bulbosa (Muhl.) BSP. Prel. Cat. N. Y. 4. 1888.

Perennial herbs from a swollen bulbous base and from tuber-bearing rootstocks; stems erect, usually unbranched, to 50 cm tall, glabrous; basal leaves oblong to nearly orbicular, usually cordate at the base, glabrous, the blade to 3 cm long, the glabrous petiole usually longer than the blade; cauline leaves ovate to lanceolate, subacute at the apex, rounded or subcuneate at the base, entire to dentate, to 4.5 cm long, glabrous, all but the lowest few sessile; racemes terminal, the flowers several, to 1.5 cm broad, on strongly ascending, glabrous pedicels to 2 cm long; sepals 4, elliptic to lanceolate, 2.5–5.0 mm long, green with a white margin; petals 4, white, to 1.5 cm long; siliques nearly straight, erect, linear to narrowly lanceoloid, to 2.5 cm long, the style persistent as a beak 2–4 mm long, tipped by a prominent stigma, the seeds several, oval. March–June.

Wet woods, along streams, fens, around springs, occasionally in standing water.

IA, IL, IN, KS, KY, MO, NE, OH (OBL).

Spring cress; bulbous cress.

This species, *C. douglassii*, and *C. rotundifolia* are the only species of the mustard family that sometimes may be found in standing water that have simple leaves. It differs from *C. douglassii* by its smooth stems, green sepals, and white flowers. The stems of *C. rotundifolia* are decumbent.

5. **Cardamine rotundifolia** Michx. Fl. Bor. Am. 2:30. 1803. Fig. 134.

Weak, decumbent perennials, with fibrous roots, forming long stolons; stems branched, glabrous; leaves simple, alternate, simple or occasionally with a pair of tiny leaflets near the base, ovate to orbicular, obtuse at the apex, rounded cordate, or tapering at the base, sinuate-toothed, glabrous, the lower leaves long-petiolate, the uppermost leaves sessile; flowers in axillary and terminal racemes, actinomorphic, on pedicels up to 2 cm long; sepals 4, free, green, oblong, acute, 2–4 mm long; petals 4, free, white, 5–10 mm long; stamens 6; fruit linear, pointed at the apex, glabrous, 12–15 mm long, up to 2 mm wide; seeds oblong. May–June.

Swampy woods; in running water; springs.

KY, MO, OH (OBL).

Fig. 133. *Cardamine rhomboidea*
(Spring cress).

a. Habit.
b. Flower (with one sepal
and one petal removed).
c. Sepal.

d. Petal.
e. Fruiting raceme.
f. Fruit.
g. Seed.

Round-leaved water-cress.

This sprawling plant is distinguished by its white petals 5–10 mm long and its usually simple, sessile cauline leaves.

Fig. 134. *Cardamine rotundifolia* (Round-leaved watercress). Habit. Flower (next to upper right). Fruit (upper right).

2. **Nasturtium** R. Br.—Water Cress

Perennial herbs; leaves pinnately compound; racemes terminal; flowers actino-morphic, bractless; sepals 4, free; petals 4, white, free; stamens 6; pistil 1, the ovary superior, the style stout; siliques cylindrical, the valves nerveless, with several wing-less seeds arranged in two rows in each cell.

Only the following species comprise the genus. Some botanists place these species in the genus *Rorippa*, a view not followed here.

1. Fruits 1.0–1.5 mm wide; lateral leaflets petiolulate; seeds in 1 row in each locule, with 100–175 depressions on each face ....1. *N. microphyllum*
1. Fruits 1.5–2.5 mm wide; lateral leaflets sessile; seeds in 2 rows in each locule, with 25–60 depressions on each face .... 2. *N. officinale*

1. **Nasturtium microphyllum** Boenn. ex Reichenb. Fl. Germ. Excurs. 683. 1832. Fig. 135.
Nasturtium officinale R. Br. var. *microphyllum* (Boenn. ex Reichenb.) Thell. Mem. Soc. Nat. Cherb. ser. 4, 38:280. 1911.
Rorippa microphylla (Boenn. ex Reichenb.) Hyl. ex A. Love & D. Love, Iceland Univ. Inst. Univ. Inst. Appl. Sci. Dept. Agr. Rep. 3:109. 1948.

Perennial herbs from fibrous roots; stems creeping or floating, rather fleshy, glabrous, rooting at many of the nodes; leaves pinnately compound, petiolate, the leaflets 3–11, oblong to oval to nearly orbicular, obtuse at the apex, rounded at the petiolulate base, entire to dentate-undulate, glabrous, sometimes fleshy; racemes terminal or from the uppermost axils on arched-recurving or ascending pedicels,

the flowers 4–6 mm broad, bractless; sepals 4, elliptic to narrowly oblong, obtuse, green, glabrous, 1.5–2.5 mm long; petals 4, white, 3–6 mm long; siliques linear, cylindric, arcuate to straight, glabrous, to 2.7 cm long, 1.0–1.5 mm broad, the pedicels arched-recurving to ascending, with several seeds in 1 row in each cell, with 100–175 depressions on each face. April–October.

Springs, brooks. NE (not listed by the U.S. Fish and Wildlife Service). Lesser water cress.

This species, often considered a variety of *N. officinale*, differs by its narrower siliques, its petiolulate leaflets, and the seeds in one row in each locule.

Fig. 135. *Nasturtium microphyllum* (Lesser water cress). Habit (left). Flower (bottom right). Fruiting branch (far upper right). Fruits (near upper right).

2. **Nasturtium officinale** R. Br. in Ait. Hort. Kew. ed., 2, 4. 109. 1812. Fig. 136.
Sisymbrium nasturtium-aquaticum L. Sp. Pl.657. 1753.
Nasturtium siifolium Reichenb. Fl. Germ. Excurs. 683. 1830.
Radicula nasturtium-aquaticum (L.) Britten & Rendle, Brit. Seed Plants 3. 1907.

Perennial herbs from fibrous roots; stems creeping or floating, rather fleshy, glabrous, rooting at many of the nodes; leaves pinnately compound, petiolate, the leaflets 3–11, oblong to oval to nearly orbicular, obtuse at the apex, rounded at the usually sessile base, entire to dentate-undulate, glabrous, sometimes fleshy; racemes terminal and from the uppermost axils on spreading to ascending pedicels, the flowers 4–6 mm broad, bractless; sepals 4, elliptic to narrowly oblong, green, glabrous, 1.5–2.5 mm long; petals 4, white, 3–6 mm long; siliques linear, cylindric, arcuate to

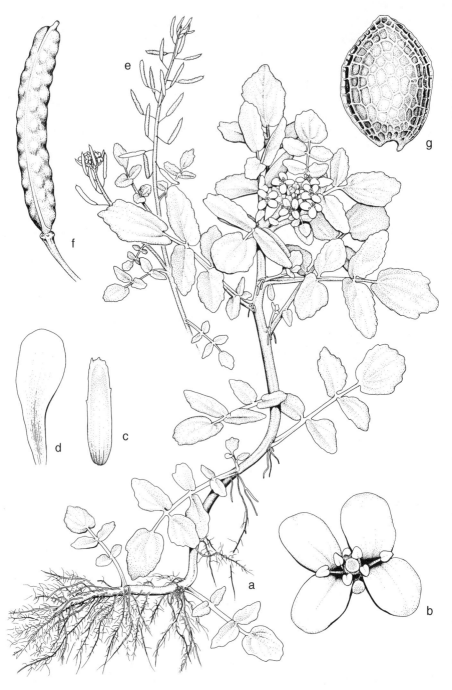

Fig. 136. *Nasturtium officinale*
(Water cress).

a. Habit.
b. Flower.
c. Sepal.
d. Petal.

e. Fruiting branch.
f. Fruit.
g. Seed.

straight, glabrous, to 1.8 cm long, 1.5–2.5 mm broad, the beak absent or up to 1.5 mm long, the pedicels spreading to ascending, glabrous, with several seeds arranged in two rows in each cell, with 25–60 depressions on each face. April–October.

Springs, brooks.

IA, IL, IN, KS, KY, MO, NE, OH (OBL).

Water cress.

Although found in most freshwater springs in the central Midwest, this species is probably not native. Its leaves are a delicacy in salads. The ornamental known as nasturtium is in a different family and different genus.

3. **Neobeckia** Greene—Lake Cress

Aquatic perennials; basal leaves immersed, capillary to filiform; emersed leaves broader; flowers racemose; sepals 4, free, green; petals 4, free, white; stamens 6; ovary superior; fruit a 1-locular silicle; seeds wingless.

Only one species comprises this genus. Several botanists include this species in the genus *Armoracia*, but the aquatic nature of this plant with its submerged capillary leaf divisions seems to distinguish these genera, just as *Megalodonta* is distinguished from *Bidens* in the Asteraceae.

1. **Neobeckia aquatica** (Eat.) Greene, Pittonia 3:95. 1896. Fig. 137.
Cochlearia aquatica Eat. Man. Bot. ed. 5, 181. 1829.
Nasturtium natans DC. var. *americanum* Gray, Ann. Lyc. N. Y. 3:223. 1836.
Nasturtium lacustre Gray, Gen. Fl. Am. Ill. 1:132. 1848.
Rorippa americana (Gray) Britt. Mem. Torrey Club 5:169. 1894.
Radicula aquatica (Eat.) Robins. Rhodora 10:32. 1908.
Armoracia aquatica (Eat.) Wieg. Rhodora 27:186. 1925.
Rorippa aquatica (Eat.) Palmer & Steyerm. Ann. Mo. Bot. Gard. 22:550. 1935.

Aquatic perennial herbs from slender rhizomes; stems much branched to simple, glabrous, to 75 cm long; basal submersed leaves to 15 cm long, long-petiolate, 1- to 5-pinnate, the ultimate segments capillary to filiform, entire, glabrous; emersed leaves lanceolate to oblong, obtuse to acute at the apex, cuneate at the sessile base, to 8 cm long, glabrous, entire to serrate; racemes lax, spreading or pendulous, few- to several-flowered, the flowers 5–10 mm broad, bractless; sepals linear to linear-lanceolate, green, glabrous, 2–4 mm long; petals 4, white, 5–9 mm long; silicles ovoid, not flat, glabrous, 1-locular, to 5 mm long, tipped by the persistent elongated style and the bilobed stigma, on spreading to ascending, glabrous pedicels to 1 cm long, with few small, plump, wingless seeds. May–August.

Standing water.

IA, IL, IN, KY, MO, OH (OBL), as *Armoracia aquatica*.

Lake cress.

The emersed leaves may be oblong and entire or merely serrulate, although they may also be finely divided like the submersed leaves.

4. **Rorippa** Scop.—Yellow Cress

Annual, biennial, or perennial herbs; leaves alternate or basal or both, pinnately compound, pinnatifid, or simple and unlobed; inflorescence racemose, the flowers

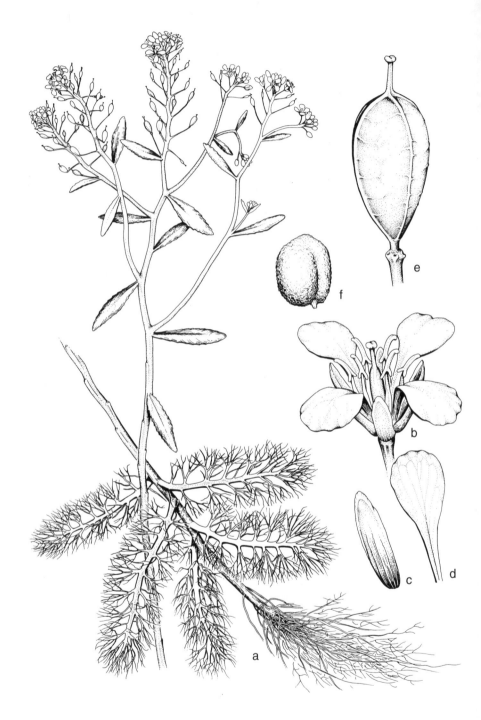

Fig. 137. *Neobeckia aquatica*
(Lake cress).

a. Habit with submersed leaves below.
b. Flower.
c. Sepal.

d. Petal.
e. Fruit.
f. Seed.

yellow, actinomorphic, bractless; sepals 4, free; petals 4, free; stamens usually 6; silicles or siliques terete, the valves nerveless, the seeds turgid, numerous, wingless, arranged in two rows.

There are about eighty species in this genus, found worldwide.

1. Petals more than 3 mm long, longer than the sepals; perennials.
  2. Leaves without auricles at base; seeds up to 0.8 mm long ....5. *R. sylvestris*
  2. Leaves with auricles at base; seeds about 1 mm long .... 4. *R. sinuata*
1. Petals absent, or up to 2.5 mm long, never longer than the sepals; annuals.
  3. Pedicels 3–7 mm long; fruits shorter than to up to two times longer than the pedicels ....
  ....2. *R. palustris*
  3. Pedicels up to 3 (–4) mm long; fruits 2–4 times longer than the pedicels.
    4. Pedicels strongly recurved or spreading at right angles to the axis..............1. *R. curvipes*
    4. Pedicels mostly ascending, sometimes spreading, never recurved.
      5. Petals absent or up to 0.5 mm long; seeds 150–200 per fruit............... 3. *R. sessiliflora*
      5. Petals 0.7–1.2 mm long; seeds up to 80 per fruit.
        6. Petals 0.9–1.2 mm long; fruits glabrous, with 30–80 seeds.................7. *R. truncata*
        6. Petals 0.7–0.9 mm long; fruits papillate, with 20–40 seeds..............6. *R. tenerrima*

1. **Rorippa curvipes** Greene, Pittonia 3:97. 1896. Fig. 138.

Annual or perennial herbs from a taproot; stems erect to decumbent or even prostrate, up to 50 cm long, glabrous; leaves alternate, sessile, oblong to spatulate, obtuse to acute at the apex, 2.5–8.0 cm long, 0.5–2.0 cm wide, pinnatifid or merely irregularly toothed, glabrous or sometimes hirsute above; racemes up to 10 cm long; sepals 1.0–1.5 mm long, green; petals 0.7–1.2 mm long, yellow; fruits on curved or nearly straight pedicels up to 3 (–4) mm long, the fruits up to 5 mm long, 0.8–2.0 mm wide, glabrous; seeds 20–50 per fruit, 0.5–0.7 mm long. May–September.

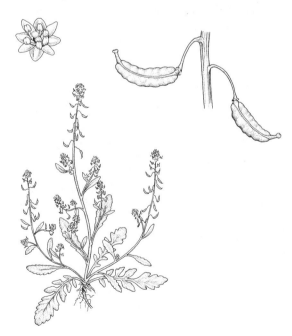

Muddy shores, wet meadows, sometimes in shallow water.

KS, MO (NI).

Curved yellow cress.

The curved pedicels are distinctive for this western species.

Fig. 138. *Rorippa curvipes* (Curved yellow cress). Habit (center, below). Flower (upper left). Fruits (upper right).

2. **Rorippa palustris** (L.) Bess. Enum. Pl. 77. 1822. Fig. 139.
Sisymbrium amphibium L. var. *palustre* L. Sp. Pl. 657. 1753.
Sisymbrium islandicum Oeder, Fl. Dan. 3:8, t. 409. 1768.
Radicula palustris (L.) Moench, Meth. 263. 1794.
Brachylobus hispidus Desv. Journ. Bot. 3:183. 1814.
Nasturtium palustre (L.) DC. Syst. 2:191. 1821.
Nasturtium hispidum (Desv.) DC. Syst. 2:201. 1821.
Nasturtium palustre (L.) DC. var. *hispidum* (Desv.) Gray, Man. ed. 2, 30. 1856.
Rorippa hispida (Desv.) Britt. Mem. Torrey Club 5:169. 1894.
Rorippa islandica (Oeder) Borbas, Bal. Tav. Partm. 392. 1900.
Radicula hispida (Desv.) Britt. Torreya 6:30. 1906.
Radicula palustris (L.) Moench var. *hispidum* (Desv.) B. L. Robins. Rhodora 10:32. 1908.
Roripppa islandica (Oeder) Borbas var. *hispida* (Desv.) Butt. & Abbe, Rhodora 42:26. 1940.
Rorippa islandica (Oeder) Borbas var. *fernaldiana* Butt. & Abbe, Rhodora 42:28. 1940.
Rorippa palustris (L.) Moench var. *elongatum* Stuckey, Sida 4:357, f. 7. 1972.

Annual or biennial herbs from thickened roots; stems erect, simple or branched, glabrous or hirsutulous, to 1.2 m tall; leaves alternate, oblong or oblanceolate, pinnately compound, pinnatifid, or only toothed, firm or membranaceous, glabrous to hirsutulous, the lower leaves usually petiolate, the middle leaves usually auriculate and clasping, the upper leaves usually sessile or nearly so; racemes terminal and from the uppermost leaf axils, several-flowered, the flowers 2–4 mm broad, on spreading or curved-ascending pedicels 3 mm long or longer; sepals 4, lanceolate, green, glabrous or pubescent, 1.5–2.0 mm long; petals 4, oblanceolate, yellow, 1.5–2.0 mm long; stamens 6; siliques ellipsoid to subglobose, sometimes curved, glabrous, 2–10 mm long, short-beaked, on spreading or curved-ascending pedicels 3 mm long or longer, the seeds up to 75 per silique. May–October.

Along rivers and streams, around lakes and ponds, bogs, sometimes in shallow water.

IA, IL, IN, KS, KY, MO, NE, OH (OBL).

Bog yellow cress.

This species is recognized by its fruits that are shorter than or only up to twice as long as the pedicels. The pedicels are 3–7 mm long.

In the past, this species has been known frequently as *R. islandica*. There is considerable variation in the pubescence or lack of it on the stems and leaves and by the size and shape of the fruits. Plants that have leaves completely glabrous on the lower surface are var. *palustris*. Plants with the lower leaf surfaces and the stems hirsute, and with fruits nearly globose or up to twice as long as broad, may be known as var. *hispida*.

Other plants with the lower leaves and the stems hirsute, and with fruits cylindrical, at least twice as long as broad, may be called var. *elongatum*.

3. **Rorippa sessiliflora** (Nutt.) Hitchc. Spring Fl. Manhattan, Kan. 18. 1894. Fig. 140.
Nasturtium sessiliflorum Nutt. ex Torr. & Gray, Fl. N. Am. 1:73. 1838.
Radicula sessiliflora (Nutt.) Greene, Leaflets 1:113. 1905.

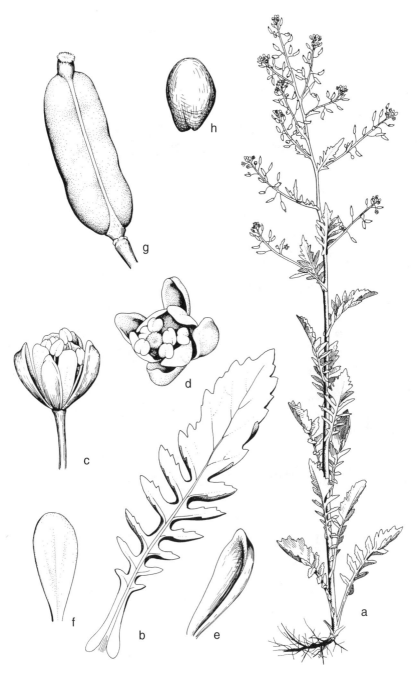

Fig. 139. *Rorippa palustris*
(Bog yellow cress).

a. Habit.
b. Leaf.
c. Flower.
d. Flower (top view).

e. Sepal.
f. Petal.
g. Fruit.
h. Seed.

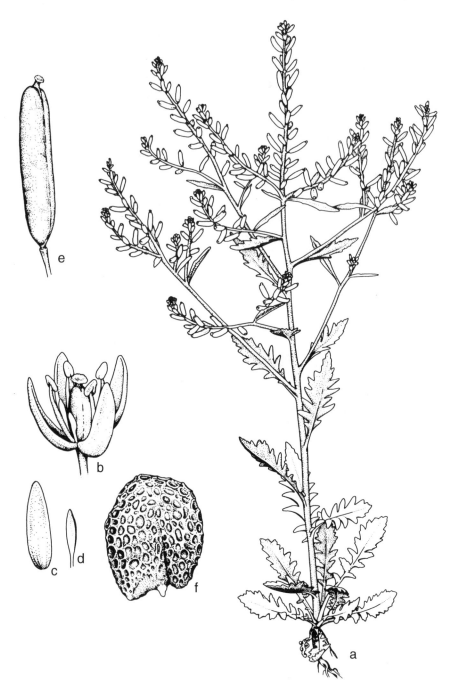

Fig. 140. *Rorippa sessiliflora*
(Sessile yellow cress).

a. Habit.
b. Flower.
c. Sepal.

d. Petal.
e. Fruit.
f. Seed.

Annual or biennial herbs from thickened roots; stems erect, simple or branched, glabrous, to 35 cm tall; basal leaves oblong, to 10 cm long, petiolate, lyrate-pinnatifid, glabrous; lower cauline leaves alternate, similar to the basal leaves; upper cauline leaves alternate, oblong to obovate, crenate, short-petiolate to nearly sessile, glabrous; racemes terminal and from the uppermost leaf axils, several-flowered, the flowers 1.5–2.0 mm broad, sessile or on pedicels to 1 mm long; sepals 4, lanceolate, green, glabrous, 0.5–1.0 mm long; petals 4, yellow, to 0.5 mm long or even absent; stamens 4; siliques narrowly oblongoid to oval, glabrous, 6–12 mm long, minutely beaked, on ascending pedicels to 1 mm long, the seeds minute, at least 175 per silique. May–October.

Along rivers and streams, around lakes and ponds, marshes.

IA, IL, IN, KS, KY, MO, NE, OH (OBL).

Sessile yellow cress.

Of the four species of *Rorippa* with petals less than 3 mm long and pedicels less than 3 mm long, *R. sessiliflora* is the most common. Its minute or absent petals and its 150 or more seeds per fruit distinguish it from *R. tenerrima* and *R. truncata*.

4. **Rorippa sinuata** (Nutt.) Hitchc. Spring Fl. Manhattan, Kans. 18. 1894. Fig. 141.
Nasturtium sinuatum Nutt. ex Torr. & Gray, Fl. N. Am. 1:73. 1838.
Radicula sinuata (Nut.) Greene, Leaflets 1:113. 1905.

Perennial herbs from rhizomes; stems clustered from a basal rosette, ascending, branched, glabrous, to 35 cm tall; leaves alternate, oblong to broadly lanceolate, pinnatifid, to 7.5 long, the lobes linear to oblong, obtuse, entire or sparsely dentate, glabrous, at least the middle and upper cauline leaves auriculate at the base; racemes terminal and from the uppermost leaf axils, several-flowered, the flowers 5–7 mm broad, on ascending pedicels; sepals 4, linear-lanceolate, green, glabrous, 2.5–4.0 long; petals 4, yellow, 6–8 mm long; siliques linear, cylindric, falcate, glabrous, to 1.5 cm long, the style persistent as a slender beak to 3 mm long, the pedicels ascending, to 1 cm long, the seeds about 1 mm long. April–July.

Wet ground, occasionally in standing water.

IA, IL, MO (FACW), KS, NE (FACW–), KY (NI).

Western yellow cress.

The leaves with auricles at the base distinguish this species from *R. sylvestris*, the only other *Rorippa* with petals more than 3 mm long and with all the leaves lobed or deeply pinnatifid.

5. **Rorippa sylvestris** (L.) Bess. Enum. 27. 1821. Fig. 142.
Sisymbrium sylvestre L. Sp. Pl. 657. 1753.
Nasturtium sylvestre (L.) R. Br. in Ait. Hort. Kew., ed. 2, 4:110. 1812.
Radicula sylvestris (L.) Druce, List Brit. Plants 4. 1908.

Perennial herbs from rhizomes; stems ascending, branched, glabrous, to 50 cm tall; leaves alternate, pinnately compound into 5–11 leaflets, petiolate, but without auricles, glabrous, to 10 cm long, the leaflets oblong to lanceolate, dentate; racemes terminal and from the uppermost leaf axils, several-flowered, the flowers 6–9 mm broad, on spreading to ascending pedicels; sepals 4, lanceolate, green, glabrous,

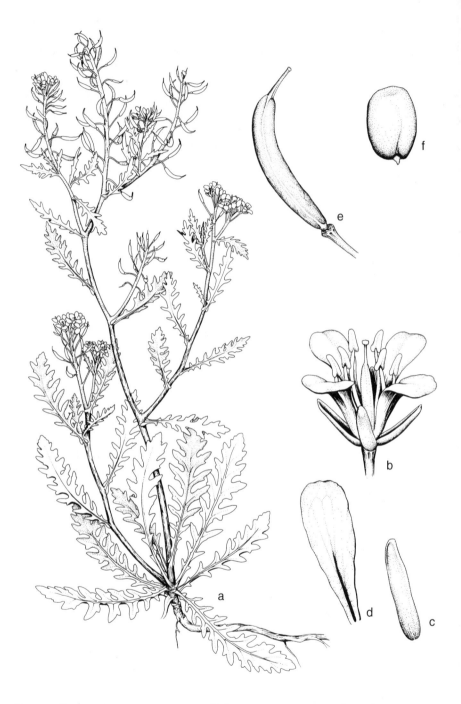

Fig. 141. *Rorippa sinuata*
(Western yellow cress).

a. Habit.
b. Flower.
c. Sepal.

d. Petal.
e. Fruit.
f. Seed.

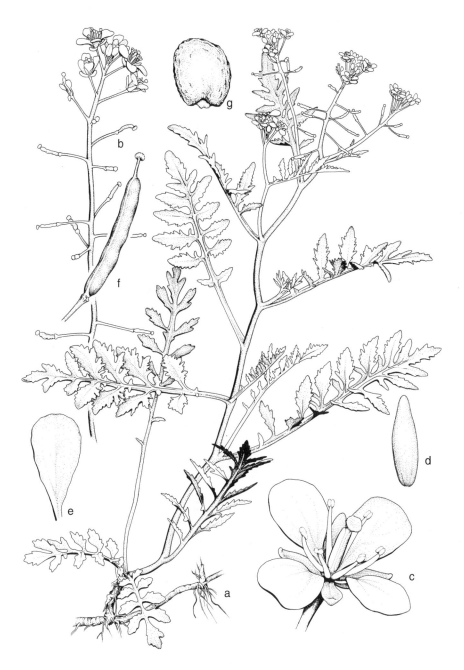

Fig. 142. *Rorippa sylvestris*
(Creeping yellow cress).

a. Habit.
b. Flowering and
fruiting raceme.
c. Flower.

d. Sepal.
e. Petal.
f. Fruit.
g. Seed.

2.5–4.0 mm long; petals 4, yellow, 6–8 mm long; siliques linear, cylindric, glabrous, to 2.5 cm long, the style persistent as a slender beak to 1 mm long, the pedicels spreading to ascending, about as long as the flower, the seeds up to 0.8 mm long. May–September.

Wet, disturbed areas.

IA, IL, IN, MO (OBL), KY, OH (FACW).

Creeping yellow cress.

The absence of auricles at the base of the leaves distinguishes this species from *R. sinuata*.

6. **Rorippa tenerrima** Greene, Erythea 3:46. 1895. Fig. 143.

Annual or biennial herbs with a taproot; stems decumbent to prostrate, to 20 cm long, glabrous; leaves alternate, oblanceolate to spatulate, obtuse at the apex, pinnatifid, glabrous, to 6 cm long, 5–15 mm wide; racemes up to 12 cm long; sepals 4, green, 0.8–1.5 mm long; petals 4, yellow, 0.7–0.9 mm long; siliques usually slightly curved, 4–6 (–9) mm long, 1.0–1.5 mm wide, 4–5 times longer than the pedicels; seeds 20–40 per fruit, 0.6–0.8 mm long. May–August.

Along river banks, rarely in standing water.

IL, MO, NE (NI).

Smooth yellow cress.

This completely glabrous yellow cress has the fewest seeds per fruit of any *Rorippa*.

Fig. 143. *Rorippa tenerrima* (Smooth yellow cress). Habit. Fruit (upper left).

7. **Rorippa truncata** (Jepson) Stuckey, Sida 2:414. 1966. Fig. 144.
Radicula sinuata (Nutt.) Greene var. *truncata* Jepson, Man. Fl. Pl. Cal. 424. 1925.
Rorippa curvipes Greene var. *truncata* (Jepson) Rollins, Harvard Pap. Bot. 1:48. 1993.

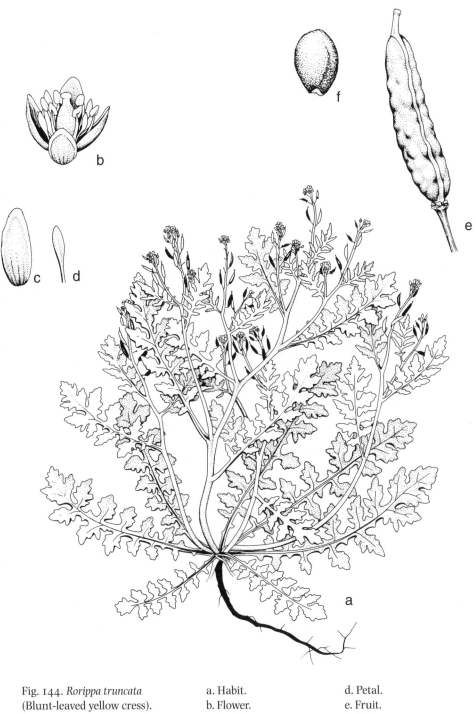

Fig. 144. *Rorippa truncata*
(Blunt-leaved yellow cress).

a. Habit.
b. Flower.
c. Sepal.

d. Petal.
e. Fruit.
f. Seed.

Annual or biennial herbs from a taproot; stems decumbent to prostrate, to 35 cm long, glabrous; leaves alternate, oblanceolate, obtuse at the apex, to 10 cm long, 7–18 mm wide, pinnate to pinnatifid, glabrous; racemes up to 10 cm long; sepals 4, green, 1.0–1.5 mm long; petals 4, yellow, 0.9–1.2 mm long; siliques cylindrical, 3.5–5.5 mm long, 1.2–1.8 mm wide, up to 4 times longer than wide, smooth or somewhat roughened, with 30–80 seeds per fruit; seeds 0.5–0.6 mm long. May–September.

Along rivers and streams, around lakes and ponds, sometimes in shallow water.
IL, MO (FAC), KS, NE (NI).

Blunt-leaved yellow cress.

This species is similar in appearance to *R. sessiliflora* from which it differs by its longer pedicels and fewer seeds per silique. It differs from *R. palustris* by its shorter petals and shorter pedicels.

51. CABOMBACEAE—WATER SHIELD FAMILY

Aquatic perennial herbs with rhizomes; stems often covered with mucilage; leaves floating or immersed, alternate, opposite, or whorled, sometimes dissected into filiform segments, some or all of them often peltate; flower solitary, axillary, perfect, actinomorphic; sepals 3 (–4), free, green; petals 3 (–4), free, white, yellow, or purple; stamens 3–18, free; pistils 2–18, free, each with 2–3 ovules; fruit a cluster of 2 or more indehiscent, 1- to 3-seeded follicles.

This family is closely related to the Nymphaeaceae and Nelumbonaceae, differing from both by its definite number of perianth parts and its follicular fruits. Several botanists combine these three families into one. This family consists of two genera and eight species. Both genera occur in the central Midwest.

1. All leaves floating, usually elliptic, entire; flowers purple; stamens 12–18 .... 1. *Brasenia*
1. Most leaves submersed, finely dissected; flowers white; stamens 3–6 .... 2. *Cabomba*

1. **Brasenia** Schreb.—Water Shield

Only the following species comprises the genus.

1. **Brasenia schreberi** J. F. Gmel. Syst. Veg. 1:853. 1796. Fig. 145.
Brasenia peltata Pursh, Fl. Am. Sept. 389. 1814.

Aquatic perennial herbs from a slender rhizome; leaves floating, oval to nearly orbicular to elliptic, obtuse at either end, to 10 cm long, up to two-thirds as wide, entire, green and glabrous above, usually purple below, both surfaces covered by a viscid gelatinous material; flower solitary, to 1.5 cm across, on long, glabrous, axillary peduncles; sepals 3 (–4), linear, green; petals 3 (–4), linear, purple; stamens 12–18, free, with long, filiform filaments; fruit a cluster of 4–18 oblongoid follicles to 1 cm long, tipped by the persistent style, 1- to 2-seeded. June–September.

In ponds, lakes, and quiet streams.
IA, IL, IN, KS, KY, MO, OH (OBL).

Water shield.

This species is widely distributed in much of the world. Although in the same family as *Cabomba*, it is easily distinguished by its uniform, undivided leaves, 12–18 stamens, 4–18 pistils, and purple flowers. The leaves are usually covered by a gelatinous material.

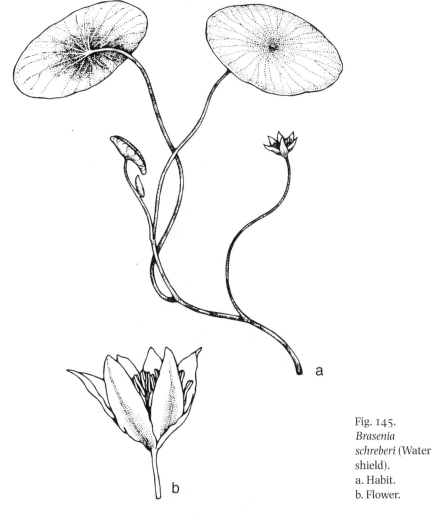

Fig. 145.
Brasenia schreberi (Water shield).
a. Habit.
b. Flower.

2. Cabomba Aubl.—Carolina Water Shield

Aquatic perennial herbs; leaves dimorphic, the submersed ones opposite or whorled, dissected into filiform segments, not peltate, the floating ones becoming alternate, entire, centrally peltate; flower solitary, axillary; sepals 3, free, green; petals 3, white or yellow, free; stamens 3–6, free; pistils 2–3, free, each with 3 pendulous ovules; fruit a cluster of 2–3 three-seeded, coriaceous, indehiscent ripened carpels.

Seven species native to the tropical and temperate regions of the New World comprise this genus. Only the following species occurs in the central Midwest.

1. **Cabomba caroliniana** Gray, Ann. Lyc. N. Y. 4:47. 1837. Fig. 146.

Aquatic perennial herbs from a slender rhizome; submersed leaves opposite or whorled, not peltate, palmately dissected into many filiform segments, the floating leaves becoming alternate, linear-oblong, entire, peltate, to 2 cm long; flower solitary, to 1.8 cm broad, on a long, glabrous, axillary peduncle; sepals 3, linear, green, free; petals 3, obovate, white with usually yellow bases, free; stamens 3–6, free; fruit a cluster of 2–3 flask-shaped, pubescent follicles to 1 cm long, with 3 seeds. May–September.

Standing water.

IL, IN, KS, KY, MO, OH (OBL).

Fanwort; Carolina water shield.

This species is readily distinguished by its dimorphic leaves, 3 sepals, 3 petals, 3–6 stamens, and a cluster of 2–3 follicles.

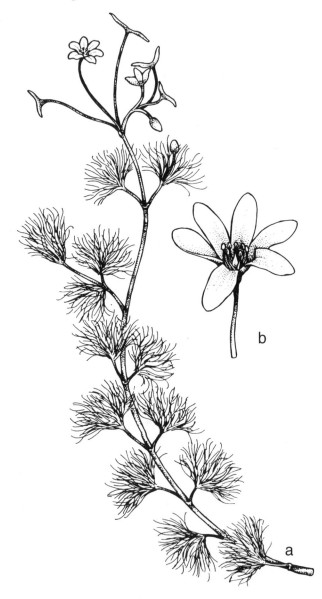

Fig. 146. *Cabomba caroliniana* (Fanwort).
a. Habit.
b. Flower.

52. CAESALPINIACEAE—CAESALPINIA FAMILY

Herbaceous or woody plants; leaves alternate, usually compound, with stipules; flowers perfect or unisexual, zygomorphic or more or less actinomorphic, usually perigynous; sepals (3–) 5, the segments equal or unequal; petals (0–3) 5, equal or unequal, usually free; stamens 5–10; ovary superior; fruit a legume.

If separated from the Fabaceae, this family consists of about 150 genera and approximately two thousand species, found worldwide.

Only the following genus has a species that occurs in standing water in the central Midwest.

1. Gleditsia L.—Locust

Trees; stems often with thorns; leaves alternate, once- to thrice-pinnate, with numerous leaflets; flowers in spikes, greenish yellow, some of the spikes entirely staminate, the others with perfect flowers; sepals 3–5; petals 3–5, free; stamens 5–8, attached to the base of the sepals; legumes woody, 1- to several-seeded, with or without pulp between the seeds.

There are about fifteen species in the genus, including the common honey locust (*G. triacanthos*) and the species described below.

1. Gleditsia aquatica Marsh. Arb. Am. 54. 1785. Fig. 147.
Gleditsia monosperma Walt. Fl. Car. 254. 1788.

Trees to 25 m tall, usually with simple or 3-pronged thorns, the base of the thorn more or less flat in cross-section; leaves mostly once-pinnate, occasionally twice- or thrice-pinnate, with 8–many pairs of leaflets, the leaflets oblong to narrowly ovate, crenulate, 1–3 cm long, up to 1.5 cm broad, glabrous, the petiolules glabrous; sepals 3–5, yellowish green, 3–4 mm long; petals 3–5, yellowish green, 4–5 mm long; stamens often 8; legume ovoid, flat, dark brown, 1-seeded, without pulp surrounding the seed, 3–5 cm long, 2–3 cm broad. April–May.

Swamps, swampy woods, often in standing water.

IL, IN, KY, MO (OBL).

Water locust; one-seeded locust.

This species strongly resembles *G. triacanthos* vegetatively, but differs by its short, 1-seeded, pulpless legumes. The leaves tend to be less commonly bi- or tri-pinnate than in *G. triacanthos*. The base of the thorns is usually flat in cross-section, while it is usually terete in *G. triacanthos*.

53. CALLITRICHACEAE—WATER STARWORT FAMILY

Only the following genus comprises the family.

1. Callitriche L.—Water Starwort

Mostly aquatic annuals; leaves opposite, entire, small; flowers axillary, minute, unisexual, monoecious, subtended by a pair of bracteoles; sepals 0; petals 0; stamen usually 1; ovary more or less compressed, 2- or 4-lobed; styles 2; fruit a cluster of 4 nutlets.

Approximately thirty-five species found worldwide comprise this genus.

Fig. 147. *Gleditsia aquatica* (Water locust). Leaves and twig (center). Fruits (below).

1. Flower not subtended by bracts; leaves uniform.................................... .1. *C. hermaphroditica*
1. Flower subtended by bracts; submersed leaves usually linear, floating leaves usually obovate.
    2. Fruit 0.8–1.0 mm long, unwinged.. 2. *C. heterophylla*
    2. Fruit 1.0–1.5 mm long, narrowly winged...3. *C. verna*

1. **Callitriche hermaphroditica** L. Cent. Pl. I, 31. 1755. Fig. 148.
Callitriche autumnalis L. Fl. Suec. ed. 2, 2. 1755.

Submersed aquatic annual or perennial herbs; stems prolonged, to 40 cm long, with numerous leaves; leaves uniform, simple, opposite, dark green, broadly linear, obtuse to retuse at the apex, broadened at the base, 1-nerved, glabrous, 4–12 mm long, 2–6 mm wide; flower solitary in the leaf axils, not subtended by bracts; sepals 0; petals 0; stamen 1; pistil 1, the ovary superior, 4-locular, with 2 reflexed styles;

Fig. 148. *Callitriche hermaphroditica* (Water starwort). Habit (right). Fruit (lower left).

capsules spherical in outline, flattened, sessile, deeply notched, 1.5–2.5 mm in diameter, each lobe broadly winged. August–September.

Slow-moving water.

NE (OBL).

Water starwort.

This species differs from other aquatic species of *Callitriche* in the central Midwest by its uniform leaves and by the absence of bracts at the base of the flower.

2. **Callitriche heterophylla** Pursh, Fl. Am. Sept. 3. 1814. Fig. 149.

Annual herbs, rooting at the lower nodes if stranded on land; stems up to 18 (–20) cm long, glabrous; submersed leaves linear, up to 15 mm long, up to 3 mm wide; floating or emersed leaves obovate to spatulate, to 15 mm long, to 5 mm broad, 3-nerved, glabrous; flowers axillary, usually one staminate and one pistillate per axil; bracteoles white, 0.5–1.5 wide; fruits 0.8–1.0 mm long, compressed, widest above the middle, as broad as long, not winged. May–October.

Shallow water, or occasionally stranded on muddy soil.

IA, IL, IN, KS, KY, MO, NE, OH (OBL).

Water starwort.

The fruit that is rounded at the base and as broad as long distinguishes this species from the following.

Fig. 149. *Callitriche heterophylla* (Water starwort). Habit. Fruit (below).

3. **Callitriche verna** L. Fl. Suec., ed. 2, 4. 1755. Fig. 150.
Callitriche palustris L. Sp. Pl. 2:969. 1753.

Annual herbs, rooting at the lower nodes if stranded on land; stems up to 15 (–20) cm long, glabrous; submersed leaves linear, up to 15 mm long, up to 3 mm broad; floating or emersed leaves obovate to spatulate, to 15 mm long, to 5 mm broad, 3-nerved, glabrous; flowers axillary, usually one staminate and one pistillate flower per axil; bracteoles white, 0.5–1.5 mm wide; fruits 1.0–1.5 mm long, compressed, widest above the middle, narrowed to the base, longer than broad, narrowly winged. May–August.

Shallow water, less commonly stranded on muddy soil.

IA, IL, OH (OBL).

Water starwort.

This species differs from *C. heterophylla* by its slightly larger fruits that are longer than broad and narrowly winged.

Fig. 150. *Callitriche verna* (Water starwort). Habit. Fruit (bottom left).

54. CAMPANULACEAE—BELLFLOWER FAMILY

Herbs (in the United States), usually with milky sap; leaves alternate; flowers usually in cymes or racemes, usually perfect, actinomophic to zygomorphic; sepals usually 5, attached to each other; petals 5, attached to each other; stamens 5, attached to a nectariferous disk or to the base of the corolla tube; ovary inferior; fruit usually a capsule, with numerous seeds.

With the genus *Lobelia* included within the Campanulaceae, this family consists of about seventy genera and approximately two thousand species, found worldwide.

The following genera have species that sometimes occur in standing water.

1. Corolla actinomorphic; carpels 3–5.. 1. *Campanula*
1. Corolla zygomorphic; carpels 2... 2. *Lobelia*

1. **Campanula** L.—Bellflower

Annual or perennial herbs; leaves alternate or basal, simple, not clasping; flowers perfect, actinomorphic, solitary or in racemes or panicles; calyx 5-lobed; corolla campanulate or rotate, 5-lobed; stamens 5, free from each other; ovary inferior, 3- to 5-locular, dehiscing by 3–5 lateral valves or pores, with numerous seeds.

There are about three hundred species in the north temperate and boreal regions of the world. Two similarly appearing species may be found in standing water habitats in the central Midwest.

1. Leaves linear-lanceolate to lanceolate, averaging 6 times longer than broad; corolla 4–10 mm long, on spreading pedicels .... 1. *C. aparinoides*
1. Leaves linear, averaging 12 times longer than broad; corolla 5–13 mm long, on ascending pedicels... 2. *C. uliginosa*

1. **Campanula aparinoides** Pursh, Fl. Am. Sept. 159. 1814. Fig. 151.

Perennial herbs from slender rootstocks; stems weak and reclining, branched, retrorsely hispid, to 50 cm long; leaves linear-lanceolate to lanceolate, averaging six times longer than wide, acute at the apex, cuneate at the sessile base, crenulate to denticulate, pubescent, to 4.5 mm long; flowers borne in leafy panicles from filiform, spreading pedicels; calyx 1.5–4.0 mm long, 5-parted, the lobes lanceolate, about as long as the calyx tube; corolla campanulate, pale lavender, 4–10 mm long, deeply 5-lobed; capsule subglobose, ascending, ribbed, to 2 mm long, opening by basal pores. June–August.

Wet meadows, marshes, sometimes in shallow water.

IA, IL, IN, MO, OH (OBL).

Marsh bellflower.

This weak-stemmed plant usually reclines on other vegetation in wet meadows and marshes. It differs from the similar *C. uliginosa* by its broader leaves, smaller flowers, and smaller capsules. The denticulate leaves give this species a scabrous feeling when touched.

2. **Campanula uliginosa** Rydb. in Britt. Man. Fl. N. States 885. 1901. Fig. 152.
Campanula aparinoides Pursh var. *grandiflora* Holz. Bull. Geol. & Nat. Hist. Surv. Minn. 9:566. 1896.
Campanula aparinoides Pursh var. *uliginosa* (Rydb.) Gl. Phytologia 4:25. 1952.

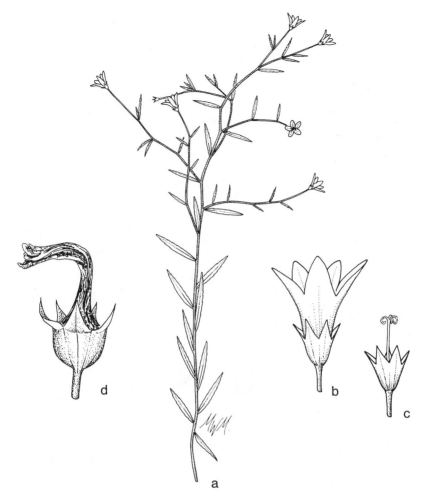

Fig. 151. *Campanula aparinoides* (Marsh bellflower).

a. Habit.
b. Flower.
c. Flower (corolla removed).
d. Developing fruit.

Perennial herbs from slender rootstocks; stems weak and reclining, branched, retrorsely hispid, to 50 cm long; leaves linear, averaging 12 times longer than wide, acuminate at the apex, cuneate to the sessile base, denticulate, pubescent, to 5 cm long; flowers borne on nearly bractless, ascending pedicels; calyx 3–7 mm long, 5-parted, the lobes usually a little longer than the calyx tube; corolla campanulate, bluish, 5–13 mm long, 5-lobed, the lobes about equaling the corolla tube; capsule subglobose, ascending, ribbed, 3–5 mm long, opening by basal pores. June–August.

Wet meadows, marshes, often in standing water habitats.

IA, IL, IN, OH (the U.S. Fish and Wildlife Service equates this species with *C. aparinoides*).

Larger-flowered marsh bellflower.

This bellflower is sometimes considered synonymous with *C. aparinoides* or as a variety of it.

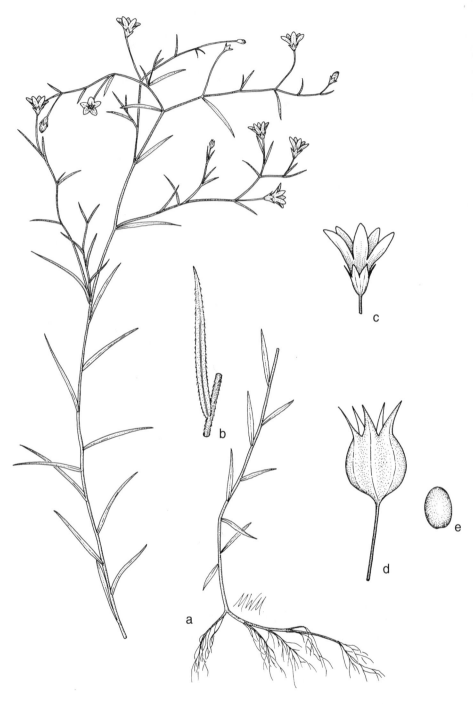

Fig. 152. *Campanula uliginosa*
(Large-flowered marsh bellflower).

a. Habit.
b. Leaf.
c. Flower.

d. Fruit.
e. Seed.

2. Lobelia L.—Lobelia

Herbs (in the central Midwest) or shrubs, with milky sap; leaves alternate or basal, simple; flowers perfect, zygomorphic, borne in spikes, racemes, or panicles; calyx tubular, 5-lobed; corolla tubular, split down one side, 2-lipped; stamens 5, the anthers united in a ring around the style, at least 2 of the anthers bearded at the tip; pistil 1, the ovary inferior, 2-locular; stigma 2-lobed; capsule 2-valved, dehiscing at the top, with numerous seeds.

This genus, with about three hundred species found worldwide, is sometimes placed in its own family.

1. Flowers 1.5 cm long or longer.
  2. Flowers red..1. *L. cardinalis*
  2. Flowers blue.
    3. Auricles at base of calyx less than 2 mm long; stems densely puberulent throughout; flowers 1.5–2.5 cm long .... 3. *L. puberula*
    3. Auricles at base of calyx 1–5 mm long; stems glabrous or sparsely hirsute; flowers 2.0–3.3 cm long... 4. *L. siphilitica*
1. Flowers up to 1.5 cm long.
  4. Cauline leaves linear to narrowly lanceolate, usually less than 3 mm wide; lower lip of corolla glabrous...2. *L. kalmii*
  4. Cauline leaves oblong to lanceolate, more than 3 mm wide; lower lip of corolla usually pubescent .... 5. *L. spicata*

1. Lobelia cardinalis L. Sp. Pl. 930. 1753. Fig. 153.

Perennial herbs with basal offshoots; stems erect, simple, glabrous or nearly so, to 1.5 m tall; leaves alternate, simple, lanceolate to ovate-lanceolate, acute or acuminate at the apex, cuneate at the base, serrulate and usually with white-tipped teeth, glabrous or hirtellous, to 15 cm long, 1.5–3.5 cm wide, the lowermost petiolate, the upper sessile; flowers in racemes, borne on pedicels shorter than the glandular, leafy bracts; calyx deeply 5-lobed, the lobes linear, glabrous or pubescent; corolla scarlet (rarely white), 2.5–4.0 cm long, 2-lipped, 5-lobed; stamens 5, the filament tube exserted; capsule glabrous; seeds wrinkled. July–September.

Wet ground, occasionally in standing water.

IA, IL, IN, MO (OBL), KS, KY, NE, OH (FACW+).

Cardinal-flower.

The bright scarlet flowers not only are distinctive for this species but make it one of the most beautiful wetland plants in the United States. The tiny teeth of the leaves are usually white-tipped.

2. Lobelia kalmii L. Sp. Pl. 930. 1753. Fig. 154.

Perennial herbs with basal offshoots; stems erect, branched, glabrous to pubescent, to 60 cm tall; basal leaves spatulate, obtuse at the apex, cuneate to the petiolate base, usually very shallowly toothed to entire, pubescent, to 2.5 cm long; upper leaves alternate, linear to narrowly lanceolate, acute to obtuse at the apex, cuneate to the sessile base, very shallowly toothed or entire, pubescent to nearly glabrous, to 2 cm long, 1–3 mm wide; flowers in loose racemes, on slender pedicels to 2.5 cm long, subtended by a pair of bracteoles midway on the pedicels; calyx campanulate,

Fig. 153. *Lobelia cardinalis*
(Cardinal-flower).

a. Habit.
b. Leaves.
c. Flower.

d. Developing fruit.
e. Seed.

deeply 5-lobed, the lobes linear-lanceolate, glabrous or hirtellous, the sinuses not appendaged; corolla pale blue, 0.5–1.5 cm long, the lower lip glabrous; stamens 5; capsule subglobose, to 4 mm in diameter. July–September.

Fens, marshes, occasionally in shallow water. IA, IL, IN, OH (OBL). Kalm's lobelia.

This species is distinguished by its linear to linear-lanceolate leaves and its small pale blue flowers with the lower lip of the corolla glabrous.

3. Lobelia puberula

Michx. Fl. Bor. Am. 2:152. 1803. Fig. 155.
Lobelia puberula Michx. var. *simulans* Fern. Rhodora 49:184. 1947.
Lobelia puberula Michx. var. *mineolana* E. Wimm. Rep. Spec. Nov. Regni Veg. 26:4. 1929.

Perennial herbs with basal offshoots; stems erect, unbranched or branched, densely puberulent, to 1.5 m tall; leaves basal and alternate, simple, lanceolate to oblong to obovate, subacute to acute at the apex, tapering to the base, denticulate to crenulate, spreading to ascending, glabrous or short-pilose on the lower surface, to 10 cm long, to 4 cm wide, the lowest leaves on elongated petioles, the upper ones sessile or nearly so; flowers in usually secund, spikelike racemes up to 30 cm long, borne on pedicels 2–5 mm long, subtended by leafy, glandular, linear-lanceolate to lanceolate bracts; calyx deeply 5-lobed, the lobes lanceolate, pubescent, with reflexed auricles less than 2 mm long in the sinuses; corolla blue, 1.5–2.5 cm long, 5-lobed, the lobes about as long as the tube; stamens 5; capsules glabrous; seeds wrinkled. August–October.

Depressions in swampy woods; wet meadows.
IL, IN, KY, MO, OH (FACW–).

Fig. 154. *Lobelia kalmii* (Kalm's lobelia).

a. Habit.
b. Flower.
c. Developing fruit.
d. Seed.

Fig. 155. *Lobelia puberula*
(Downy lobelia).

a, b. Habit.
c. Leaf base.
d. Leaf variation.

e. Flower.
f. Fruit.

Downy lobelia.

This species resembles *L. siphilitica* because of its large blue flowers but differs by its densely puberulent stems and the short auricles at the base of the calyx.

Most of our plants belong to var. *simulans*. In southeastern Missouri, plants with short pilose leaves and longer bracts have been referred to as var. *mineolana*.

4. Lobelia siphilitica L. Sp. Pl. 931. 1753. Fig. 156.
Lobelia siphilitica L. var. *ludoviciana* A. DC. in DC. Prodr. 7:377. 1839.

Perennial herbs with basal offshoots; stems erect, usually unbranched, glabrous or sparsely hirsute, to 1 m tall; leaves lanceolate to oblong to ovate, acute or acuminate at the apex, cuneate at the base, serrulate to denticulate, usually with white-tipped teeth, glabrous or strigose, to 15 cm long, the lowermost petiolate, the upper sessile; flowers densely racemose, borne on short pedicels subtended by leafy bracts; calyx deeply 5-lobed, the lobes lanceolate, usually hirsute, with reflexed auricles 2–5 mm long in the sinuses; corolla blue, except for the white base of the lower lip, 2.0–3.3 cm long, 5-lobed, the lobes about as long as the tube; stamens 5; capsule glabrous; seeds wrinkled. August–October.

Wet ground, sometimes in shallow depressions in swampy woods.

IA, IL, IN, KY, MO, OH (FACW+), KS, NE (OBL).

Blue cardinal-flower; great blue lobelia.

This species shows considerable variation in the degree of pubescence on the stems, leaves, and calyces. Plants that are nearly glabrous may be known as var. *ludoviciana*.

5. Lobelia spicata Lam. Encycl. 3:587. 1789. Fig. 157.
Lobelia leptostachya DC. Prodr. 7:376. 1839.
Lobelia spicata Lam va. *hirtella* Gray, Syn. Fl. 2:6. 1878.
Lobelia spicata Lam. var. *leptostachya* (DC.) Mack. & Bush, Fl. Jackson Co., Mo. 183. 1902.
Lobelia spicata Lam. var. *campanulata* McVaugh, Rhodora 38:316. 1936.

Perennial herbs with basal offshoots; stems erect, usually unbranched, pubescent above, glabrous or pubescent below, to 1 m tall; leaves alternate, simple, lanceolate to oblong to obovate, obtuse to acute at the apex, tapering to the short-petiolate or sessile base, denticulate, pubescent on both surfaces, to 10 cm long, to 4 cm wide; flowers densely crowded in spikelike racemes to 60 cm long, borne on pedicels 2–5 mm long, subtended by linear to lanceolate, acute bracts; calyx deeply 5-lobed, the lobes lanceolate, hirsute to glabrous, with or without reflexed auricles in the sinuses; corolla blue or whitish, 7–10 mm long, the lower lip pubescent; capsules glabrous; seeds wrinkled. June–August.

Wet meadows; a variety of dry habitats.

IA, IL, IN, KS, MO, NE (FAC), KY, OH (FAC–).

Spiked lobelia.

Typical var. *spicata* has stems that are pubescent in the upper part and glabrous or nearly so in the lower half. Var. *hirtella* has stems that are pubescent throughout. Var. *leptostachys* has auricles, between the calyx lobes up to 5 mm long.

Fig. 156. *Lobelia siphilitica*
(Blue cardinal-flower).

a. Habit.
b. Leaf.
c. Flower.

d. Developing fruit.
e. Seed.

Fig. 157. *Lobelia spicata*
(Spiked lobelia).

a. Habit.
b. Flower.

c. Fruit, with corolla persisting.
d. Seed.

55. CAPRIFOLIACEAE—HONEYSUCKLE FAMILY

Woody plants or perennial herbs; leaves opposite, simple or pinnately compound, with or without stipules; flowers perfect, usually in cymes, actinomorphic or sometimes zygomorphic; sepals united, 3- to 5-lobed, the tube attached to the ovary; petals united, mostly 5-lobed; stamens 5, inserted on the corolla tube; ovary inferior, 1- to 6-locular, with 1–many ovules; fruit a berry, drupe, or capsule.

This family consists of fifteen genera and about four hundred species found in north temperate and boreal regions of the world; also in South America and Australia.

Only the following genera have species that sometimes occur in shallow water, bogs, or fens in the central Midwest.

1. Leaves pinnately compound ....2. *Sambucus*
1. Leaves simple.
  2. Vines .... 1. *Lonicera*
  2. Erect shrubs.
    3. Flowers in flat-topped clusters; leaves toothed or lobed (entire in *V. nudum*)...................
    ....3. *Viburnum*
    3. Flowers in pairs; leaves entire.. 1. *Lonicera*

1. **Lonicera** L.—Honeysuckle

Woody vines or shrubs; leaves opposite, simple, entire; flowers in clusters or racemes, actinomorphic or 2-lipped; calyx teeth 5, minute; corolla 5-lobed, tubular, sometimes gibbous at base; stamens 5; ovary inferior; fruit a several-seeded berry.

There are approximately 180 species in this genus, most of them in the Northern Hemisphere. The following may occur in standing water, particularly in cedar swamps.

1. Vines; uppermost 1–4 pairs of leaves connate; flowers several in clusters ....2. *L. dioica*
1. Shrubs; none of the leaves connate; flowers in pairs.
  2. Leaves glaucous or pale on the lower surface, rather thick, strongly reticulate-veined.
    3. Leaves ciliate; berries blue; peduncles shorter than the flowers.................... 4. *L. villosa*
    3. Leaves not ciliate; berries red; peduncles as long as the flowers............. 3. *L. oblongifolia*
  2. Leaves green on the lower surface, thin, not strongly reticulate-veined...... 1. *L. canadensis*

1. **Lonicera canadensis** Marsh. Arb. 81. 1785. Fig. 158.
Lonicera ciliata Muhl. Cat. 23. 1813.

Shrubs to 1.8 m tall; branchlets glabrous; leaves opposite, the uppermost not connate, thin, ovate, acute at the apex, rounded to subcordate at the base, not strongly reticulate-veined, glabrous or nearly so on the upper surface, villous on the lower surface, not glaucous, entire, to 10 cm long, to 5 cm wide; flowers in pairs on filiform peduncles up to 4 cm long; calyx teeth 5, minute; corolla greenish yellow, 5-lobed, the lobes about equal, glabrous, tubular, to 2 cm long; stamens 5; ovary inferior; berries 2, free at base, red, glabrous or nearly so, subglobose. May–June.

Wet woods, cedar swamps.

IN, OH (FACU).

American fly honeysuckle.

This is the only wetland shrubby *Lonicera* with thin leaves that are neither glaucous nor strongly reticulate-veined.

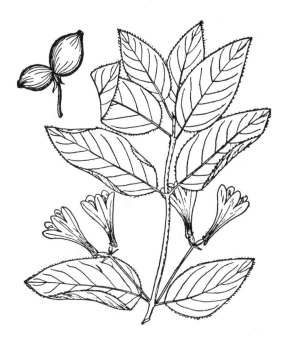

Fig. 158. *Lonicera canadensis* (American fly honeysuckle). Habit. Fruits (upper left).

2. **Lonicera dioica** L. Syst. ed. 12, 165. 1767. Fig. 159.
Lonicera glaucescens Rydb. Bull. Torrey Club 24:90. 1897.
Lonicera dioica L. var. *glaucescens* (Rydb.) Britt. Minn. Trees & Shrubs 289. 1912.

Climbing or twining woody vines; twigs glabrous; leaves opposite, entire, oblong to elliptic, obtuse at the apex, rounded at the nearly sessile base, the uppermost 1–4 pairs of leaves connate at the base, strongly glaucous on the lower surface, glabrous or densely pubescent on the lower surface, to 10 cm long, to 5 cm broad; flowers in clusters on peduncles up to 3 cm long; calyx teeth 5, minute; corolla 5-lobed, 2-lipped, tubular, yellow to red to purple, to 2.5 cm long, hirsute on the inner side,

Fig. 159. *Lonicera dioica* (Glaucous honeysuckle). Habit (center). Stamen (upper left). Fruits with bracts (upper right).

somewhat gibbous at the base; stamens 5; ovary inferior, glabrous or glandular; berries red, smooth or glandular. May–July.

Woods, sometimes in cedar swamps.

IA, IL, IN, KS, KY, MO, NE, OH (FACU).

Glaucous honeysuckle.

This species is readily recognized by its viny habit and its opposite, entire leaves that are strongly glaucous on the lower surface. Specimens that have the lower surface of the leaves densely hairy may be known as var. *glaucescens*.

3. **Lonicera oblongifolia** (Goldie) Hook. Fl. Bor. Am. 1:284. 1833. Fig. 160.
Xylosteum oblongifolium Goldie, Edinb. Phil. Journ. 6:323. 1822.

Shrubs to 1.5 m tall; twigs puberulent; leaves opposite, entire, the uppermost not connate, entire, rather thick, strongly reticulate-veined, obtuse to subacute at the apex, more or less rounded at the base, glabrous on the upper surface, downy-pubescent on the lower surface when young, not ciliate, to 5 cm long, to 3 cm wide; flowers in pairs on filiform peduncles up to 4 cm long, as long as the flowers; calyx teeth 5, minute; corolla 5-lobed, 2-lipped, tubular, yellow, gibbous at the base, pubescent, 1.0–1.5 cm long; stamens 5; ovary inferior; berries 2, free or united at the base, red to orange, subglobose. May–June.

Swamps, bogs.

OH (OBL).

Swamp fly honeysuckle.

Lonicera oblongifolia is similar in appearance to *L. villosa*, but it lacks cilia on its leaves and has red berries.

Fig. 160. *Lonicera oblongifolia* (Swamp fly honeysuckle). Habit. Fruits (lower left). Flowers (upper left). Seed (lower right).

4. **Lonicera villosa** (Michx.) Roem. & Schultes, Syst. Veg. 5:256. 1819. Fig. 161.
Xylosteum villosum Michx. Fl. Bor. Am. 1:106. 1803.

Ascending shrubs to 1 m tall; twigs pubescent, at least when young; leaves opposite, simple, entire, oblong, green on both sides, obtuse at the apex, rounded at the nearly sessile base, pubescent and strongly ciliate, to 4.5 cm long, to 2.5 cm wide; flowers in pairs, on peduncles as long as the flowers; calyx teeth 5, minute; corolla 5-lobed, not 2-lipped, tubular, yellow, to 1.5 cm long; stamens 5; ovary inferior; berries blue, oblongoid, up to 7 mm broad. May–July.

Woods, bogs, cedar swamps.

OH (FACW+), as *L. caerulea*.

Mountain fly honeysuckle.

This species is distinguished by its very short stature, its strongly ciliate, non-glaucous leaves, and its blue berries. The berries are edible.

2. **Sambucus** L.—Elderberry

Shrubs or trees; leaves opposite, pinnately compound, serrate; flowers numerous in cymes, perfect, actinomorphic; sepals 5, united below; petals 5, united below; stamens 5; ovary inferior, 3- to 5-locular; drupes berrylike, 1-seeded.

About twenty species, found worldwide, comprise this genus.

1. Cymes umbel-like, flat-topped; twigs with white pith; fruit purple................. 1. *S. canadensis*
1. Cymes paniculate, ovoid; twigs with orange pith; fruit usually red.................. 2. *S. racemosa*

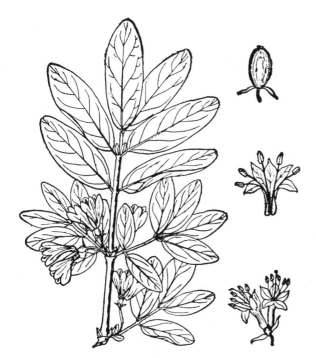

Fig. 161. *Lonicera villosa* (Mountain fly honeysuckle). Habit (left). Fruit (upper right). Flower (center right). Flowers (lower right).

1. **Sambucus canadensis** L. Sp. Pl. 269. 1753. Fig. 162.

Shrub with soft wood; stems to 4 m tall, glabrous, strongly lenticellate, with white pith; leaves opposite, pinnately compound; leaflets 5–11, lanceolate to elliptic to ovate, acuminate at the apex, more or less rounded at the short-petiolulate base, glabrous above, usually somewhat pubescent beneath, sharply serrate, to 10 cm long, to 6 cm wide; flowers numerous in flat-topped or convex cymes, actinomorphic, perfect, 2–4 mm across; sepals 5, green, 1–2 mm long, united below; petals 5, white, 2–3 mm long, united below, fragrant; stamens 5; ovary inferior; drupes purple-black, 4–5 mm in diameter, 1-seeded. June–July.

Floodplain woods, swampy woods, mesic woods, thickets, along streams.

IA, IL, IN, KY, MO, OH (FACW–), KS, NE (FAC+).

Elderberry.

The large inflorescence of white flowers makes this plant a candidate for ornamental use.

The soft wood with conspicuous lenticels on the twigs is distinctive, as is the white pith and purple-black drupes.

2. **Sambucus racemosa** L. Sp. Pl. 270. 1753. Fig 163.
Sambucus pubens Michx. Fl. Bor. Am. 1:181. 1803.
Sambucus racemosa L. ssp. *pubens* (Michx.) House, Bull. N. Y. State Mus. 244:50. 1923.

Shrub with soft wood; stems to 3 m tall, glabrous to puberulent, lenticellate, with orange pith; leaves opposite, pinnately compound; leaflets 5–7, lanceolate to oblong to ovate, acuminate at the apex, rounded at the short-petiolulate base, more or less glabrous above, usually short-hairy beneath, sharply serrate, to 10 cm long, to 6 cm wide; flowers numerous in paniculate cymes longer than broad, actinomorphic, perfect; sepals 5, green, 1–2 mm long, united below; petals 5, yellow-white, 2–3 mm long, united below, with an unpleasant odor; stamens 5; ovary inferior; drupes red, 4–5 mm in diameter. April–July.

Mesic woods, rocky woods, swampy woods.

IA, IL, IN, MO (FACU+), OH (FACU).

Red elderberry.

This plant is distinguished by its bright orange pith and its red berries. The flowers have an unpleasant odor.

Although primarily a shrub of mesic or rocky woods, it sometimes occurs in swampy woods.

3. **Viburnum** L.—Nannyberry; Arrowwood

Shrubs or trees; buds with or without bud scales; leaves opposite, simple, usually with stipules; flowers perfect, in compound cymes; sepals 5, united below to form a tube; petals 5, united at base, rotate or campanulate; stamens 5, attached to the corolla tube; ovary inferior, 1- to 3-locular, each locule with one ovule; fruit a 1-seeded drupe.

Nearly 250 species comprise this genus. The following may be found occasionally in shallow water, bogs, or fens in the central Midwest.

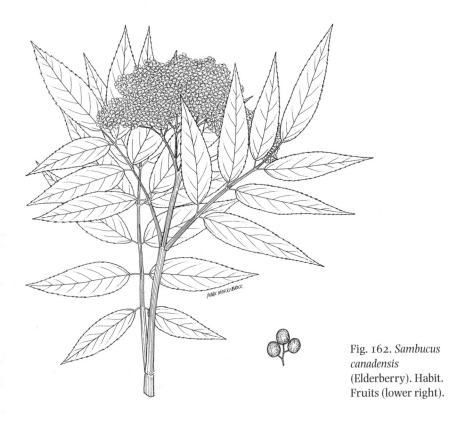

Fig. 162. *Sambucus canadensis* (Elderberry). Habit. Fruits (lower right).

Fig. 163. *Sambucus racemosa* (Red elderberry). Habit (center). Fruit (upper left). Flower (lower left).

1. Leaves entire or crenulate, never lobed.
  2. Leaves crenulate; peduncles shorter than the cymes..................................... 1. *V. cassinoides*
  2. Leaves entire; peduncles as long as or longer than the cymes .... 2. *V. nudum*
1. Leaves dentate, sometimes lobed.
  3. Some of the outer flowers of the cyme sterile and enlarged; leaves 3-lobed; drupes red.
    4. Glands on petiole saucer-shaped, with a concave summit............................. 3. *V. opulus*
    4. Glands on petiole round-topped, club-shaped... 5. *V. trilobum*
  3. None of the flowers sterile and enlarged; leaves unlobed; drupes blue-black......................
    .... 4. *V. recognitum*

1. **Viburnum cassinoides** L. Sp. Pl. ed. 2, 384. 1762. Fig. 164.

Viburnum nudum L. var. *cassinoides* (L.) Torr. & Gray, Fl. N. Am. 2:14. 1841.

Shrubs to 4 m tall; bud scales 2, scurfy-yellow; branchlets gray, scurfy, becoming glabrous; leaves opposite, simple, ovate to oblong-lanceolate, acute at the apex, tapering to the base, crenulate, glabrous or nearly so, up to 15 cm long, up to 6 cm wide; flowers foul-smelling, in cymes, the peduncles shorter than the cymes; sepals 5, united; petals 5, united, white; stamens 5; ovary inferior; drupes pink, becoming blue-black, ellipsoid to ovoid, 6–10 mm long, glaucous, with a single flat, oblong stone. May–July.

Swamps, damp thickets.

IN, OH (FACW).

Withe-rod.

This species has attractive white flowers borne in cymes, followed by pinkish drupes that turn blue-black when mature. It is distinguished from *V. nudum* by its numerous crenulate teeth on the leaves.

Fig. 164. *Viburnum cassinoides* (Withe-rod). Habit (right).
Fruit (upper left). Flower (lower left).

2. **Viburnum nudum** L. Sp. Pl. 268. 1753. Fig. 165.

Shrubs or small trees to 6 m tall; bud scales 2, brown; branchlets usually not scurfy, greenish; leaves opposite, simple, coriaceous, elliptic to narrowly ovate, acute or subacute at the apex, tapering to the base, entire or nearly so, glabrous or scurfy on the upper surface, glabrous on the lower surface, up to 15 cm long, up to 7.5 cm wide; flowers in cymes, the peduncles as long as or longer than the cymes; sepals 5, united; petals 5, united, white; stamens 5; ovary inferior; drupes blue-black, subglobose, up to 10 mm in diameter, with a single flat, obovate stone. June–August.

Swamps.

KY (OBL).

Raisin tree; possum haw.

This species differs from *V. cassinoides* by its entire or nearly entire leaves. It is the only large-leaved wetland tree in the central Midwest with leathery, entire, opposite leaves. The dried fruits resemble raisins.

Fig. 165. *Viburnum nudum* (Raisin tree). Habit (right). Flower (lower left).

3. Viburnum opulus L. Sp. Pl. 268. 1753. Fig. 166.

Shrubs or small trees to 6 m tall; stems and branchlets glabrous; leaves opposite, simple, palmately 3-lobed, dentate, ovate, rounded or truncate at the base, glabrous on the upper surface, pubescent on the lower surface, at least on the veins, to 8 cm long, the petioles with saucer-shaped, concave glands; outer flowers of the cyme enlarged, sterile, whitish, up to 2 cm across; sepals 5, united; petals 5, united, white; stamens 5; ovary inferior; drupes red, subglobose, 10–15 mm long and nearly as broad, with a single flat, suborbicular stone. June–July.

Wet ground, sometimes invasive in bogs and fens.

IA, IL, IN, OH (FACW).

European highbush cranberry.

This non-native shrub or small tree from Europe and Asia is extremely invasive in wetlands, often replacing native species. The white sterile outer flowers and the red drupes make this species an attractive ornamental. The plant is dispersed by birds who eat the drupes and pass the seeds through their digestive tracts unharmed.

This species is very similar in appearance to the native *V. trilobum* but is readily distinguished by its saucer-shaped, concave petiolar glands.

Fig. 166. *Viburnum opulus* (European highbush cranberry). Habit. Fruit (upper right). Flower (lower right).

4. Viburnum recognitum Fern. Rhodora 43:647. 1941. Fig. 167.

Shrubs to 5 m tall; branchlets gray, glabrous; leaves opposite, simple, coarsely dentate, ovate to orbicular, acute at the apex, rounded or subcordate at the base, glabrous on both surfaces, the petioles glabrous; flowers in cymes, the branches of the cymes glabrous; sepals 5, united; petals 5, united, white; stamens 5, exserted; ovary inferior; drupes blue-black, more or less globose, to 10 mm in diameter, containing a single grooved stone. May–July.

Wet ground, seldom in standing water.

IL, IN, OH (FACW–).

Smooth arrowwood.

This species is very similar to *V. dentatum*, but this latter species usually has pubescence on the twigs, leaves, and petioles.

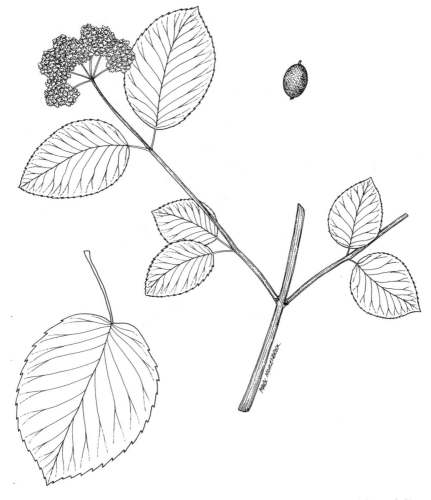

Fig. 167. *Viburnum recognitum* (Smooth arrowwood). Habit (center). Leaf (lower left). Fruit (upper right).

5. **Viburnum trilobum** Marsh. Arb. Am. 168. 1785. Fig. 168.
Viburnum americanum Mill. Gard. Dict., ed. 8, 8. 1768, misapplied.
Viburnum opulus L. var. *americanum* Ait. Hort. Kew. 1:280. 1789.

Shrubs or small trees to 5 m tall; stems and branchlets glabrous; leaves opposite, simple, palmately 3-lobed, dentate, ovate, rounded to truncate at the base, glabrous on the upper surface, pubescent on the lower surface, at least on the veins, to 8 cm long, the petioles with round-topped, club-shaped glands; outer flowers of the cyme enlarged, sterile, whitish; sepals 5, united; petals 5, united; stamens 5; ovary inferior; drupes red, subglobose, 10–15 mm long and nearly as broad, with a single flat, suborbicular stone. June–July.

Wet ground, bogs.

IA, IL, IN, MO, OH (FACW).

Highbush cranberry.

This native species is very similar to *V. opulus*, differing by its round-topped, club-shaped petiolar glands.

Fig. 168. *Viburnum trilobum* (Highbush cranberry). Flowers and leaves (upper right). Cluster of fruits (lower left). Base of leaf with gland on petiole (lower right).

56. CARYOPHYLLACEAE—PINK FAMILY

Annual, biennial, or perennial herbs; leaves simple, opposite or less commonly whorled, entire (rarely minutely serrulate), with or without stipules; inflorescence cymose or paniculate, terminal or axillary, or flower solitary in the leaf axils, with or without bracts; flowers actinomorphic, usually perfect; sepals (4–) 5, free or united into a tubular or urn-shaped calyx; petals (4–) 5, free, occasionally absent, sometimes with basal appendages; stamens 5 or 10, rarely 4 or 8; ovary superior, 1-locular; styles 2, 3, 5, or rarely 4; fruit a 1-seeded utricle or a capsule; seeds often tuberculate or roughened.

There are approximately eighty-five genera and twenty-four thousand species of Caryophyllaceae in the world, particularly in arctic and alpine areas of the Northern Hemisphere.

The following genera are represented in wetlands in the central Midwest.

1. Petals absent .... 2. *Sagina*
1. Petals present.
    2. Leaves oblong to ovate, cordate at the base, pubescent; styles 5 .... 1. *Myosoton*
    2. Leaves linear, tapering to the base, glabrous; styles 3 .... 3. *Stellaria*

1. **Myosoton** Moench—Giant Chickweed

Perennial herbs; leaves opposite, simple, entire, without stipules; inflorescence cymose or reduced to a single axillary flower; flowers perfect, actinomorphic; sepals 5, united below; petals 5, free; stamens 10; styles 5; ovary superior, 1-locular, with many ovules; fruit a capsule, 5-toothed; seeds not flattened, roughened.

The only species that comprises this genus is often merged into the genus *Stellaria*. Reasons for segregating *Myosoton* from *Stellaria* are its five styles that are opposite the petals rather than three styles that are opposite the sepals in *Stellaria*.

1. **Myosoton aquaticum** (L.) Moench, Meth. 225. 1794. Fig. 169.
Cerastium aquaticum L. Sp. Pl. 439. 1753.
Stellaria aquatica (L.) Scop. Fl. Carn. ed. 2, 1:319. 1772.
Alsine aquatica (L.) Britt. Bull. Torrey Club 5:356. 1894.

Perennial herbs from extensive rhizomes; stems decumbent to ascending, glandular-puberulent, 4-angled, branched, up to 75 cm tall; leaves opposite, simple, ovate-lanceolate to ovate, acute at the apex, rounded to subcordate at the base, usually puberulent, up to 5.5 (–7.0) cm long, the median and upper leaves sessile or barely clasping, the lowermost leaves usually short-petiolate; flowers to 1.0 (–1.2) cm across, in open cymes from the upper axils, with ovate-lanceolate, green bracts, with glandular-viscid pedicels up to 2.5 cm long, becoming deflexed in fruit; sepals 5, united below, ovate, acute to obtuse, glandular-puberulent, green, up to 9 mm long in fruit; petals 5, free, 2-cleft at the apex, longer than the sepals in flower; stamens 10; styles 5, free, opposite the petals; capsules ovoid to oblongoid, with five 2-cleft valves; seeds more or less orbicular, 0.7–0.8 mm in diameter, dark brown, roughened. May–October.

Moist soil or wet soil, particularly along streams; sometimes in pastures.

Native of Europe; IA, IL, IN, MO (FAC+), KY, OH (FACW), KS, NE (not listed).

Fig. 169. *Myosoton aquaticum*
(Giant chickweed).

a. Habit.
b. Node.
c. Flowering branch.

d. Flower.
e. Fruit.
f. Seed.

Giant chickweed.

The five styles that are opposite the petals are the most distinctive features of this species.

2. **Sagina** L.—Pearlwort

Delicate annuals or perennials; leaves opposite, simple, connate at base, entire, without stipules; flowers solitary or in cymes; sepals 4–5, free; petals 0, 4, 5, free; stamens 4, 5, 8, 10; ovary superior; styles 5; capsules with numerous seeds.

Fifteen species in the Northern Hemisphere comprise this genus of small herbs.

1. **Sagina fontinalis** Short & Peter, Transyl. Journ. Med. 7:600. 1834. Fig 170.
Stellaria fontinalis (Short & Peter) Robins. Proc. Am. Acad. 29:286. 1894.
Alsine fontinalis (Short & Peter) Britt. Mem. Torrey Club 5:356. 1894.

Weak annuals with fibrous roots; stems spreading to ascending, very slender, branched, glabrous, to 15 cm long; stems opposite, simple, linear, obtuse to acute at the apex, tapering to the base, entire, glabrous, to 2 cm long, to 2 mm wide, the upper sessile; flowers 1–3 from the axils of the leaves, on filiform pedicels up to 25 mm long; sepals 4, free, 5-nerved, green, linear, 1–2 mm long; petals absent; stamens 4; styles 4, very short; ovary superior; capsules ovoid; seeds numerous, shining, papillate, 0.5–0.6 mm long. March–May.

Seepage areas on wet cliffs.

KY (FACW).

Kentucky pearlwort.

This small species occurs in wet seepage areas on a few cliffs in central Kentucky. It is apetalous. Some botanists have put this species in the genus *Stellaria*.

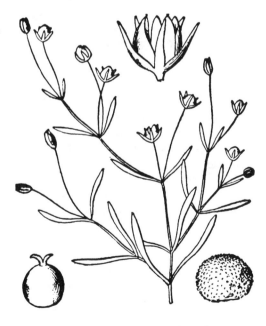

Fig. 170. *Sagina fontinalis* (Kentucky pearlwort). Habit (center). Fruit (lower left). Seed (lower right). Flower (upper).

3. **Stellaria** L.—Chickweed

Annual or perennial herbs; leaves opposite, entire, without stipules; inflorescence cymose or reduced to a single axillary flower; flowers perfect, actinomorphic; sepals (4–) 5, free; petals 5, free, rarely absent; stamens (2, 8) 10; styles 3 (–4); ovary superior, 1-locular, with many ovules; fruit a capsule, usually with 6 teeth; seeds compressed to globose, smooth or variously roughened.

About ninety species throughout most of the Northern Hemisphere comprise this genus.

1. **Stellaria longifolia** Muhl. ex Willd. Enum. Hort. Ber. 479. 1809. Fig. 171. *Alsine longifolia* (Muhl.) Britt. Mem. Torrey Club 5:150. 1894.

Perennial herbs from white rhizomes; stems decumbent to ascending, weak, glabrous, 4-angled, branched above, sometimes scabrous on the angles, up to 45 cm tall; leaves opposite, simple, linear to linear-lanceolate, acute at the apex, tapering to the sessile base, glabrous except for ciliate margins near the base, up to 4 cm long, up to 5 mm wide; flowers to 8 mm across, numerous in much branched lateral cymes, with scarious, ciliate, lanceolate bracts up to 3 mm long, with spreading to ascending to erect pedicels; sepals 5, free, lanceolate to ovate-lanceolate, acute to obtuse, usually glabrous, green with somewhat scarious margins, without conspic-uous nerves, 2.5–4.0 mm long; petals 5, free, 2-cleft at the apex, white, as long as or a little longer than the sepals; stamens usually 10; capsules ovoid to oblongoid, 5–8 mm long; seeds oblongoid to ovoid, 0.7–1.0 mm long, reddish brown, smooth or nearly so. May–July.

Moist ground, including bogs and floodplains.

IA, IL, IN, MO (FACW+), KY, OH (FACW), KS, NE (not listed).

Long-leaved stitchwort.

This species is distinguished by its nerveless sepals and its mostly smooth seeds.

57. CERATOPHYLLACEAE—HORNWORT FAMILY

Only the following genus comprises the family.

1. **Ceratophyllum** L.—Hornwort

Herbaceous aquatic herbs with much branched stems; leaves whorled, several-forked; flowers unisexual, sessile, solitary in the axils of the leaves, each flower sub-tended by a several-forked involucre, the plants monoecious; calyx absent; corolla absent; stamens 10–numerous, free; pistil one, the ovary superior, 1-locular, with 1 pendulous ovule; fruit an achene beaked by the persistent style.

The whorled leaves and sessile, axillary, unisexual flowers distinguish this genus.

There are three species in the genus, with two of them occurring in the central Midwest.

1. Achenes with two basal spines; ultimate leaf segments toothed on the margins ....
....1. *C. demersum*
1. Achenes with several spines both lateral and basal; ultimate leaf segments not toothed on the margins .... 2. *C. echinatum*

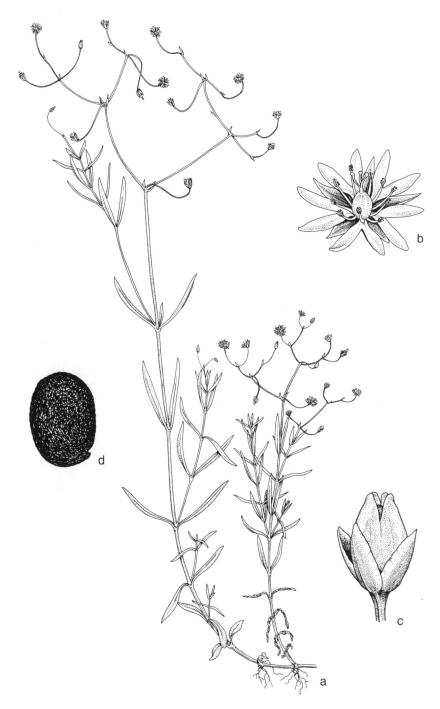

Fig. 171. *Stellaria longifolia*
(Long-leaved stitchwort).

a. Habit.
b. Flower.

c. Fruit.
d. Seed.

1. **Ceratophyllum demersum** L. Sp. Pl. 992. 1753. Fig. 172.

Herbaceous aquatics; stems much branched, glabrous, varying in length in accordance with the depth of the water; leaves whorled, sessile, to 3 cm long, divided usually into 3-forked, filiform segments, the segments flattened and serrated; flower solitary in the axils, unisexual, subtended by an 8- to 12-fid involucre; stamens 10–20, free, the anthers sessile or nearly so; fruit ellipsoid, compressed, glabrous, wingless, the body 4–5 mm long, the 2 basal spines 2–5 mm long, the beak 4–6 mm long. July–September.

Quiet waters.

IA, IL, IN, KS, KY, MO, NE, OH (OBL).

Coontail; hornwort.

This species is often very abundant in quiet waters. The length and texture of the stems, the degree of serrations of the ultimate leaf segments, and characters of the fruit are all variable.

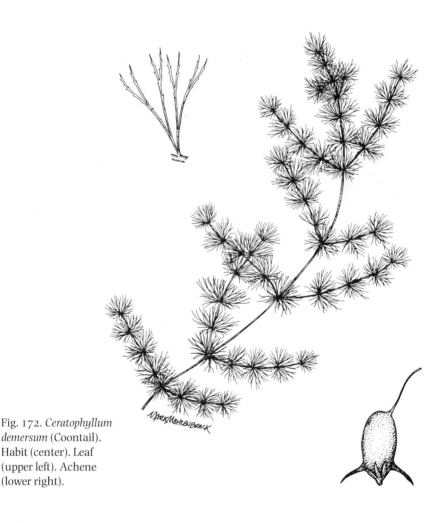

Fig. 172. *Ceratophyllum demersum* (Coontail). Habit (center). Leaf (upper left). Achene (lower right).

2. Ceratophyllum echinatum Gray, Man. ed. 1, 401. 1848. Fig. 173.

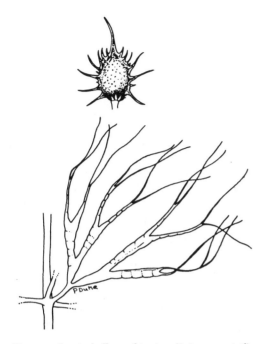

Herbaceous aquatics; stems much branched, glabrous, varying in length in accordance with the depth of the water; leaves whorled, sessile, to 3 cm long, glabrous, divided usually into 3-forked, filiform segments, the segments flattened and not serrated; flower solitary in the axils, unisexual, subtended by an 8- to 12-fid involucre; stamens 10–20, free, the anthers sessile or nearly so; fruit ellipsoid, compressed, minutely tuberculate, narrowly winged, the body 5–7 mm long, the several lateral and basal spines 2–5 mm long, the beak 5–10 mm long. July–September. Quiet waters.

IL, IN, KS, KY, MO, OH (OBL). Spiny coontail; spiny hornwort. This species often occurs with *C. demersum*, although usually less common. It is distinguished by its numerous spines on the fruits and by its lack of serrations on the leaves.

Fig. 173. *Ceratophyllum echinatum* (Spiny coontail). Leaf (below). Fruit (above).

58. CONVOLVULACEAE—MORNING-GLORY FAMILY

Herbs, shrubs, or trees, sometimes twining, sometimes with latex; leaves alternate, simple, without stipules; inflorescence cymose, or the flower solitary; flowers perfect, actinomorphic, bracteate; sepals 5, usually free; corolla mostly tubular, 5-toothed or entire; stamens 5, free, attached to the base of the corolla tube; disk often present; pistil 1, the ovary superior, 2-locular, with 1–2 ovules per locule on axile placentae; fruit a capsule.

This family consists of about fifty genera and thirteen hundred species, many of them in tropical America and tropical Asia.

Only the following genus has a species that sometimes occurs in wetlands.

1. Calystegia R. Br.—Bindweed

Perennial herbs; stems trailing, twining, or erect; leaves alternate, simple, entire or basally lobed; inflorescence cymose, axillary, 1- to 4-flowered; flowers perfect, actinomorphic, subtended by a pair of foliar bracts; calyx deeply 5-parted, or the sepals free; corolla funnelform, shallowly 5-lobed or entire; stamens 5, attached to the base of the corolla; pistil 1, the ovary superior, 2-locular, the stigmas ovoid to ellipsoid; fruit a 2-locular capsule.

The genus contains approximately twenty-five species. Only the following may sometimes be found in wetlands in the central Midwest. .

1. **Calystegia sepium** (L.) R. Br. ssp. **americana** (Sims) Brummitt, Ann. Mo. Bot. Gard. 52:216. 1965. Fig. 174.
Convolvulus sepium L. Sp. Pl. 153. 1753.
Convolvulus sepium L. var. *americanus* Sims, Bot. Mag. 190:pl. 732. 1804.
Convolvulus americanus (Sims) Greene, Pittonia 3:328. 1898.
Calystegia sepium (L.) R. Br. var. *americana* (Sims) Kitagawa, Rep. Inst. Sci. Res. Man-choukuo 3 App. 1:365. 1939.

Perennial herbs; stems twining or trailing, branched, glabrous or sparsely pubescent, to 4 m long; leaves alternate, simple, oblong to ovate to deltoid, obtuse to acute to acuminate at the apex, mostly hastate at the base, with a V-shaped sinus, entire or sparsely dentate, glabrous or pubescent, to 10 cm long, on glabrous to pubescent petioles to 5 cm long; flowers 1–2 per axil, on peduncles longer than the petioles, subtended by a pair of ovate, cordate, foliaceous bracts to 3.5 cm long; sepals 5, free or nearly so, green, glabrous to pubescent, concealed by the bracts; corolla white to pinkish, to 8 cm long, nearly as broad, shallowly 5-lobed; stamens 5, included to barely exserted; stigmas oblongoid; capsules up to 4-seeded. June–August.

Moist soil, fields, fens, sometimes in disturbed soil.

IA, IL, IN, MO (FAC), KS, KY, NE, OH (FAC–).

American bindweed; hedge bindweed.

The V-shaped sinus between the basal lobes of the leaf and the square rather than rounded basal lobes are distinctive for this plant.

59. CORNACEAE—DOGWOOD FAMILY

Shrubs or trees, less commonly herbs; leaves simple, usually opposite, entire; flowers perfect, borne in cymes; sepals 4, tiny, attached to the ovary; petals 4, free; stamens 4, attached to the petals; ovary inferior; fruit a drupe, with 1–2 seeds.

There are about ten genera and eighty-five species in this family.

1. **Cornus** L.—Dogwood

Shrubs or trees; leaves simple, usually opposite, entire; inflorescence cymose; sepals 4, united at base; petals 4, free; stamens 4; ovary inferior, 2-locular, each locule with 1 ovule; fruit a 2-seeded drupe.

There are approximately fifty species in this genus.

1. Leaves softly pilose on the lower surface..4. *C. sericea*
1. Leaves with appressed pubescence or glabrous on the lower surface.
   2. Leaves whitened or glaucous beneath.
      3. Young twigs glabrous, red; pith white; drupes white ....4. *C. sericea*
      3. Young twigs pubescent, brownish; pith dark; drupes blue .... 3. *C. obliqua*
   2. Leaves green or rufescent beneath.
      4. Pith white; leaves glabrous beneath...2. *C. foemina*
      4. Pith brownish; leaves with reddish pubescence beneath ....1. *C. amomum*

Fig. 174. *Calystegia sepium* (American bindweed).

a. Habit.
b. Flower with bracteoles removed, showing calyx.
c. Bracteole.
d. Corolla partly cut away.

e. Pistil.
f, g, h. Leaves.
i. Flower.
j. Progression of bracteoles to sepals.

1. **Cornus amomum** Mill. Gard. Dict. ed. 8, 5. 1768. Fig. 175.

Shrubs to 4 m tall; twigs brown to purplish, usually with some pubescence, with brownish pith; leaves opposite, simple, entire, ovate to elliptic, acuminate at the apex, rounded at the base, green and glabrous on the upper surface, rufescent with reddish pubescence on the lower surface, to 12 cm long, to 8 cm wide; flowers white, in compact, flat cymes; sepals 4; petals 4, free from each other; stamens 4; ovary inferior; drupes blue, glabrous, 6–10 mm in diameter, with ridged stones. April–June.

Swamps, moist thickets.

IA, IL, IN, MO (FACW+), KY, OH (FACW).

Southern silky dogwood; southern blue-fruited dogwood.

This shrub of the southeastern United States is distinguished from *C. foemina* by its dark pith and glabrous leaves, and from *C. obliqua* by its leaves that are not whitened on the lower surface.

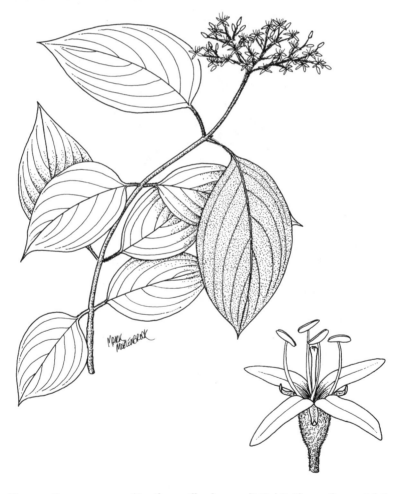

Fig. 175. *Cornus amomum* (Southern silky dogwood). Habit. Flower (lower right).

2. **Cornus foemina** Mill. Gard. Dict. ed. 8, 4. 1768. Fig. 176.
Cornus stricta Lam. Encycl. 2:116. 1786.

Shrubs to 4 m tall; twigs brown to purplish, glabrous, with white pith; leaves opposite, simple, entire, ovate to ovate-lanceolate, acute to acuminate at the apex, tapering or rounded at the base, green and glabrous on both surfaces, to 8 cm long, to 4 cm wide; flowers white, in flat-topped or convex cymes; sepals 4; petals 4, free; stamens 4; ovary inferior; drupes blue, sometimes whitish, globose, 5–8 mm in diameter, with smooth, glabrous stones. April–May.
Swamps, low woods.
IL, IN, MO (FACW–), KY (FAC).
Southern swamp dogwood.
This southeastern species is the only dogwood with blue drupes and white pith.

3. **Cornus obliqua** Raf. Am Nat. 13:1820. Fig. 177.
Cornus schuetzeana C. A. Meyer, Einige Cornus—Arten 22. 1845.

Shrubs to 4 m tall; twigs brown to purplish, with some pubescence when young, with brown pith; leaves opposite, simple, entire, ovate to elliptic, acuminate at the apex, rounded at the base, green on the upper surface, whitened and pubescent on the lower surface, to 10 cm long, to 6 cm wide; flowers white, in compact, flat cymes; sepals 4; petals 4, free; stamens 4; ovary inferior; drupes blue, glabrous, 6–10 mm in diameter, with ridged stones. May–July.
Swamps, low woods, wet prairies, fens.
IA, IL, IN, MO (FACW+), KS, KY, NE, OH (FACW) (the U.S. Fish and Wildlife Service calls this plant *C. amomum*).
Blue-fruited dogwood; silky dogwood.
The name silky dogwood is derived from the appressed hairs found on the young twigs of this species. It is the only blue-fruited dogwood with the leaves whitened on the lower surface.

4. **Cornus sericea** L. Mant. Pl. 2:199. 1771. Fig. 178.
Cornus stolonifera Michx. Fl. Bor. Am. 1:92. 1803, misapplied.
Cornus baileyi Coult. & Evans, Bot. Gaz. 15:37. 1890.
Svida interior Rydb. Bull. Torrey Club 31:572. 1904.
Cornus interior (Rydb.) Petersen, Fl. Nebr. 163. 1912.
Cornus stolonifera Michx. var. *baileyi* (Coult. & Evans) Drescher, Trans. Wisc. Acad, Sci. 28:190. 1933.
Cornus sericea L. var. *baileyi* (Coult. & Evans) Mohlenbr. Vasc. Fl. Ill. 187. 2002.

Shrubs to 5 m tall; twigs red or brownish, glabrous or sometimes pubescent or even tomentose when young, with white pith; leaves opposite, simple, entire, broadly ovate to ovate-lanceolate, acute to acuminate at the apex, rounded at the base, glabrous or appressed pubescent on the upper surface, glabrous or appressed pubescent or densely soft pilose and whitened on the lower surface, to 10 cm long, to 7.5 cm broad; flowers white, in flat or slightly convex cymes; sepals 4; petals 4, free; stamens 4; ovary inferior; drupes white, globose, glabrous, 7–10 mm in diameter, with rounded stones. May–September.

Fig. 176. *Cornus foemina*
(Southern swamp
dogwood). Habit.
Seed (upper left).
Flower (lower right).

Fig. 177. *Cornus obliqua*
(Blue-fruited dogwood).
Habit. Flower (upper
left). Seed (lower right).

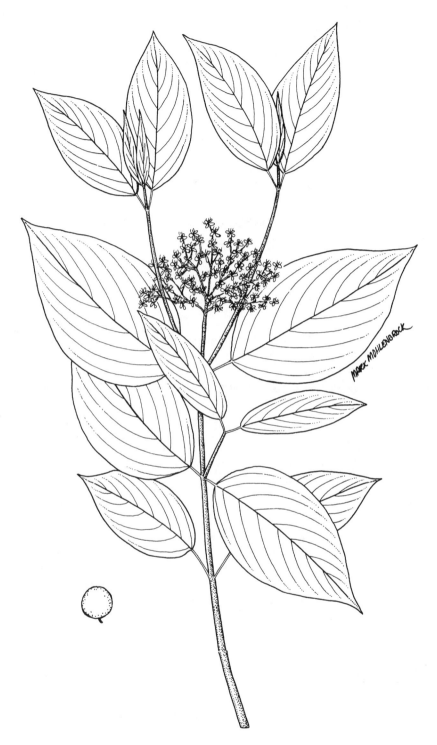

Fig. 178. *Cornus sericea* (Red-osier dogwood). Habit. Fruit (lower left).

Fens, marshes, wet ground.

IA, IL, IN (FACW), KY, OH (FACW+), as *C. stolonifera*.

Red-osier dogwood; red twig dogwood; red osier.

This species has long been known as *C. stolonifera*. In its typical variety, the plants have bright red twigs and leaves that are either glabrous or appressed on the lower surface. Plants with more brownish twigs and densely soft pilose twigs and lower leaf surfaces may be called var. *baileyi*. Plants of the western part of our range with densely tomentose twigs have been called *C. interior*.

60. CORYLACEAE—HAZELNUT FAMILY

Trees or shrubs; leaves alternate, simple, stipulate; flowers unisexual, the staminate usually in elongated, pendulous catkins, the pistillate pendulous or erect, short; calyx absent in the staminate flowers, 2- to 4-parted in the pistillate flowers; petals absent; stamens more than 4; ovary inferior; ovules 2; fruit a nut, enclosed in an involucre.

The Corylaceae consists of four genera and about eighty-five species.

1. **Carpinus** L.—Hornbeam

Trees with smooth, thin, ashy-gray bark; twigs slender, terete, tough, reddish brown or gray; buds small, acute; leaves ovate, acute at the apex, serrate; staminate flowers in pendulous catkins; stamens 3–20, subtended by acute, leathery scales that exceed them; pistillate flowers in terminal catkins, in pairs, subtended by 3-lobed foliaceous bracts; fruit a nut, ovoid, acute, separating from the bracts when mature.

Carpinus is a genus of several species in the Northern Hemisphere.

1. **Carpinus caroliniana** Walt. Fl. Carol. 236. 1788. Fig. 179.
Carpinus betululus virginiana Marsh. Arb. Am. 25. 1785.
Carpinus caroliniana Walt. var. *virginiana* (Marsh.) Fern. Rhodora 37:425. 1935.

Trees up to 12 m tall, to 2 dm in diameter, the crown open and rounded; bark thin, ashy-gray, fluted, marked with broad, dark stripes and appearing twisted; twigs reddish brown, slender, terete, with a zigzag appearance; buds small, acute, chestnut-brown; leaves alternate, simple, ovate-oblong, thin, acute at the apex, doubly serrate, glabrous except for tufts of hairs in the vein axils, up to 12 cm long; petioles short, slender, hairy, terete; staminate flowers in catkins up to 5 cm long, the flowers subtended by broadly ovate scales; pistillate flowers in catkins up to 2 cm long, in pairs, subtended by ovate, acute scales; fruit a nut, ovoid, to 8 mm long, subtended by a 3-lobed, foliaceous bract. April–May.

Wet woods, mesic woods.

IA, IL, IN, KY, MO, OH (FAC).

Musclewood, blue beech, American hornbeam, ironwood.

This species is distinguished from America beech by shorter buds, more teeth per leaf, and a fluted bark. It differs from *Ostrya virginiana* by its smooth bark and its less pubescent and usually narrower leaves.

Fig. 179. *Carpinus caroliniana*
(Musclewood).

a. Fruiting branch.
b, c. Staminate catkins.

d. Bract and nutlet.
e. Nutlet.

61. CUSCUTACEAE—DODDER FAMILY

Twining, annual parasitic herbs, without chlorophyll, attached to hosts by means of haustoria; stems threadlike, orange or yellow; leaves scalelike, much reduced, without chlorophyll; flowers in cymes, sometimes densely crowded and resembling a head or spike, perfect, actinomorphic; calyx 4- or 5-parted; corolla 4- or 5-parted, white, with a diversity of fringed or cleft scales on the tube beneath each stamen; stamens 4–5, alternating with the lobes of the corolla; ovary superior, 2-locular, with 2 free or united styles; fruit a capsule.

This family is sometimes merged with the Convolvulaceae. The Cuscutaceae consists of approximately 150 species in one genus. All species are noxious and harmful weeds that severely parasitize their hosts.

1. **Cuscuta** L.—Dodder

Parasitic annual herbs; stems twining, with minute suckers, or haustoria, orange or yellow or red; leaves reduced to alternate scales, or absent; flowers in cymes, perfect, actinomorphic; calyx 4- to 5-parted, or the sepals free; corolla campanulate to urceolate, 4- to 5-parted; stamens 4–5, attached to the corolla; pistil 1, the ovary superior, 2-locular, with 2 ovules per locule; styles 2, usually free; capsule circumscissile, 1- to 4-seeded, the seeds glabrous.

The seeds of *Cuscuta* germinate in the soil. Upon emerging from the ground, the seedling immediately attaches itself to the host plant by means of minute haustoria.

Although the species of this family do not themselves grow in water, some species are parasitic upon plants that are in standing water. The following have been observed on plants growing in water.

1. Sepals free to base.
  2. Bracts at base of sepals appressed; lobes of corolla obtuse; seeds 2.5–2.6 mm long .... 2. *C. compacta*
  2. Bracts at base of sepals with recurved tips; lobes of corolla acute; seeds 1.7–1.8 mm long . ... 3. *C. glomerata*
1. Sepals united, at least at base.
  3. Flowers with narrowly oblong scales at base...1. *C. cephalanthi*
  3. Flowers with obsolete scales at base...4. *C. polygonorum*

1. **Cuscuta cephalanthi** Engelm. Am. Journ. Sci. 43:336. 1842. Fig. 180.

Stem coarse, yellow; flowers pedicellate in glomerules; bracts absent; calyx usually 4-lobed, the lobes ovate, obtuse, glabrous, much shorter than the corolla tube; corolla cylindric-campanulate, white, with 4 ovate, obtuse, spreading lobes, about half as long as the tube, with 5 narrowly oblong, toothed scales; stamens 5, included; styles about as long as the ovary; stigmas capitate; capsules globose, with the withered corolla persistent; seeds 1.6–1.7 mm long. August–October.

Parasitic primarily on *Cephalanthus* and various shrubs and herbs in wet ground. IA, IL, IN, KS, KY, MO, NE, OH (not rated by the U.S. Fish and Wildlife Service). Dodder.

The distinguishing features of this species are the 4-lobed, erect corollas with toothed scales.

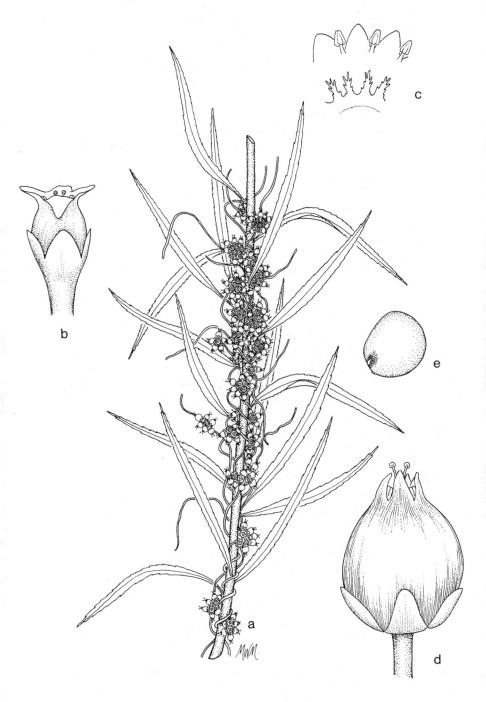

Fig. 180. *Cuscuta cephalanthi*
(Dodder).

a. Habit, on leafy host.
b. Flower.
c. Corolla cut open.

d. Fruit.
e. Seed.

2. Cuscuta compacta Juss. ex Choisy, Mem. Soc. Gen. 9:281, pl. 4, f. 2. Fig. 181.

Stems rather white, yellowish or whitish; flowers sessile in dense clusters; bracts 3–5, orbicular, obtuse, serrulate, glabrous; sepals 5, free, orbicular, obtuse, glabrous, to 3 mm long; corolla salverform to urceolate, white, with 5 oblong to ovate, obtuse to subacute, spreading lobes less than half as long as the tube, with 5 narrow, fimbriate scales shorter than the corolla tube; stamens 5, reaching the base of the sinuses between the corolla lobes; styles slightly shorter than the ovary; stigmas capitate; capsules oblongoid, with the withered corolla persistent; seeds 2.5–2.6 mm long. July–October.

Parasitic mostly on shrubby plants in low ground.

IA, IL, IN, KS, KY, MO, NE, OH (not rated).

Dodder.

This species is similar to *C. glomerata* in its sessile flowers, but the flowers do not form the dense clusters as the flowers of *C. glomerata* do. *Cuscuta compacta* has the largest seeds of any species of *Cuscuta* in the central Midwest.

3. Cuscuta glomerata Choisy, Mem. Soc. Gen. 9:280, pl. 4, f. 1. 1842. Fig. 182.
Lepidanche compositorum Engelm. Am. Journ. Sci. 43:344. 1842.

Stems slender, yellow or whitish; flowers sessile in dense clusters, usually completely concealing the stem of the host; bracts 8 or more, lanceolate, acute and recurved at tip, serrulate, glabrous; sepals 4, free, oblong, obtuse, serrulate, glabrous, to 2.5 mm long; corolla oblong-cylindric, white, with 5 oblong to lanceolate, subacute, spreading lobes less than half as long as the tube, with 5 narrow, fimbriate scales shorter than the corolla tube; stamens 5, reaching the base of the sinuses between the corolla lobes; styles at least twice as long as the ovary; stigmas capitate; capsules oblongoid, with the withered corolla persistent; seeds 1.7–1.8 mm long. July–October.

Parasitic in low areas, primarily on Asteraceae.

IA, IL, IN, KS, KY, MO, NE, OH (not rated).

Dodder.

The sessile flowers grow in compact clusters so densely that the stem of the host is usually completely covered. The flowers are somewhat fragrant. The long styles are also distinctive.

4. Cuscuta polygonorum Engelm. Am. Journ. Sci. 43:342. 1842. Fig. 183.
Cuscuta chlorocarpa Engelm. in Gray, Man. Bot. 350. 1848.

Stems slender, orange; flowers pedicellate in glomerules; bracts absent; calyx usually 4-lobed, the lobes ovate, obtuse, glabrous, about as long as the corolla tube; corolla campanulate, white, with 4 deltoid, acute, erect lobes about as long as the tube, with minute toothlike scales, or the scales absent; stamens 5, arising from the sinuses between the corolla lobes; styles shorter than the ovary; stigmas capitate; capsules globose, with the withered corolla persistent; seeds 1.3–1.4 mm long. July–September.

Parasitic on various herbs in wet soil.

IA, IL, IN, KS, KY, MO, NE, OH (not rated).

Dodder.

This species occurs in low areas, particularly on species of *Polygonum*. This is the only species of *Cuscuta* in the central Midwest with obsolete scales.

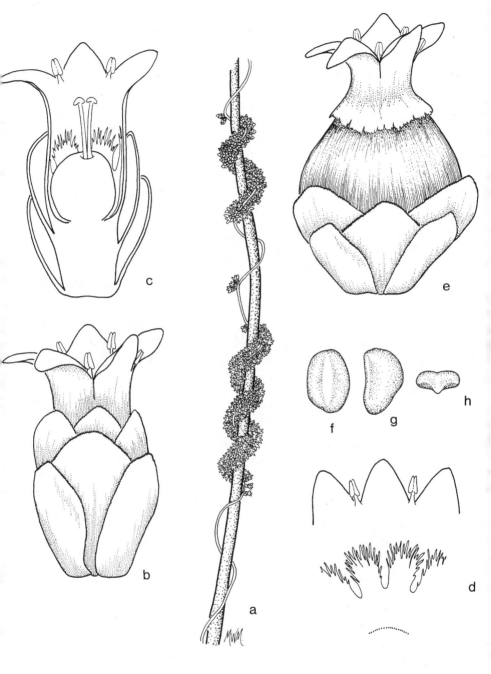

Fig. 181. *Cuscuta compacta*
(Dodder).

a. Habit, on host.
b. Flower.
c. Flower partly cut away.
d. Corolla opened.

e. Developing fruit.
f. Seed, face view.
g. Seed, side view.
h. Seed, top view.

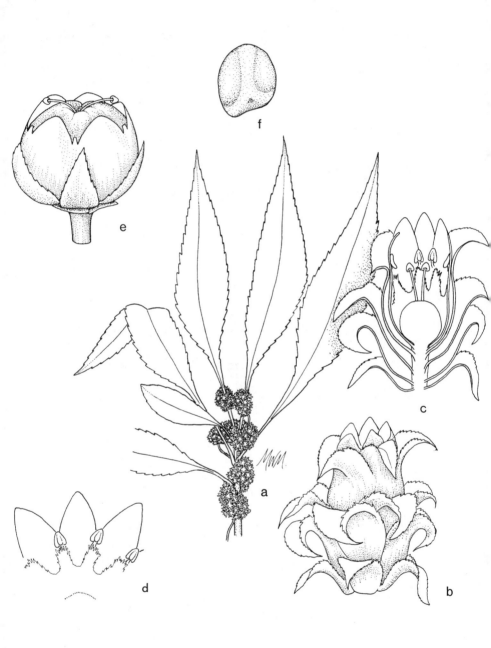

Fig. 182. *Cuscuta glomerata* (Dodder).

a. Habit, on leafy host.
b. Flower.
c. Side view of flower.

d. Corolla opened out.
e. Fruit.
f. Seed.

Fig. 183. *Cuscuta polygonorum*
(Dodder).

a. Habit, on leafy host.
b. Flower.
c. Corolla opened out.
d. Scale.

e. Developing fruit.
f. Seed, face view.
g. Seed, top view.

243

62. DROSERACEAE—SUNDEW FAMILY

Carnivorous biennial or perennial herbs; leaves basal, bearing gland-tipped hairs; flowers in 1-sided racemes, perfect, actinomorphic; sepals 5; petals 5, free; stamens 5; ovary superior, the styles 3 or 5; fruit a capsule.

This family consists of four genera and about one hundred species, found in many parts of the world.

1. **Drosera** L.—Sundew

Carnivorous herbs; leaves basal, simple, bearing gland-tipped hairs, with stipules; flowers in a 1-sided raceme, perfect, actinomorphic; sepals 5; petals 5, free; stamens 5; ovary superior; styles 3; fruit a capsule.

Approximately ninety-five species are in this genus.

1. Leaves suborbicular, broader than long ....3. *D. rotundifolia*
1. Leaves obovate to spatulate, longer than broad.
   2. Scapes glandular-pubescent .... 1. *D. brevifolia*
   2. Scapes not glandular-pubescent ....2. *D. intermedia*

1. **Drosera brevifolia** Pursh, Fl. Am. Sept. 1:211. 1814. Fig. 184.

Perennial herbs; leaves basal, forming a rosette up to 3.5 cm across, obovate, obtuse at the apex, tapering to the petiolate base, glandular-hairy, to 15 mm long, to 6 mm wide, without stipules; scape filiform, to 8 cm tall, stipitate-glandular, with 1–2 flowers; flowers in 1-sided racemes to 1.5 cm across; sepals 5, glandular-pubescent, 3–4 mm long; petals 5, free, white, obovate, 4–7 mm long; stamens 5; capsule with 3 valves, containing numerous black, obovoid, pitted seeds. May–June.

Wet meadows.

KY (OBL).

Short-leaved sundew.

This is the only *Drosera* with glandular-pubescent scapes. Primarily a species of the southern United States, *D. brevifolia* enters our area in Kentucky where it grows in wet meadows.

The flowers are open only during morning hours.

2. **Drosera intermedia** Hayne in Schrad. Journ. Bot. 1:37. 1800. Fig. 185.
Drosera longifolia Michx. Fl. Bor. Am. 1:186. 1803.

Perennial herbs; leaves basal, forming a rosette, spatulate to obovate, subacute at the apex, tapering to long, glabrous petioles, glandular-hairy, to 20 mm long, to 8 mm wide, with setaceous stipules; scape to 20 cm tall, glabrous, with up to 20 flowers; flowers in 1-sided racemes to 1.2 cm long; sepals 5, glabrous or sparsely pubescent, 3–4 mm long; petals 5, free, white, obovate, 5–8 mm long; stamens 5; capsule with 3 valves, containing numerous reddish brown, oblongoid, papillose seeds. June–August.

Bogs, pools in excavated sandy areas.

IL, IN, KY, OH (OBL).

Narrow-leaved sundew.

This species differs from other sundews in our area by its spatulate leaves and glabrous scapes and petioles.

Fig. 184. *Drosera brevifolia*
(Short-leaved sundew).

a. Habit.
b. Leaf.
c. Flower, face view.

d. Calyx and capsule.
e. Seed.

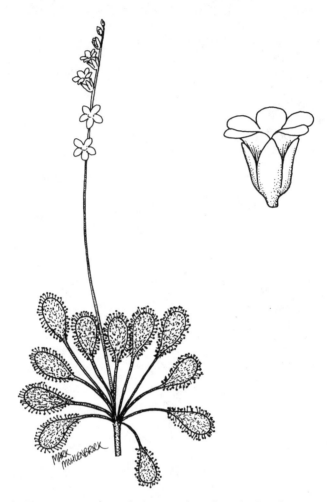

Fig. 185. *Drosera intermedia* (Narrow-leaved sundew). Habit. Flower (right).

3. **Drosera rotundifolia** L. Sp. Pl. 281. 1753. Fig. 186.

Perennial herbs; leaves basal, forming a rosette 5–12 cm across, suborbicular, broadly rounded at the apex, rounded at the petiolate base, glandular-hairy, to 10 mm long, 10–12 mm wide, with fimbriate stipules, the petioles with non-glandular heads; scape to 30 cm long, glabrous, with up to 15 flowers; flowers in 1-sided racemes, to 10 mm across; sepals 5, glabrous or pubescent, 4–5 mm long; petals 5, free, obovate, pink; stamens 5; capsule with 3 valves, containing numerous pale brown, shiny, fusiform, striate seeds. June–August.

Bogs, pools in excavated sandy areas.

IA, IL, IN, KY, OH (OBL).

Round-leaved sundew.

The rounded leaf blades are distinctive for this species.

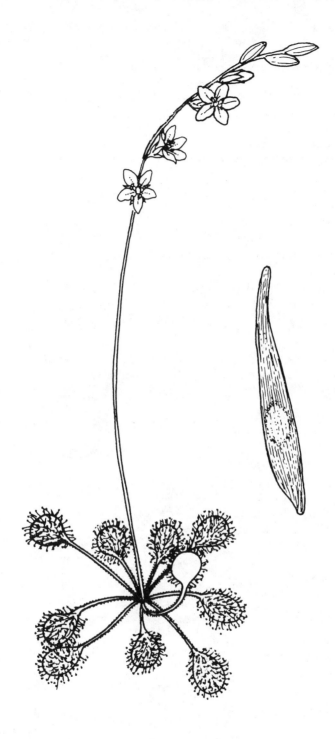

Fig. 186. *Drosera rotundifolia* (Round-leaved sundew).
Habit. Seed (right).

Herbs; leaves opposite, simple, with stipules; flowers axillary or in cymes, perfect, actinomorphic; sepals 2, 3, 4, or 5, usually free; petals 2, 3, 4, or 5, free; stamens 3, 4, 5, 6, 8, or 10; ovary superior; styles 2–5; fruit a capsule with many seeds.

There are two genera and about forty species in this family.

1. Sepals 5; petals 5; stamens 5 or 10; plants glandular-pubescent .... 1. *Bergia*
1. Sepals 2 or 3; petals 2 or 3; stamens 2 or 3; plants glabrous .... 2. *Elatine*

1. **Bergia** L.—Bergia

Small herbs; leaves alternate, toothed; flowers perfect; sepals 5, free; petals 5, free; stamens 5 or 10; ovary superior; fruit a capsule.

There are approximately twenty species in this genus.

1. **Bergia texana** (Hook.) Seub. in Walp. Rep. 1:285. 1842. Fig. 187.
Merimea texana Hook. Icon. Pl. pl. 278. 1840.

Annual; stems much branched, to 25 cm tall, glandular-pubescent; leaves alternate, simple, elliptic to oblong to spatulate, 1–3 cm long, 3–8 mm broad, denticulate, glandular-pubescent; stipules lanceolate, glandular-ciliate; flowers 1–3 in the axils of the leaves, on pedicels 1–3 mm long; sepals 5, lanceolate, acuminate, 2.5–3.0 mm long; petals 5, oblong, acute, 2.0–2.5 mm long, shorter than the sepals; stamens 5 (–10); capsule globose, 2 mm in diameter; seeds 0.3–0.5 mm long. June–August.

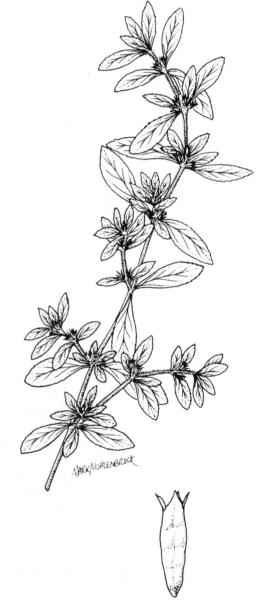

Fig. 187. *Bergia texana* (Bergia). Habit. Fruit (below).

Muddy banks, wet fields, sometimes in standing water.

IL, KS, MO, NE (OBL).

Bergia.

This usually obscure species is distinguished by its glandular-pubescent leaves, small axillary flowers, and globose fruits.

2. Elatine L.—Waterwort

Annuals; leaves alternate, simple; flowers axillary, perfect, actinomorphic; sepals 2, 3, or 4, free; petals 2, 3, or 4, free; stamens 2, 3, or 4; ovary superior; fruit a capsule.

There are approximately twenty species in this genus, found worldwide.

1. Petals 3; stamens 3; leaves oblong to spatulate, retuse at the apex......................3. *E. triandra*
1. Petals 2; stamens 2; leaves obovate, not retuse at the apex.
    2. Seeds with 9–15 areoles in each longitudinal row................................2. *E. brachysperma*
    2. Seeds with 16–25 areoles in each longitudinal row.....................................1. *E. americana*

1. **Elatine americana** (Pursh) Arn. Edinb. Journ. Sci. 1:430. 1813. Fig. 188.

Peplis americana Pursh, Fl. Am. Sept. 238. 1814.

Elatine triandra Schkuhr var. *americana* (Pursh) Fassett, Rhodora 41:374. 1939.

Matted annual herbs; stems creeping, spreading, or sometimes erect or floating, glabrous; leaves alternate, simple, obovate, not retuse, up to 8 mm wide, glabrous; flowers axillary; sepals 2, free; petals 2, free; stamens 2; capsules globose, about 1 mm in diameter, composed of checkered seeds with 16–25 areoles in each longitudinal row. June–August.

Fig. 188. *Elatine americana* (Waterwort). Habit.

In water, on muddy banks.

KS, MO (OBL).

Waterwort.

The flowers of this species are similar to those of *E. brachysperma*, but the seeds have 16–25 areoles in each longitudinal row.

2. **Elatine brachysperma** Gray, Proc. Am. Acad. 13:361. 1878. Fig. 189.

Elatine triandra Schkuhr var. *brachysperma* (Gray) Fassett, Rhodora 41:374. 1939.

Matted annual herbs, sometimes submerged, glabrous; leaves alternate, simple, obovate, not retuse at the apex, to 8 mm wide, glabrous; flowers axillary; sepals 2, free; petals 2, free; stamens 2; capsules globose, about 1 mm in diameter, composed of checkered seeds with 9–15 areoles per longitudinal row. June–August.

Shallow water, muddy banks.

IL, OH (OBL), NE (FACW).

Waterwort.

The seeds of this species have only 9–15 areoles in each longitudinal row.

Fig. 189. *Elatine brachysperma* (Waterwort). Habit (center). Seed (upper left).

3. **Elatine minima** (Nutt.) Fisch. & C. A. Meyer, Linnaea 10:75. 1836. Fig. 190.

Annual herbs; stems creeping or floating, glabrous; leaves alternate, simple, oblong to spatulate, retuse at the apex, up to 8 mm wide, glabrous; flowers axillary; sepals 3, free; petals 3, free; stamens 3; capsules globose, about 1 mm in diameter with numerous seeds, the seeds checkered with narrowly elliptic areoles in each longitudinal row. June–August.

Shallow water, muddy banks.

IL (OBL).

Waterwort.

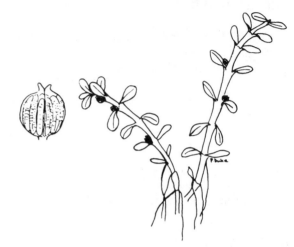

Fig. 190. *Elatine minima* (Waterwort). Habit (right). Fruit (left).

This species differs from the other two species in the central Midwest by its 3-parted flowers and its usually retuse leaves.

64. ERICACEAE—HEATH FAMILY

Perennial herbs, shrubs, or small trees; leaves mostly alternate or basal, often evergreen, frequently coriaceous; flowers perfect, borne variously; sepals (2–) 4–5 (–6), often persistent; petals (4–) 5, free or united; stamens (8–) 10 (–12), dehiscing by vertical slits or apical pores; ovary superior or inferior, (2–) 5- (10-) locular; fruit a capsule, berry, or drupe.

This family is comprised of about 125 genera and approximately thirty-five hundred species.

1. Trees; leaves 10–15 cm long .... 5. *Oxydendrum*
1. Shrubs or shrublike herbs; leaves up to 10 cm long.
  2. Plants trailing; corolla lobes 4; fruit a red berry.. 6. *Vaccinium*
  2. Plants erect or, if the stems trailing, the flowering branches ascending; corolla lobes 5; fruit a capsule or blue-black berry (red in *Gaultheria*).
    3. Plants with the scent of wintergreen... 3. *Gaultheria*
    3. Plants not wintergreen-scented.
      4. Leaves scurfy on the lower surface; flowers in 1-sided racemes......... 2. *Chamaedaphne*
      4. Leaves pubescent, not scurfy; flowers in dense racemes or in umbels.
        5. Leaves up to 6 mm wide .... 1. *Andromeda*
        5. Leaves more than 6 mm wide.
          6. Flowers in umbels, more than 3 cm long; fruit a capsule .... 4. *Rhododendron*
          6. Flowers in racemes, up to 1 cm long; fruit a berry........................... 6. *Vaccinium*

1. Andromeda L.—Bog Rosemary

Low slender shrubs with a creeping base and ascending branches; leaves alternate, simple, entire; inflorescence umbellate; flowers perfect; calyx with nearly 5 free sepals; corolla 5-lobed, urceolate; stamens 10, the anthers opening by terminal pores, awned; ovary superior, 5-locular; fruit a 5-valved capsule, with numerous seeds attached to the columella.

There are two species in this genus, both found in boreal parts of the world.

1. Andromeda glaucophylla Link, Enum. Pl. Hort. Berol. 1:394. 1821. Fig. 191.

Creeping shrubs with ascending stems to 60 cm tall, the branches glaucous, glabrous; leaves evergreen, linear to narrowly oblong, to 6 cm long, 4–6 mm wide, coriaceous, more or less glabrous above, paler and puberulent below, entire, revolute; petioles up to 2 mm long; inflorescence an umbel of 2–6 flowers; pedicels thick, glabrous, 8–12 mm long; calyx whitish, spreading, the segments 1.5–2.5 mm long; corolla urceolate, to 8 mm long, 5-lobed, white or pink; stamens 10, included, the anther with a single, slender awn; style included; capsule depressed-globose, glaucous, 5-valved, 3–5 mm across; seeds numerous, ellipsoid, lustrous, attached to the summit of the columella. May–June.

Bogs.

IL, IN, OH (OBL).

Bog rosemary.

This is a characteristic species of acid bogs in the extreme northern part of the central Midwest. Its evergreen, linear, revolute leaves that have a dense covering of white pubescence on the lower surface are distinctive.

2. Chamaedaphne Moench—Leatherleaf

Only the following species comprises this genus.

1. Chamaedaphne calyculata (L.) Moench var. angustifolia (Ait.) Rehder in Bailey, Cycl. Am. Hort. 1:287. 1900. Fig. 192.
Andromeda calyculata ß angustifolia Ait. Hort. Kew. 2:70. 1900.

Shrub to 1 m tall, the stems much branched, light brown, glabrous; leaves usually evergreen, oblong to oblong-lanceolate, to 5 cm long, less than half as wide, coriaceous, glabrous or scurfy above, scurfy beneath, entire, the petioles up to 3 mm long; inflorescence one-sided, racemose; pedicels to 2 mm long; sepals 5, free, acute, to 2 mm long; corolla urceolate, white, to 6 mm long, 5-lobed, the lobes recurved; stamens 10, included; style slightly exserted; capsule depressed-globose, to 4 mm in diameter; seeds numerous, flat. April–June.

Bogs.

IL, IN, OH (OBL).

Leatherleaf.

This bog-inhabiting shrub often forms dense, impenetrable thickets. It is distinguished by its evergreen, flat leaves that are scurfy on the back.

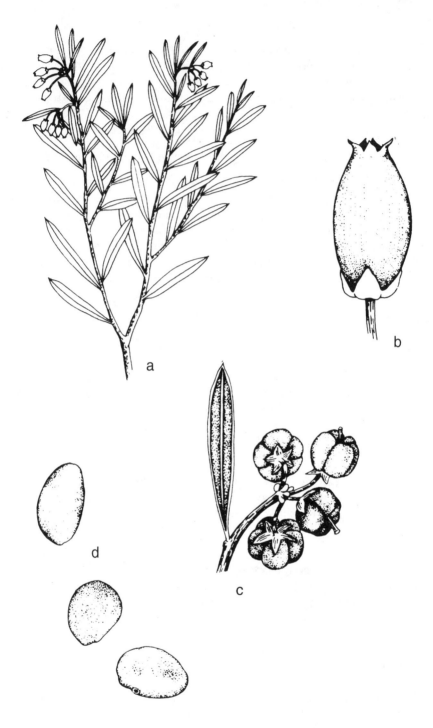

Fig. 191. *Andromeda glaucophylla*   a. Habit.   c. Fruits with leaf.
(Bog rosemary).   b. Flower.   d. Seeds.

Fig. 192. *Chamaedaphne calyculata* (Leatherleaf). Habit, with fruit (left). Habit, with flowers (right).

3. Gaultheria L.—Winterberry

Low shrubs; leaves alternate, simple, evergreen; flowers in racemes or panicles, actinomorphic, perfect; sepals 4–5, united below, subtended by 2 bracteoles; petals 4–5, united below; stamens 10; ovary superior; fruit a dry berry.

This genus consists of about 150 species found in much of the world, particularly South America.

1. Gaultheria procumbens L. Sp. Pl. 395. 1753. Fig. 193.

Perennial aromatic herb from creeping rhizomes and stolons; stems bearing leaves and flowers erect, to 15 cm tall, glabrous or nearly so; leaves simple, alternate but crowded at the top of the stem, evergreen, elliptic to oblong to obovate, acute at the apex, tapering or somewhat rounded at the base, entire or crenulate, sometimes slightly revolute, glabrous or nearly so, dark green above, paler beneath, to 5 cm long, to 2.5 cm wide, the petioles up to 5 mm long; flower solitary from the axils of the leaves, on recurved pedicels up to 10 mm long, subtended by 2 bracteoles; calyx saucer-shaped, 5-parted; corolla urceolate, 5-toothed, white, 7–10 mm long; stamens 10, with white-pubescent filaments; berry fleshy but dry, globose, bright red, 7–11 mm in diameter, the withered corolla often persistent. July–August.

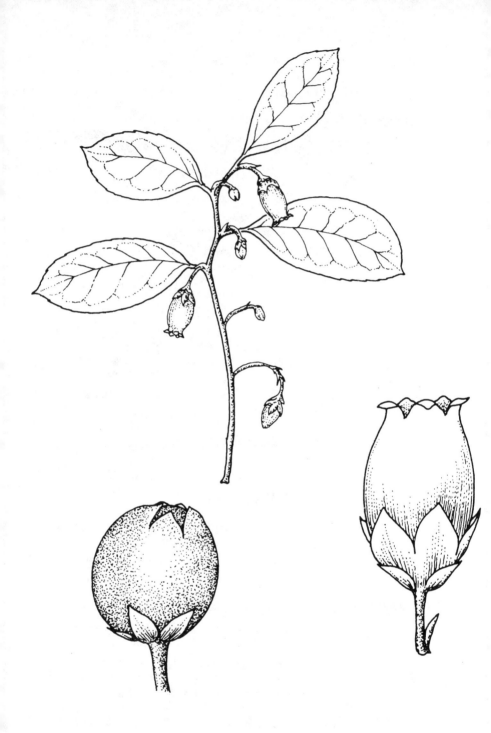

Fig. 193. *Gaultheria procumbens* (Wintergreen). Habit (center). Fruit (lower left). Flower (lower right).

Edge of marshes and bogs, in bogs, mesic woodlands.

IA, IN, KY, OH (FACU).

Wintergreen; checkerberry.

Despite the designation of FACU for this species, it is a frequent inhabitant of bogs. The wintergreen scent when the leaves are crushed is distinctive and pleasant.

4. Oxydendrum DC.—Sourwood

Tree; leaves simple, alternate, deciduous; flowers white, in 1-sided racemes, forming a panicle; calyx deeply 5-parted; corolla ovoid, 5-toothed; stamens 10; ovary superior, 5-locular; capsules angular, the seeds elongated, reticulate.

Only the following species is in the genus.

1. **Oxydendrum arboreum** (L.) DC. Prodr. 7:601. 1839. Fig. 194.

Andromeda arborea L. Sp. Pl. 394. 1753.

Tree to 20 m tall; twigs glabrous, branched; leaves simple, alternate, deciduous, elliptic to oblong-lanceolate, acuminate at the apex, rounded or somewhat tapering to the base, serrate to entire, glabrous, to 15 cm long, to 7 cm wide, the petioles 1.0—1.5 cm long; flowers in 1-sided erect to curving racemes to 15 cm long, on pubescent pedicels up to 8 mm long, with a bracteole near the middle; sepals 5, united at base, 1–2 mm long; corolla white, 5-lobed at the summit, 6–7 mm long; stamens 10; ovary superior, angular; capsules 5-angled, the style persistent, 5–7 mm long, with several elongated, reticulate seeds. June–July.

Swampy woods, mesic woods, dry woods.

IN, KY, OH (NI); introduced in IL.

Sourwood.

This tree has attractive white flowers on curved racemes. The leaves turn a brilliant red in the autumn. It occasionally occurs in swampy woods in our area.

The leaves contain a very sour substance when chewed.

5. Rhododendron L.—Rhododendron; Azalea

Shrubs or small trees; leaves simple, alternate, deciduous or evergreen; flowers showy, in umbel-like clusters or in racemes, actinomorphic, perfect; sepals 5, united below; corolla usually funnelform, 5-lobed; stamens 5 or 10; ovary superior, 5-locular; capsules many-seeded.

Approximately 850 species comprise the genus. If the leaves are deciduous, the plants are called azaleas; if evergreen, they are called rhododendrons.

Only the following species is generally found in wetland habitats.

1. **Rhododendron canescens** (Michx.) Porter, Bull. Torrey Club 16:220. 1889. Fig. 195.

Azalea canescens Michx. Fl. Bor. Am. 1:150. 1803.

Shrub to 5 m tall; stems much branched, glabrous or pubescent; leaves alternate, simple, often crowded near the ends of the branchlets, oblanceolate to oblong, acute at the apex, rounded or tapering to the base, serrulate-ciliate, soft-pubescent on both surfaces, green on the upper surface, paler on the lower surface, to 10

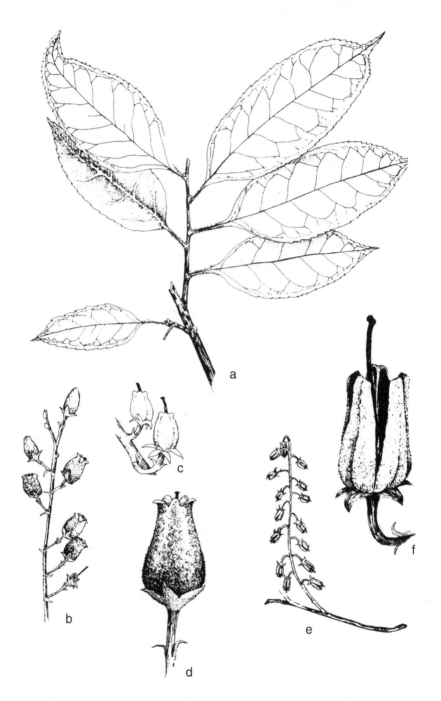

Fig. 194. *Oxydendrum arboreum*
(Sourwood).

a. Branch, with leaves.
b. Upper part of inflorescence.
c. Flowers.
d. Flower.

e. Fruits.
f. Fruit,
longitudinal
view.

cm long, to 4 cm wide, on pubescent petioles up to 1 cm long; flowers in umbel-like clusters, showy, opening before or as the leaves expand, very fragrant; sepals 5, united below, without glandular hairs; petals 5, united below to form an elongated, densely glandular-pubescent tube, to 6 cm long; stamens 10, exserted; ovary superior, pubescent; capsules narrowly oblongoid, usually somewhat pubescent but not glandular, to 1.5 cm long. April–May.

Swampy woods, along streams, dry woods.

KY, OH (FACW).

Hoary azalea; piedmont azalea.

This handsome shrub is distinguished by its soft-hairy leaves that are clustered near the tips of the branchlets.

This is primarily a species of the southeastern United States where it is of common occurrence in the coastal plain and the piedmont.

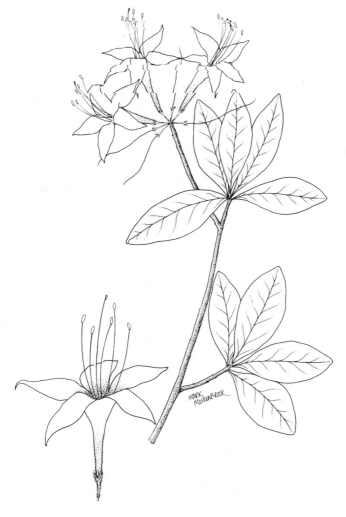

Fig. 195.
*Rhododendron
canescens*
(Hoary azalea).
Habit. Flower
(lower left).

6. **Vaccinium** L.— Blueberry

Shrubs or small trees with alternate, simple, often coriaceous leaves; inflorescence various; flowers perfect; calyx 4- to 5-lobed; corolla 4- to 5-lobed, often campanulate; stamens 8 or 10, the anthers opening by terminal pores, awned or awnless; ovary adnate to the calyx tube, 4- to 5-locular; fruit a many-seeded berry.

More than four hundred species comprise this worldwide genus. The two trailing species are sometimes segregated into the genus *Oxycoccus*.

1. Stems, or at least the flowering ones, ascending or erect; corolla 5-lobed; berries blue-black.
    2. Low shrubs up to 1.5 m tall; corolla 4–5 mm long..................................... 3. *V. myrtilloides*
    2. Tall shrubs; corolla 5–10 mm long .... 2. *V. corymbosum*
1. Stems trailing; corolla 4-lobed; berries red.
    3. Bracteoles leaflike, green, 2–4 mm long, borne at the middle of the pedicel; leaves obtuse.
    .... 1. *V. macrocarpon*
    3. Bracteoles scalelike, red, up to 1.5 mm long, borne below the middle of the pedicel; leaves
    acute .... 4. *V. oxycoccos*

1. **Vaccinium corymbosum** L. Sp.Pl. 350. 1753. Fig. 196.
Vaccinium albiflorum Hook. Bot. Mag. 62, pl. 3428. 1835.
Vaccinium corymbosum L. var. *glabrum* Gray, Man. Bot. N. U.S. ed. 2, 250. 1856.
Vaccinium corymbosum L. var. *albiflorum* (Hook.) Fern. Rhodora 51:104. 1949.

Shrubs to 3 m tall, often growing in clumps, the brown to gray-brown; branchlets glabrous; leaves alternate, simple, elliptic to narrowly ovate, to 8 cm long, about half as wide, subcoriaceous, green on both sides, entire or rarely serrulate, glabrous above, pubescent along the nerves below; inflorescence in dense terminal or axillary clusters; pedicels not jointed,

Fig. 196. *Vaccinium corymbosum* (High-bush blueberry). Habit. Flower (lower left).

3–6 mm long; calyx 0.5–2.0 mm long; corolla urceolate, white or tinged with pink, 5-lobed, 5–10 mm long; stamens 10, included, the anthers awnless; style included; berry globose, blue-black, glaucous, up to 12 mm in diameter, sweet, with several seeds. April–July.

Swampy woods, bogs, sometimes in upland woods and old fields.

IL, IN (FACW), KY, OH (FACW–).

High-bush blueberry.

This is the only wetland *Vaccinium* in the central Midwest that attains heights greater than 1.5 meters.

Rarely there are specimens with serrulate leaves. Var. *albiflorum* has green leaves, and var. *glabrum* has glaucous and nearly glabrous leaves.

2. **Vaccinium macrocarpon** Ait. Hort. Kew. 2:13, pl. 7. 1789. Fig. 197.
Oxycoccus macrocarpus (Ait.) Pers. Syn. 1:419. 1805.

Stems creeping, much branched, 2–4 mm in diameter, pale to dark brown, glabrous; leaves alternate, simple, evergreen, oblong to elliptic, obtuse at the apex, to 1.5 (–1.7) cm long, less than half as wide, coriaceous, green above, somewhat paler beneath, entire, slightly revolute, glabrous; inflorescence racemose, to about 3 cm long, the flowering rachis terminated by a vegetative shoot; pedicels to 12 mm long, not jointed, with a pair of bracteoles near the tip; calyx 1–2 mm long; corolla deeply 4-lobed, the lobes reflexed, 6–10 mm long, very pale pink; stamens 10, exserted, the anthers awnless; style exserted; berry ellipsoid, red to purple to bluish, to 2 cm in diameter, acid, with several seeds. June–August.

Bogs.

IL, IN, OH (OBL).

Large cranberry.

The leaves, which are rarely revolute, are usually obtuse at the apex. The berries average slightly larger than those of *V. oxycoccos*.

3. **Vaccinium myrtilloides** Michx. Fl. Bor. Am. 1:234. 1803. Fig. 198.
Vaccinium canadense Kalm ex Richards. Frankl. Journ. 736. 1823.

Shrubs to 1.0 (–1.5) m tall, the dark brown branchlets densely pubescent; leaves alternate, simple, lanceolate to elliptic, to 4 cm long, about one-half as wide, subcoriaceous, scarcely shiny above, entire, downy on the lower surface and sometimes on the upper surface; inflorescence a dense, terminal, bractless raceme; pedicels not jointed, 3–6 mm long; calyx 1–3 mm long; corolla urceolate, greenish with an occasional purplish tinge, 5-lobed, 4–6 mm long; stamens 10, included, the anthers awnless; style included; berry globose, blue, strongly glaucous, 5–8 mm in diameter, sour, with several seeds. May–June.

Bogs; sandy or rocky slopes.

IA, IL, IN (FACW), OH (FAC).

Velvet-leaved blueberry.

This species is similar to the upland *V. pallidum* from which it differs by its densely pubescent branchlets and lower leaf surfaces and its sour berries.

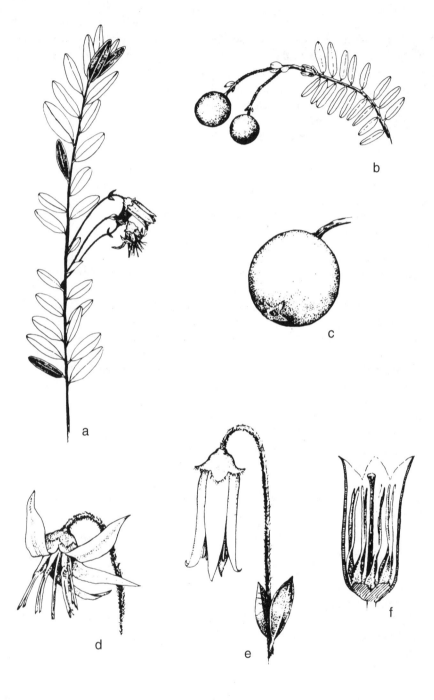

Fig. 197. *Vaccinium macrocarpon*
(Large cranberry).

a. Flowering branch.
b. Fruiting branch.
c. Fruit.

d, e. Flowers.
f. Flower,
longitudinal view.

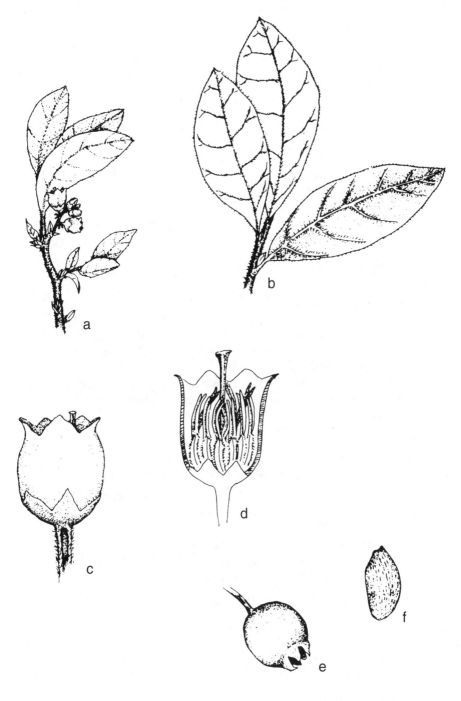

Fig. 198. *Vaccinium myrtilloides*
(Velvet-leaved blueberry).

a. Branch.
b. Leaves.
c. Flower.

d. Flower, longitudinal view.
e. Fruit.
f. Seed.

4. Vaccinium oxycoccos L. Sp. Pl. 35331. 1753. Fig. 199.

Oxycoccus palustris Pers. Syn. 1:419. 1805.

Oxycoccus microcarpus Turcz. ex Rupr. Beitr. Pflanzenk. Russ. Reiches 4:56. 1845.

Stems creeping, slender, rooting at the nodes, to 15 cm long, the flowering branches ascending; leaves simple, alternate, evergreen, ovate, subacute at the apex. rounded or cordate at the sessile base, entire, pale on the lower surface, to 8 mm long, to 3 mm wide, more or less revolute; flowers 1–6, erect to nodding, with 2 bracteoles below the middle of the pedicel; sepals 4; petals 4, 5–6 mm long, pink, attached at the very base; stamens 8 or10; berries red, speckled at first, globose, 5–8 mm in diameter.

Bogs.

IL, IN, OH (OBL).

Lesser cranberry.

The leaves of this species are more or less revolute and usually subacute, thus distinguishing it from *V. macrocarpon*.

Fig. 199. *Vaccinium oxycoccos* (Lesser cranberry). Habit (right). Fruiting branch (upper left). Leaf (next to left).

65. ESCALLONIACEAE—ESCALLONIA FAMILY

Mostly shrubs or small trees; leaves alternate, simple; flowers often in racemes, perfect, actinomorphic, usually 5-merous, with a hypanthium; ovary superior; fruit a capsule.

This family is composed of seven genera and 150 species. Its members are sometimes merged in the Saxifragaceae and sometimes in the Grossulariaceae.

Only the following genus occurs in the United States.

1. Itea L.—Sweetspire

Shrubs; pith chambered; leaves alternate, simple; flowers perfect, actinomorphic, in racemes; sepals 5, united below to form a hypanthium; petals 5, free; stamens 5; ovary superior, 2-locular; fruit a capsule.

There are approximately twenty species in this genus, most of them in eastern Asia.

1. Itea virginica L. Sp. Pl. 199. 1753. Fig. 200.

Shrubs to 3 m tall; stem glabrous; leaves alternate, simple, elliptic to oblong to broadly lanceolate, to 8 cm long, to 4 cm wide, acute to acuminate at the apex, cuneate, serrulate, more or less glabrous; racemes up to 20 cm long, erect to spreading to pendulous; flowers on pedicels 2–3 mm long; sepals 5, green, united below, 2–3 mm long; petals 5, free, white, 5–6 mm long, narrowly lanceolate; stamens 5, shorter than the petals; capsules reflexed at maturity, oblongoid, 7–10 mm long, the style persistent as an abrupt short beak, containing several seeds. May–June.

Swamps, usually in standing water.

IL, IN, KY, MO (OBL).

Virginia sweetspire; Virginia willow; tassel-white.

Fig. 200. *Itea virginica* (Virginia sweetspire). Habit, in fruit. Flower (lower right).

This handsome shrub, now available commercially as an ornamental, is distinguished by its serrulate, alternate leaves and its racemes of white flowers. The strongly beaked fruits persist throughout the winter and well into the next summer.

66. EUPHORBIACEAE—SPURGE FAMILY

Herbs, shrubs, or trees, sometimes succulent and cactuslike, often with latex; leaves usually alternate, simple, stipulate; flowers unisexual, monoecious or dioecious, variously arranged; calyx present or absent, deeply 3- to 6-parted when present; corolla often absent, the parts usually 3–6 when present; stamens 1–several, the filaments free or united; ovary superior, mostly 3-locular, with 1–2 ovules per locule, the placentation axile; flowers often arranged in a cyathium with one pistillate flower and one staminate flower composed of 1 stamen, surrounded by an involucre; fruit a capsule, berry, or drupe.

This diverse family consists of about three hundred genera and over eight thousand species, in most parts of the world.

Only the following genus has a species that may occur in wetlands.

1. Phyllanthus L.—Leaf-flower

Herbs (in the central Midwest), shrubs, or trees, with watery sap; leaves alternate, simple, entire, stipulate; flowers unisexual, usually monoecious, solitary or arranged in axillary cymes; staminate flower with a 4- to 6-lobed calyx, no corolla, 2–15 free or connate stamens, and a segmented disk; pistillate flower with a 4- to 6-lobed calyx, no corolla, a superior, 3-locular ovary with two ovules per locule, and a segmented disk; fruit an explosively dehiscent capsule.

This genus consists of about seven hundred species, mostly in the tropics.

1. **Phyllanthus caroliniensis** Walt. Fl. Carol. 228. 1788. Fig. 201.

Annual or perennial herbs; stems erect, simple or branched, glabrous, to 30 cm tall; leaves alternate, simple, elliptic to oblong to obovate, obtuse and apiculate at the apex, tapering to the base, entire, mostly glabrous, to 20 (–30) mm long, to 10 (–15) mm wide, nearly sessile; flowers borne in axillary cymules, the cymules with one staminate and 2–3 pistillate flowers; staminate flower with a 6-lobed calyx, the lobes oblong to suborbicular, obtuse at the apex, pale yellow, 0.5–0.7 mm long, with 6 disk segments, with 3 stamens; pistillate flower with a 6-lobed calyx, the lobes linear-lanceolate, acute, green, 0.7–1.4 mm long, with an entire, cupular disk and a glabrous ovary; capsules reddish green, 1.6–2.0 mm wide, with gray-brown seeds about 1 mm long. June–October.

Wet soil, often in roadside ditches.

IL, IN, MO (FAC), KS, KY, OH (FAC+).

Leaf-flower.

This species is distinguished by its small, obtuse, alternate, entire leaves and its tiny 6-parted axillary flowers.

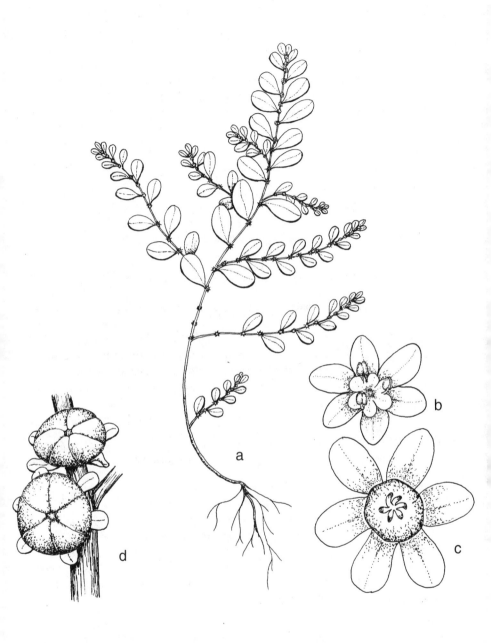

Fig. 201. *Phyllanthus caroliniensis*
(Leaf-flower).

a. Habit.
b. Staminate flower.

c. Pistillate flower.
d. Fruits.

67. FABACEAE—PEA FAMILY

Herbaceous or woody plants, sometimes climbing; leaves alternate, usually compound, with stipules; flowers in racemes, spikes, or capitate clusters, perfect, zygomorphic; sepals usually 5, united below; petals usually 5, of unequal sizes, two of them often united to form a keel; stamens usually 10, with 9 of them united for a part of their length and one of them free; ovary superior; fruit usually a legume.

Excluding the Caesalpiniaceae and Mimosaceae, the Fabaceae consists of approximately 440 genera and twelve thousand species, found in all parts of the world.

1. Plants woody.
  2. Shrubs or small trees .... 1. *Amorpha*
  2. Vines .... 7. *Wisteria*
1. Plants herbaceous.
  3. Plants erect; flowers deep blue, 2 cm long or longer .... 4. *Baptisia*
  3. Plants vines; flowers pale blue or maroon or purple.
    4. Leaves terminating in a tendril; leaflets even in number .... 6. *Lathyrus*
    4. Leaves without a tendril; leaflets uneven in number.
      5. Leaflets 5, 7, or 9; flowers maroon .... 3. *Apios*
      5. Leaflets 3; flowers pale blue or purple.
        6. Flowers of 2 kinds, the ones with petals not subtended by bracteoles ....
        .... 2. *Amphicarpaea*
        6. Flowers all alike, subtended by a pair of bracteoles .... 5. *Galactia*

1. **Amorpha** L.—False Indigo

Shrubs or small trees; leaves alternate, pinnately compound, the leaflets glandular-punctate, stipellate; flowers crowded in racemes, blue or purple; sepals 5, united below to form a short tube; petal 1; stamens 10, exserted; ovary superior, 2-locular; fruit dry, indehiscent, 1- to 2-seeded.

Amorpha consists of approximately fifteen species, all in North America.

1. Leaflets and branches pubescent; fruits glandular-dotted .... 1. *A. fruticosa*
1. Leaflets and branches glabrous or nearly so; fruits not glandular-dotted .... 2. *A. nitens*

1. **Amorpha fruticosa** L. Sp. Pl. 713. 1753. Fig. 202.

Amorpha fruticosa L. var. *angustifolia* Pursh, Fl. Am. Sept. 2:466. 1814.
Amaorpha tennesseensis Shuttlew. ex Kunze, Del. Sem. Hort. Lips. 1848:1. 1848.
Amorpha croceolanata P. W. Wats. Dendrol. Britt. pl. 139. 1825.
Amorpha fruticosa L. var. *croceolanata* (P. W. Wats.) Schneid. Ill. Handb. Laubholzk. 2:73. 1907.
Amorpha fruticosa L. var. *tennesseensis* (Shuttlew.) Palmer, J. Arn. Arb. 12:192. 1931.
Amorpha fruticosa L. var. *oblongifolia* Palmer, J. Arn. Arb. 12:192. 1931.

Shrubs or small trees to 5 m tall; branchlets pubescent, sometimes with tawny or orange pubescence; leaves alternate, pinnately compound, not blackened when dry, the petioles up to 5 cm long; leaflets 13–35, oval, elliptic, oblong, or obovate, usually obtuse and mucronate at the apex, rounded or tapering to the base, entire, pubescent, not shiny, 2–5 cm long, to 2.5 cm wide, on short, pubescent petiolules; flowers densely crowded in racemes up to 20 cm long, on pedicels 3–5 mm long; calyx 5-toothed, 2–3 mm long; petal 1, purple, 5–6 mm long; stamens 10; ovary superior; pod usually slightly curved, glabrous, glandular, 5–8 mm long, 2–4 mm wide, indehiscent, 1- to 2-seeded. May–June.

Along rivers and streams, swampy woods, floodplain forests.

IA, IL, IN, MO (FACW+), KS, KY, NE, OH (FACW).

False indigobush.

False indigo is distinguished by its pinnately divided leaves with 13–35 mucronate leaflets, its densely purple-flowered racemes, and its short, curved, glandular fruits.

Much variation of the leaves and pubescence occurs in the central Midwest.

Typical var. *fruticosa* has 13–25 ovate to oblong leaflets rounded at the base with short, spreading grayish hairs, the leaflets are up to two times as long as broad. Var. *tennesseensis* is similar, but has 21–35 leaflets 1–2 cm long, 2–3 times longer than broad. Var. *oblongifolia* is very similar to var. *tennesseensis*, but the leaflets are 2–5 cm long and nearly glabrous. Var. *angustifolia* has elliptic to obovate leaflets that taper to the base and with appressed pubescence. Var. *croceolanata* has soft-hairy leaflets with tawny or orange hairs.

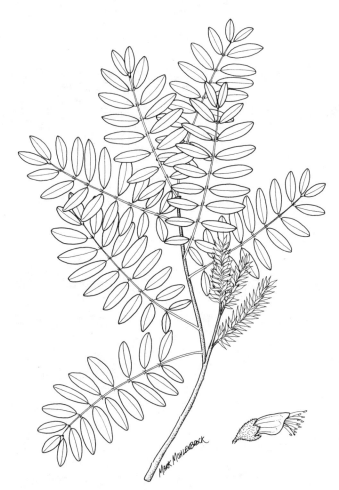

Fig. 202. *Amorpha fruticosa* (False indigobush). Habit. Flower (lower right).

2. **Amorpha nitens** Boynton, Biltmore Bot. Stud. 1:139. 1902. Fig. 203.

Shrub to 3 m tall; branchlets glabrous; leaves alternate, pinnately compound, blackened when dry, the petioles up to 5 cm long; leaflets 9–19, oblong to oblong-ovate, obtuse at the apex, rounded at the base, entire, glabrous or sparsely pubescent, shiny, 2–6 cm long, 1–2 cm wide, on short glabrous or sparsely pubescent petiolules; flowers densely crowded in 1–3 racemes up to 25 cm long, on pedicels 3–5 mm long; calyx 5-toothed, 2–3 mm long; petal 1, purple, 5–6 mm long; stamens 10; ovary superior; pod curved, glabrous, not glandular, 5–8 mm long, 2–4 mm wide, indehiscent, 1- to 2-seeded. May–June.

Swampy woods, along streams.

IL (FACW), KY (NI).

Shiny false indigo.

This rather rare species with a restricted range differs from *A. fruticosa* by its shiny leaves that turn black when dry, its glabrous branches, and its non-glandular pods.

Fig. 203. *Amorpha nitens* (Shiny false indigo). Habit (center). Fruiting branch (lower right).

2. **Amphicarpaea** Ell.—Hog Peanut

Twining non-woody herbs; leaves alternate, compound, with 3 leaflets, the leaflets stipellate; flowers in axillary racemes, of two kinds, one with petals and one without petals; calyx asymmetrical, 4-lobed, united below; corolla (when present) 5-parted; stamens 10; ovary superior, 2-locular; legumes flat, 1- or 3-seeded.

Three species in Africa, Asia, and North America comprise this genus.

1. **Amphicarpaea bracteata** (L.) Fern. Rhodora 35:276. 1933. Fig. 204.
Glycine bracteata L. Sp. Pl. 754. 1753.
Glycine comosa L. Sp. Pl. 754. 1753.
Amphicarpaea pitcheri Torr. & Gray, Fl. N. Am. 1:292. 1838.
Falcata comosa (L.) Kuntze, Rev. Gen. Pl. 182. 1891.
Falcata pitcheri (L.) Kuntze, Rev. Gen. Pl. 182. 1891.
Amphicarpaea bracteata (L.) Fern. var. *comosa* (L.) Fern. Rhodora 39:318. 1937.

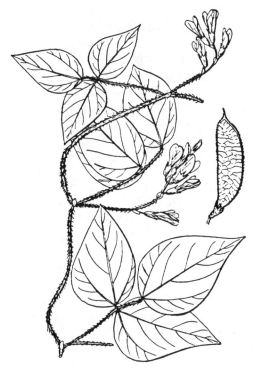

Fig. 204. *Amphicarpaea bracteata* (Hog peanut). Habit. Fruit (right).

Twining non-woody annual herbs; stems appressed-pubescent to villous-hirsute, to 1.5 m long; leaves alternate, compound, the leaflets 3; leaflets stipellate, lanceolate to ovate, acute to acuminate at the apex, rounded at the base, to 10 cm long, to 5 cm wide, strigose to softly pubescent on both surfaces; flowers of two kinds: petalous flowers bracteate, in axillary racemes, and apetalous flowers solitary at the end of stolons; calyx irregular, 4-lobed, tubular below, the lobes 1–2 mm long, the tube 4–5 mm long; petals 5, pale blue, 8–14 mm long; stamens 10; ovary superior; legumes of petaliferous flowers flat, 1.5–4.0 cm long, strigose or glabrous, with a terminal beak 1–5 mm long; legumes of apetalous flowers flat, strigose.

Swampy woods, as well as a variety of other woods and thickets.

IA, IL, IN, KY, MO, OH (FAC), KS, NE (FACU).
Hog peanut.

This very common vine has stipellate leaflets. It produces two kinds of flowers. Those that occur in short axillary racemes have petals and produce elongated legumes. Those that are at the end of stolons that trail on the ground have no petals and produce shorter legumes.

Two quite different appearing varieties occur. Plants with appressed-pubescent stems are the typical variety, while those with spreading hairs may be called var. *comosa*. This latter variety is sometimes treated as a separate species.

3. **Apios** Medic.—Groundnut

Twining perennial herbs from underground tubers; leaves alternate, pinnately compound, with 5–9 stipellate leaflets; flowers few to several in axillary racemes, each flower subtended by a pair of bractlets; calyx obscurely 5-lobed, short-tubular below; petals 5; stamens 10; ovary superior; legumes several-seeded.

About ten species in North America, Asia, and Africa comprise this genus.

1. **Apios americana** Medic. Vorles. Churpfalz. Phys.-Ocon. Ges. 2:355. 1787. Fig. 205.

Apios americana Medic. var. *turrigera* Fern. Rhodora 41: 546. 1939.

Twining perennial herbs from tuberous rootstocks; stems pubescent or glabrous, to 3 m long; leaves alternate, compound, with minute stipules; leaflets 5–9, lanceolate to ovate, acute to acuminate at the apex, rounded at the base, entire, glabrous or pubescent on the lower surface, to 6 cm long, to 4 cm wide, stipellate; flowers crowded or loosely arranged in racemes, each flower subtended by a pair of linear bracts; calyx 5-lobed, united below; petals 5, maroon, 10–12 mm long; stamens 10; ovary superior; legumes linear, straight or curved, to 10 cm long, with many seeds. July–September.

Swampy woods, wet thickets.

IA, IL, IN, KY, MO, OH (FACW), KS, NE (FAC).

Groundnut.

Plants with loosely arranged flowers have been called var. *turrigera*.

The tuberous-thickened roots contain starch which is reported to be very nutritious. The common name groundnut refers to these edible tubers.

This is the only vine in the central Midwest wetlands that has alternate leaves with 5–9 leaflets.

4. **Baptisia** Vent.—False Indigo

Perennial herbs from rhizomes; leaves alternate, with 3 leaflets that turn black when dry; flowers large, showy, in racemes; calyx bilabiate, 4- or 5-lobed; petals 5; stamens 10; ovary superior; legumes thick, curved at the tip, with many seeds.

Approximately fifteen to eighteen species are in this genus, all in the United States.

1. **Baptisia australis** (L.) R. Br. in Ait. Hort. Kew. ed. 2, 3:6. 1811. Fig. 206.

Sophora australis L. Syst. Nat. ed. 12, 2:287. 1767.

Perennial herbs from thickened rhizomes; stems erect to spreading, much branched, glabrous, to 1.5 m tall; leaves alternate, petiolate, stipulate, with 3 leaflets, the stipules up to 2 cm long; leaflets oblong to oblanceolate, obtuse at the apex, tapering to the sessile base, entire, glabrous, to 8 cm long, to 4 cm wide, turning black when dry; flowers loosely arranged in erect, terminal racemes, subtended by

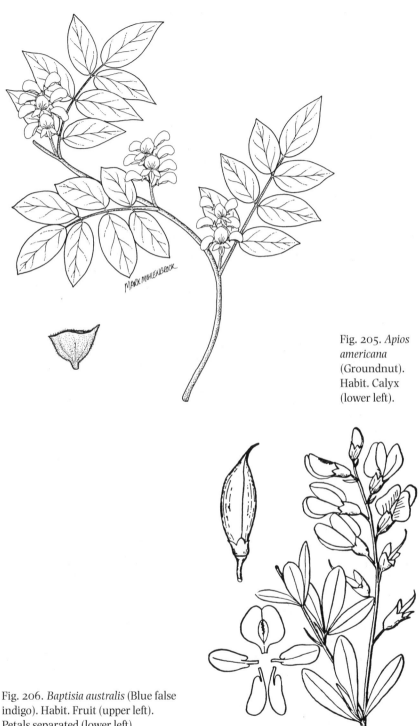

Fig. 205. *Apios americana* (Groundnut). Habit. Calyx (lower left).

Fig. 206. *Baptisia australis* (Blue false indigo). Habit. Fruit (upper left). Petals separated (lower left).

bracts 10–12 mm long, each flower 2–3 cm long, on pedicels 5–15 mm long; calyx 4- or 5-lobed, 8–10 mm long; petals 5; stamens 10; ovary superior; legumes ellipsoid to oblongoid, up to 5 cm long with a curved beak up to 1.8 cm long, glabrous, stipitate, with many seeds. May–June.

Woods, prairies, depressions in wet barrens.

IA, IL, IN, KS, KY, MO, NE, OH (UPL).

Blue false indigo.

This handsome species is recognized by its large blue flowers, much branched stems, and 3 leaflets.

This is an upland species, as indicated by the UPL status given it in all states by the U.S. Fish and Wildlife Service, but in Kentucky, it occurs in depressions in wet barrens.

5. Galactia P. Browne—Milk Pea

Twining semi-woody to herbaceous perennials; leaves alternate, minutely stipulate, with 3 leaflets; flowers subtended by a pair of bracteoles, purple or bluish; calyx 4-lobed, tubular below; petals 5; stamens 10; legumes flat, oblongoid, with few seeds.

Galactia consists of about eighty species in tropical and subtropical parts of the New World.

1. Galactia mohlenbrockii Maxwell, Castanea 44:242. 1979. Fig. 207.

Dolichos multiflora Torr. & Gray, Fl. N. Am. 1:281. 1838.
Dioclea multiflora (Torr. & Gray) C. Mohr, Contr. U.S. Natl. Herb. 6:580. 1901.

High-climbing semi-woody perennial vine; stems retrorsely pubescent, to 10 m long; leaves alternate, minutely stipulate, with 3 leaflets; leaflets ovate to suborbicular, obtuse or subacute at the apex, rounded at the base, entire, more or less glabrous above, sparsely pubescent beneath, to 15 cm long, about as broad, on pubescent petiolules; flowers crowded in a raceme to 5 cm long, on peduncles up to 6 cm long, each flower 12–15 mm long, subtended by a pair of bracteoles; calyx 4-lobed, tubular below; petals 5, bluish or purple; stamens 10; ovary superior; legumes flat, narrowly 2-winged, oblongoid, to 5 cm long, with few seeds. May–July.

Swampy woods.

IL (FACW–), KY (FAC).

Mohlenbrock's milk pea.

Fig. 207. *Galactia mohlenbrockii* (Mohlenbrock's milk pea). Habit. Fruit (right).

This plant has usually been called *Dioclea multiflora*, but Maxwell (1979) believes it should be in the genus *Galactia*, but the binomial *Galactia multiflora* already exists, hence the new specific epithet.

This is a high-climbing semi-woody vine.

6. Lathyrus L.—Sweet Pea; Vetchling

Annual or perennial herbs; stems winged or wingless; leaves alternate, pinnately compound, the terminal leaflet often modified into a tendril, often with large stipules; flowers few in clusters or several in racemes; calyx 5-lobed, the lobes similar or dissimilar in size, gibbous at base; petals 5, the wings obovate; stamens 10; ovary superior; legume flat to terete.

This genus consists of approximately 150 species, found worldwide. Only the following may occur in standing water in the central Midwest.

1. **Lathyrus palustris** L. Sp. Pl. 2:733–734. 1753. Fig. 208.

Lathyrus myrtifolius Muhl. in Willd. Sp. Pl. 3:1091. 1803.

Lathyrus palustris L. var. *myrtifolius* (Muhl.) Gray, Man. ed. 2, 104. 1856.

Perennial herbs with creeping rhizomes; stems often winged, to 1 m long; leaves alternate, pinnately compound, with 2–4 (–5) pairs of leaflets and a branched, terminal tendril, the leaflets linear to lanceolate to ovate, acute and mucronate at the tip, cuneate at the base, to 2 cm long, to 1.5 cm wide, glabrous; flowers 2–9 on a slender peduncle; flowers purple, 10–20 mm long; calyx irregular; legume linear, sessile, glabrous or somewhat pubescent, to 2 cm long, to 5 mm broad. May–July.

Wet ground, marshes, fens, bogs.

IA, IL, IN, MO (FACW), KY, OH (FACW+).

Marsh sweet pea; marsh vetchling.

The typical variety is readily recognized by its winged stems. Plants with wingless or nearly wingless stems may be called var. *myrtifolius*. These latter plants resemble *Vicia*.

Fig. 208. *Lathyrus palustris* (Marsh sweet pea). Habit. Fruit (lower right). Seed (lower left).

7. Wisteria Nutt.—Wisteria

High-climbing woody vines; leaves alternate, pinnately compound with 9–15 leaflets; flowers many in showy racemes; calyx asymmetrical, 4- or 5-lobed; petals 5, the standard petal reflexed near the middle; stamens 10; ovary superior, glabrous; legumes flattened, glabrous, with reniform seeds.

Stritch (1984) has removed a few species from *Wisteria* to *Rehsonia*, leaving only 2 species in *Wisteria*, both in the southern United States.

1. **Wisteria frutescens** (L.) Poir. in Lam. Ill. 3:674. 1823. Fig. 209.

Glycine frutescens L. Sp. Pl. 753. 1753.

Wisteria speciosa Nutt. Gen. 2:116. 1818.

Wisteria macrostachys Nutt. in Torr. & Gray, Fl. N. Am. 1:283. 1838, in synon.

Wisteria frutescens (L.) Poir. var. *macrostachys* Torr. & Gray, Fl. N. Am. 1:283. 1838.

Krauhnia frutescens (L.) Greene, Pittonia 2:175. 1891.

Krauhnia macrostachys (Torr. & Gray) Small, Bull. Torrey Club 25:134. 1898.

Fig. 209. *Wisteria frutescens* (Wisteria). Habit (above). Flower, with petals removed (lower center). Standard petal (lower right).

High-climbing perennial woody vines; stems glabrous or minutely pubescent, up to 15 m long; leaves alternate, with 9–15 leaflets; leaflets oblong to lance-ovate, acute to acuminate at the apex, more or less rounded at the base, more or less pubescent on the lower surface, to 8 cm long, to 4 cm wide; flowers densely crowded in racemes to 30 cm long, on spreading pilose to glandular-hairy pedicels, 4–10 mm long, subtended by bracts up to 1.5 cm long; calyx campanulate, finely pubescent; petals bluish purple, to 2 cm long; stamens 10; ovary superior; legumes linear, flattened, glabrous, to 10 cm long, stipitate, with large seeds. June–September.

Swampy woods.

IA, IL, IN, KY, MO, OH (FACW+).

Wisteria.

This handsome vine is distinguished by its pinnately compound alternate leaves with 9–15 leaflets.

68. FAGACEAE—BEECH FAMILY

Monoecious trees or shrubs with alternate, simple leaves and narrow, deciduous stipules; staminate flowers in pendulous or erect catkins or heads; pistillate flowers solitary or in clusters of 2–4, subtended by an involucre of bracts that persists in fruit as an indurated cupule or other envelope; ovary inferior and with 1 or 2 pendulous ovules per locule; fruit a one-seeded (by abortion) nut lacking endosperm.

This family consists of six genera and more than six hundred species.

1. Quercus L.—Oak

Trees or rarely shrubs; leaves alternate, simple, entire or variously indented, the stipules caducous; plants monoecious, the staminate flowers in pendulous catkins of 6–20 flowers, the calyx with 3–7 elements, the stamens 3–12; pistillate flowers solitary or in groups of 2 or 3, subtended by an involucre of many parts, the ovule 1-seeded by abortion, 3-locular; fruit a nut, subtended by, and partly or nearly enclosed in, an indurated involucre.

Nearly six hundred species comprise the genus.

1. Some or all the leaves lobed, the sinuses at least halfway to the midvein.
    2. Tips of lobes rounded; acorns nearly completely covered by the cap .... 2. *Q. lyrata*
    2. Tips of lobes with bristles; acorns not completely covered by the cap.
        3. Acorn cap just covering the base of the acorn; acorn 1.0–1.2 cm long, 1.0–1.2 cm wide .... 5. *Q. palustris*
        3. Acorn cap covering 1/3 of the acorn; acorn 2.2–2.5 cm long, 1.2–1.5 cm wide .... 7. *Q. texana*
1. Leaves entire, toothed, or shallowly lobed, the sinuses less than halfway to the midvein.
    4. Some or all of the leaves entire or occasionally irregularly lobed.
        5. All leaves entire, rounded at the base .... 6. *Q. phellos*
        5. Some leaves toothed or lobed, tapering to the base .... 4. *Q. nigra*
    4. Leaves regularly toothed or undulate.
        6. Not all lateral veins extending to the tip of each tooth; acorns long-stalked .... 2. *Q. bicolor*
        6. All lateral veins extending to the tip of each tooth; acorns sessile or short-stalked .... 3. *Q. michauxii*

1. **Quercus bicolor** Willd. in Muhl. & Willd. Gesell. Naturf. Freunde Berlin Neue Schr. 3:396. Fig. 210.

Quercus platanoides Lam. Encycl. Meth. Bot. 1:720. 1785, *nomen illeg.*

Quercus prinus L. var. *tomentosa* Michx Hist. Chenes Am. pl. 9. 1801.

Quercus bicolor Willd. var. *mollis* Nutt. Gen. N. Am. Pl. 2:215. 1818.

Quercus platanoides (Lam.) Sudw. U.S.D.A. Rep. Sec. Agr. 1892:327. 1893.

Medium trees to 25 m tall, with a rounded and broad crown and with a trunk diameter up to 1 m; bark gray-brown, deeply furrowed, becoming flaky, the twigs stout, gray-brown to yellow-brown, the buds ellipsoid to spherical, up to 3 mm long, yellow-brown, glabrous or with a few hairs at the tip; leaves alternate, simple, usually broadest above the middle, up to 15 cm long, up to 10 cm wide, coarsely round-toothed or sometimes with a few shallow lobes, glabrous or somewhat pubescent on the upper surface, white and softly pubescent on the lower surface, the petioles nearly 2 cm long, glabrous or slightly pubescent; staminate and pistillate flowers borne separately, but on the same tree, appearing when the leaves begin to unfold, minute, without petals, the staminate in slender, pendulous catkins, the pistillate in clusters of 2–4; acorns in pairs, on stalks 2 cm long or longer, the nut ovoid, pale brown, 2–3 cm long, enclosed about 1/3 its length by the cup, the cup thick, light brown, pubescent, roughened. March–April.

Swampy woods.

IA, IL, IN, KS, KY, MO, NE, OH (FACW+).

Swamp white oak.

The distinguishing features of this species are its leaves which are coarsely round-toothed and softly white pubescent on the lower surface and by its long-stalked acorns.

2. **Quercus lyrata** Walt. Fl. Carol. 235. 1788. Fig. 211.

Medium to large tree to 27 m tall, with a trunk diameter up to 1 m; crown rounded to oblong, the lowermost often pendulous; bark gray or grayish brown, divided into flat, sometimes squarish plates; twigs slender, smooth, buff-colored, with buds nearly round, smooth, pale brown, up to 3.5 mm long; leaves alternate, simple, divided into 5–7 rounded lobes, the sinuses shallow to deep, up to 25 cm long, up to 12 cm wide, dark green and smooth on the upper surface, pale and softly hairy to nearly smooth on the lower surface, the petioles up to 2 cm long, smooth or hairy; staminate and pistillate flowers borne separately, but on the same plant, appearing when the leaves begin to unfold, without petals, the staminate in slender, yellow, pendulous catkins, the pistillate few in a group; acorns solitary or 2 together, with or without a short stalk, the nut nearly spherical, up to 2 cm in diameter, pale brown, often nearly entirely enclosed by the cup, the cup finely pubescent, with some of the scales forming a ragged rim near the base. April.

Bottomland woods, swamps.

IL, IN, KY, MO, OH (OBL).

Overcup oak.

The acorns of this oak are the distinguishing feature, with the cup covering nearly all of the nut. The leaves are usually pale on the under surface and range from glabrous to pubescent.

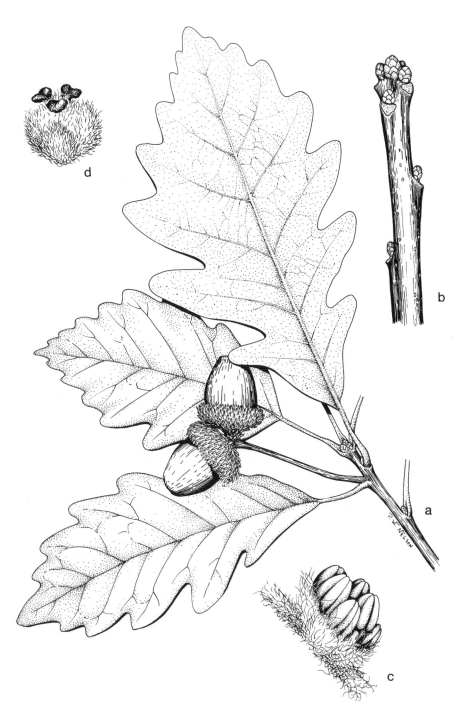

Fig. 210. *Quercus bicolor*
(Swamp white oak).

a. Branch with leaves and acorns.
b. Twig.

c. Staminate flower.
d. Pistillate flower.

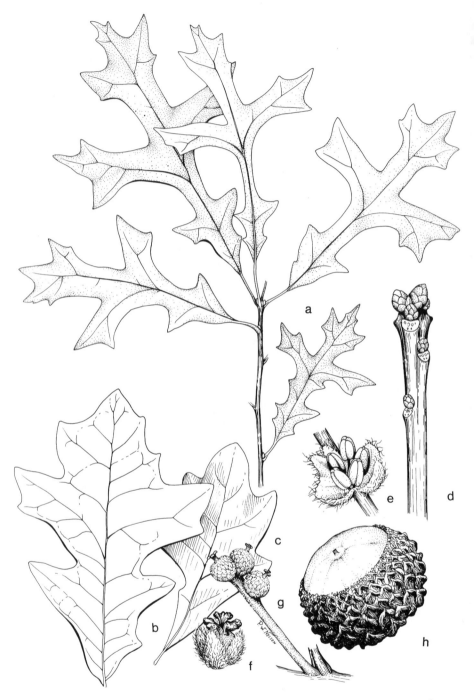

Fig. 211. *Quercus lyrata*
(Overcup oak).

a, b, c, Leaves.
d. Twig.
e. Staminate flower.

f. Pistillate flower.
g. Pistillate flowers.
h. Acorn.

3. **Quercus michauxii** Nutt. Gen. N. Am. Pl. 2:215. 1818. Fig. 212.

Medium to large tree up to 33 m tall, the trunk diameter up to 2 m, with a rounded crown; bark gray or silvery white and scaly; twigs stout, reddish brown, glabrous or nearly so; buds pointed, finely pubescent, reddish brown, up to 6 mm long; leaves alternate, simple, obovate, acute at the apex, rounded or tapering to the base, up to 25 cm long, up to 15 cm wide, coarsely scalloped along the edges, thick, green and sparsely pubescent on the upper surface, whitish and densely hairy on the lower surface, the petiole up to 3 cm long, pubescent; staminate and pistillate flowers borne separately but on the same tree, without petals, the staminate crowded into long, slender catkins, the pistillate few in a cluster; acorns solitary or paired, with or without short stalks, the nut ovoid to ellipsoid, brown, up to 3 cm long, enclosed about 1/3 the length by the cup, the cup thick, cup-shaped, pubescent, short-fringed along the rim. April.

Low woods.

IL, IN, KY, MO (FACW+).

Swamp chestnut oak; basket oak; cow oak.

This is the only coarsely toothed oak that lives in wetland habitats.

4. **Quercus nigra** L. Sp. Pl. 2:995. 1753. Fig. 213.

Tree to 25 m tall; twigs brown, becoming dark gray, glabrous or nearly so; buds ovoid, 3–6 mm long, pubescent; leaves alternate, simple, entire to variously toothed or lobed, tapering to the base, glabrous on the upper surface, glabrous or with axillary hairs beneath, sometimes bristle-tipped, to 10 cm long, to 5 cm wide, on petioles to 1 cm long; acorns to 1 cm long, ovoid to subglobose, apiculate, covered up to 1/3 by the saucer-shaped cup. April–May.

Swampy woods, upland woods.

KY (FAC), MO (FACW).

Water oak.

This oak has the most diversely shaped leaves of any oak, particularly the leaves of young shoots. Many leaves on some of the trees look like those of *Q. phellos*, but some leaves have occasional lobes and the leaves taper to the base.

5. **Quercus palustris** Muenchh. Hausv. 5:253. 1770. Fig. 214.

Medium tree up to 25 cm tall, the trunk diameter usually less than 1 m, the crown narrowly rounded or oblong, the lower branches pendulous; bark light brown or dark brown, scarcely furrowed; twigs slender, smooth, reddish brown to dark gray; buds pointed, reddish brown or dark gray, glabrous, up to 2 mm long; leaves alternate, simple, divided more than half-way to the middle into 5 (–7) bristle-tipped lobes, dark green, shiny and more or less glabrous on the upper surface, paler and with tufts of hairs along the veins on the lower surface, up to 18 cm long, up to 10 cm wide, the petiole up to 4 cm long, slender, usually glabrous; staminate and pistillate flowers borne separately but on the same tree, appearing when the leaves begin to unfold, without petals, the staminate in slender pendulous catkins, the pistillate in groups of 1–3; acorns 1–4 together, with or without short stalks, the nut hemispherical, up to 1 cm across, pale brown, frequently with darker lines,

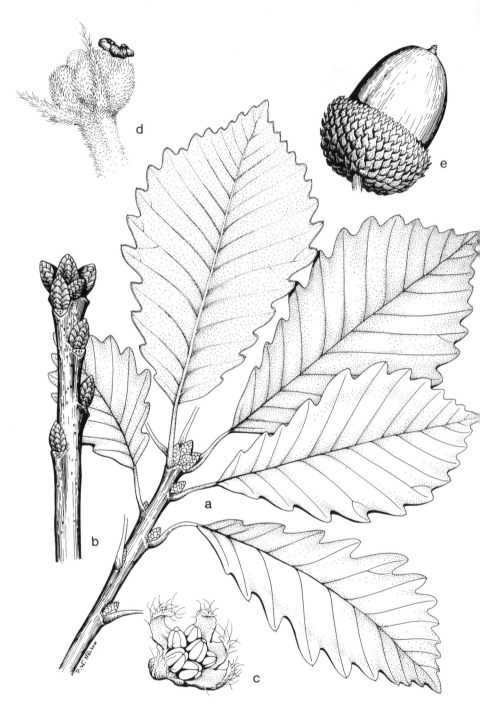

Fig. 212. *Quercus michauxii*
(Swamp chestnut oak).

a. Branch with leaves.
b. Twig.
c. Staminate flower.

d. Pistillate flower.
e. Acorn.

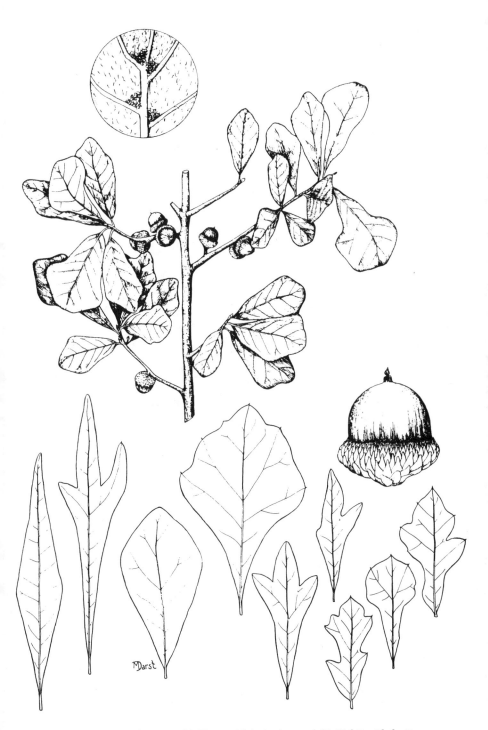

Fig. 213. *Quercus nigra* (Water oak). Veins with hairs (upper left). Habit with fruits (upper). Leaf variations (below). Acorn (center right).

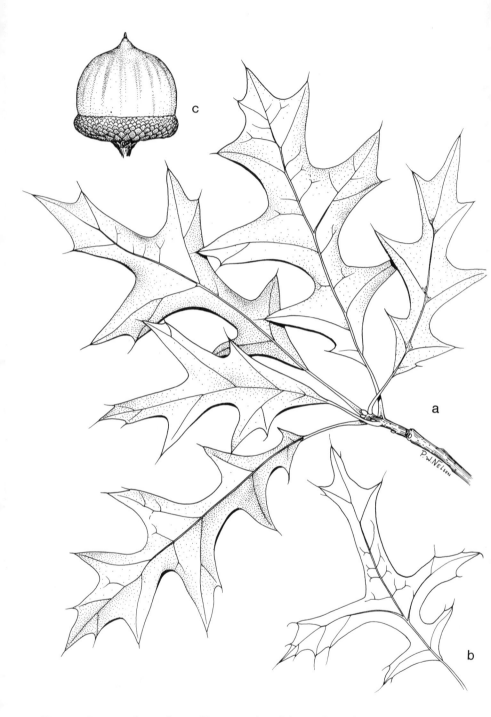

Fig. 214. *Quercus palustris* (Pin oak). a. Branch with leaves. b. Leaf. c. Acorn.

enclosed less than one quarter by the cup, the cup thin, saucer-shaped, reddish brown, finely pubescent. April.

Floodplain woods, along streams, edges of swamps and ponds.

IA, IL, IN, KY, MO, OH (FACW), KS, NE (FAC).

Pin oak.

The leaves of this oak are usually 5-lobed with the sinuses cut more than half-way to the middle. The cup of the small acorns covers less than one quarter of the nut. The lower branches of this oak usually hang low to the ground.

6. **Quercus phellos** L. Sp. Pl. 944. 1753. Fig. 215.

Medium tree to 75 m tall, with a trunk diameter up to 1 m and a narrowly round-topped crown; bark reddish brown, smooth at first, becoming irregularly and shallowly furrowed; twigs slender, glabrous, reddish brown; buds ovoid, acute, glabrous, up to 3.5 mm long; leaves alternate, simple, without lobes or teeth, narrowly lanceolate to narrowly oblong, with a bristle tip, narrowed to the base, up to 12.5 cm long, up to 2.2 cm wide, light green and glabrous on the upper surface, usually glabrous and paler on the lower surface, the petioles up to 1 cm long, glabrous or slightly pubescent; staminate and pistillate flowers borne separately but on the same tree, appearing as the leaves begin to unfold, without petals, the staminate in slender, pendulous catkins, the pistillate few in a cluster; acorns solitary or 2 together, with or without a short stalk, the nut more or less spherical, pale yellow-brown, enclosed less than a quarter its length by the cup, the cup reddish brown, finely pubescent. April.

Swampy woods.

IL, MO (FACW), KY (FAC+).

Willow oak.

This oak is readily distinguished by its narrow, toothless and lobeless leaves.

7. **Quercus texana** Buckl. Proc. Acad. Nat. Sci. Phila. 12:444. 1860. Fig. 216.
Quercus nuttallii E. J. Palmer, Journ. Arn. Arb. 8:52. 1927.

Tree to 20 m tall; twigs brown, becoming dark with age, glabrous or nearly so; buds ovoid, 2–5 mm long; leaves alternate, simple, deeply pinnately 3- to 5- (7-) lobed, the sinuses cut deep toward the midvein, the lobes bristle-tipped, glabrous on the upper surface, pubescent in the axils of the veins beneath, tapering to the base, to 15 cm long, on petioles to 5 cm long; acorns to 2.5 cm long, to 1.8 cm wide, oblong-ovoid, apiculate, covered up to one-third by the cup. April–May.

Swampy woods.

IL, MO (OBL), KY (not listed). The U.S. Fish and Wildlife Service calls this plant *Q. nuttallii*.

Nuttall's oak.

The leaves of this species are very similar to those of *Q. palustris* but often have seven lobes. The acorns, however, are very different.

This species is often called *Q. nuttallii*.

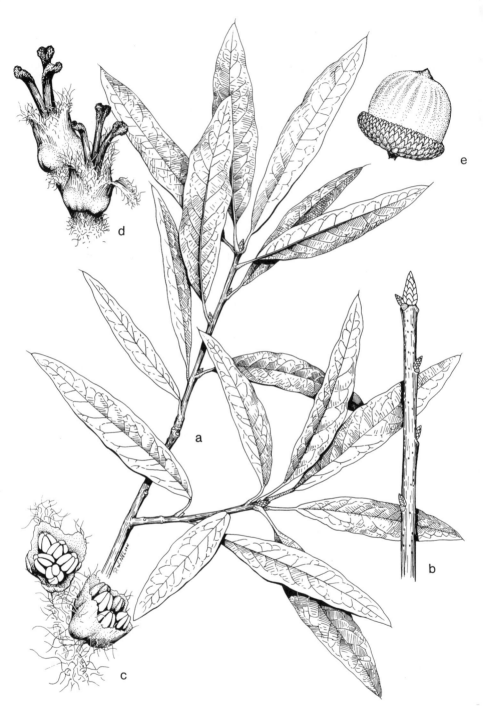

Fig. 215. *Quercus phellos*
(Willow oak).

a. Branch with leaves.
b. Twig.
c. Staminate flowers.

d. Pistillate flowers.
e. Acorn.

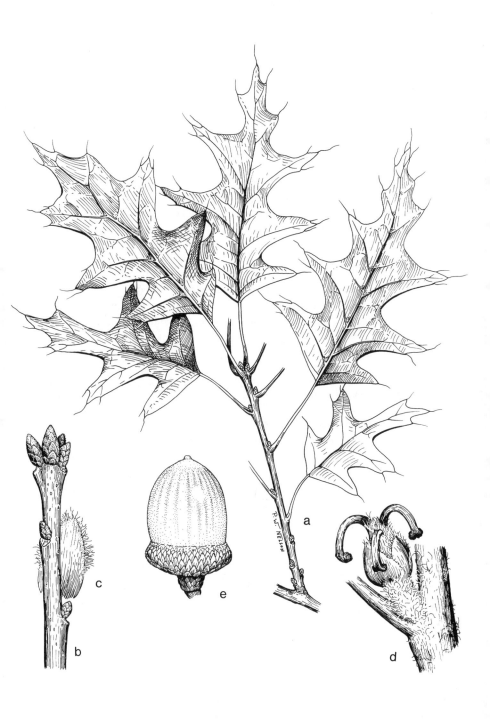

Fig. 216. *Quercus* texana (Nuttall's oak).

a. Branch with leaves.
b. Twig.
c. Bud scale.

d. Pistillate flower.
e. Acorn.

69. GENTIANACEAE—GENTIAN FAMILY

Herbs; leaves opposite, simple, entire, without stipules; flowers in various inflorescences, perfect, actinomorphic; sepals 4–5, united below; petals 4–5, united, at least at base; stamens 4–5; ovary superior, with numerous ovules; fruit a capsule.

Approximately seventy-five genera and about one thousand species comprise this family. The trend in recent years is to subdivide the genus *Gentiana* into several genera, a concept followed in this work.

1. Leaves all reduced to scales...1. *Bartonia*
1. None of the leaves reduced to scales.
   2. Lobes of the corolla longer than the tube; stems 4-angled ....4. *Sabatia*
   2. Lobes of the corolla as long as or shorter than the tube; stems terete, ridged.
      3. Corolla not fringed at the apex ....2. *Gentiana*
      3. Corolla fringed at the apex ....3. *Gentianopsis*

1. **Bartonia** Muhl.—Bartonia; Screwstem

Annual or biennial herbs; stems filiform, often spiral; leaves opposite or alternate, reduced to scales; flowers in racemes or panicles; sepals 4, united at base or free; petals 4, united at base; stamens 4; ovary superior, 1-locular; capsules ovoid, containing many seeds.

Four species, all in the United States, comprise this genus.

1. Most of the scale leaves alternate ....1. *B. paniculata*
1. Most of the scale leaves opposite ....2. *B. virginica*

1. **Bartonia paniculata** (Michx.) Muhl. Cat. Pl. Amer. Sept. 16, 1813. Fig. 217. *Centaurella paniculata* Michx. Fl. Bor. Am. 1:98. 1803.

Annual herbs with fibrous roots; stems filiform, spreading to erect, branched, flexuous, yellow-green, glabrous, to 40 cm tall; leaves scalelike, alternate, or the lowest opposite; flowers in a raceme to 15 cm long, on slender arching pedicels, each flower 2.5–4.5 mm long; sepals 4, free, lance-subulate, 2.0–2.5 mm long;

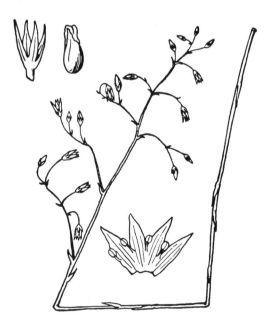

Fig. 217. *Bartonia paniculata* (Branched screwstem). Habit. Calyx (upper left). Seed (next to upper left). Flower opened out (lower right).

petals 4, united at base, yellowish, lanceolate, acuminate; stamens 4, with yellow anthers; ovary superior; capsules ovoid, flattened, with numerous minute seeds. August–October.

Swampy woods.

IL, KY, MO (OBL).

Branched screwstem.

This slender species is recognized by its alternate scale leaves.

2. **Bartonia virginica** (L.) BSP. Prel. Cat. N.Y. 36. 1888. Fig. 218.
Sagina virginica L. Sp. Pl. 128. 1753.

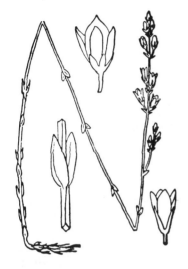

Fig. 218. *Bartonia virginica* (Yellow screwstem). Habit. Leaves (left). Fruit (center above). Flower (lower right).

Annual herbs with fibrous roots; stems filiform, spreading to erect, branched, slightly flexuous, yellow-green, glabrous, to 30 cm tall; leaves scalelike, opposite; flowers in a raceme or panicle, to 10 cm long, on ascending to erect pedicels, each flower 2.5–4.5 mm long; sepals 4, united at base; petals 4, united at base, yellowish, lanceolate to oblong, obtuse to acute; stamens 4, with yellow anthers; ovary superior; capsules ovoid, flattened, with numerous minute seeds. July–September.

Swampy woods, wet meadows, bogs.

IL, IN, MO (FACW+), KY, OH (FACW).

Yellow screwstem.

The distinguishing features of this species are its opposite scale leaves.

2. **Gentiana** L.—Gentian

Mostly glabrous herbs; leaves opposite, entire, sessile or with short petioles; flowers actinomorphic, perfect, terminal or axillary, subtended by a pair of bracts; calyx 5-cleft, usually green; corolla tubular, campanulate to funnelform, 5-lobed; stamens 5, attached to the corolla tube; ovary superior, 1-locular, with numerous ovules; fruit a capsule, the seed winged or wingless.

1. Lobes of the corolla shorter than the appendages between them .... 1. *G. andrewsii*
1. Lobes of the corolla as long as or longer than the appendages between them . 2. *G. saponaria*

1. **Gentiana andrewsii** Griseb. in Hook.Fl. Bor. Am. 2:55. 1834. Fig. 219.

Perennial herbs; leaves opposite, simple, entire, ovate to lanceolate, acute at the apex, tapering or rounded at the base, 3- to 7-nerved, to 8 cm long, to 5 cm wide, glabrous, roughened along the margins; flowers in a terminal, sessile cluster as well as a few in the upper axils, subtended by a pair of bracts; calyx 5-lobed, lanceolate to ovate, ciliolate, green; corolla tubular, to 4 cm long, blue, with 5 minute lobes, the tube nearly closed at the apex, the appendages between the lobes very broad and longer than the corolla lobes, paler in color; stamens 5, attached to the corolla tube; ovary superior; capsules lanceoloid, stipitate, with oblongoid, winged seeds. September–October.

Moist soil, bogs.

IA, IL, IN, KY, MO, NE, OH (FACW).

Closed gentian; bottled gentian.

This species is recognized by its nearly closed corolla and the corolla lobes shorter than the appendages between them.

Fig. 219. *Gentiana andrewsii* (Closed gentian). Habit (left). Flower (right).

2. Gentiana saponaria L. Sp. Pl. 228. 1753. Fig. 220.

Perennial herbs; leaves opposite, simple, entire, lanceolate to narrowly ovate, acute to acuminate at the apex, tapering to the sessile base, 3- to 7-nerved, to 10 cm long, to 6 cm wide; flowers crowded in a terminal, sessile cluster as well as a few in the upper axils, subtended by a pair of bracts; calyx 5-lobed, the lobes lanceolate to oblong, to 15 mm long; corolla tubular, to 3.7 cm long, blue, with 5 minute lobes, the tube nearly closed at the apex, the appendages between the corolla lobes a little longer or shorter than the lobes; stamens 5, attached to the corolla tube; ovary superior; capsules lanceoloid, stipitate, with oblongoid winged seeds. September–October.

Fig. 220. *Gentiana saponaria* (Soapwort Gentian). Habit. Flower partly cut away (upper left).

Moist soil, bogs.

IL, IN (FACW–), KY, OH (FACW).

Soapwort gentian.

The appendages between the corolla lobes are shorter than or barely longer than the lobes.

3. **Gentianopsis** Ma—Fringed Gentian

Annual, biennial, or perennial herbs; leaves opposite, simple, entire, glabrous, without stipules; flowers solitary or in cymes, perfect, actinomorphic; sepals 4, united below, the lobes dissimilar; petals 4, united, the lobes fringed or erose at the top; stamens 4; ovary superior; capsule with numerous seeds.

About two hundred species are in this genus, found in temperate and boreal regions of North America, Europe, and Asia.

1. Corolla lobes distinctly fringed at tip; leaves lanceolate to ovate-lanceolate.......... 1. *G. crinita*
1. Corolla lobes erose at tip; leaves linear to linear-lanceolate................................. 2. *G. virgata*

1. **Gentianopsis crinita** (Froel) Ma, Acta Phyto. Sin. 1:7. 1951. Fig 221.
Gentiana crinita Froel. Gen. 112. 1796.

Annual or biennial herbs; stems erect, branched, glabrous, to 1.2 m tall; leaves opposite, lanceolate to ovate-lanceolate, acute to acuminate at the apex, rounded at the sessile base, entire, glabrous, to 6 cm long, to 2 cm wide; flower solitary on a pedicel up to 20 cm long; calyx 4-parted, 2.5–4.0 mm long, the lobes green, lanceolate, acuminate at the tip, unequal in size; corolla 4-parted, 3.5–5.5 cm long, bright blue, the lobes obtuse and deeply fringed at the tip; stamens 4; ovary superior; capsule lanceoloid to ellipsoid, stipitate, to 3 cm long; seeds numerous, papillose. September–October.

Wet woods, bogs, fens, wet meadows.

IA, IL, IN (FACW+), OH (OBL).

Fringed gentian.

This beautiful species is distinguished from *G. virgata* by the deeply fringed corolla lobes and by its broader leaves. It has often been called *Gentiana crinita*.

2. Gentianopsis virgata (Rusby) Holub, Folia Geobot. & Phytotax. Praha 2:120. 1967. Fig. 222.

Gentiana virgata Rusby, Mem. Torrey Club 6:81. 1896.
Gentiana procera Holm, Ottawa Nat. 15:11. 1901.
Gentianopsis procera (Holm) Ma, Acta Phyto. Sin. 1:8. 1951.

Annual herbs; stems erect, branched or unbranched, glabrous, to 1.5 m tall; leaves opposite, linear to linear-lanceolate, acuminate at the apex, tapering to the sessile base, entire, glabrous, to 5 cm long, to 8 mm wide; flower often solitary, on a pedicel to 10 cm long; calyx 4-parted, 1.5–3.5 cm long, the lobes green,

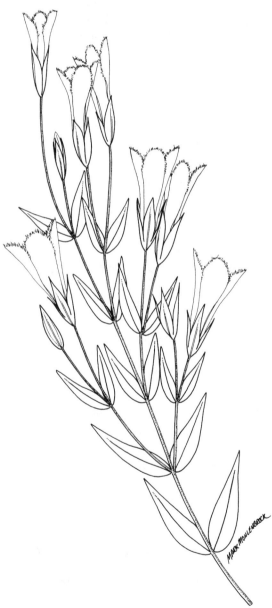

Fig. 221. *Gentianopsis crinita* (Fringed gentian). Habit.

lanceolate, acuminate at the tip, more or less all the same size, papillate; corolla 4-parted, 2–5 cm long, bright blue, erose at the tip; stamens 4; ovary superior; capsules narrowly lanceoloid, short-stipitate, to 2 cm long; seeds numerous, usually smooth. July–September.

Bogs, wet prairies, wet meadows, sloughs.

IA, IL, IN (OBL), OH (FACW+).

Lesser fringed gentian.

This species is generally smaller in all respects than *G. crinita*. The tips of the corolla lobes are merely erose, rather than deeply fringed. This species has been called *Gentiana procera* for many years.

Fig. 222. *Gentianopsis virgata* (Lesser fringed gentian). Habit (center). Lower part of plant (lower left). Corolla spread out (upper right).

4. **Sabatia** Adans.—Rose Gentian

Annual or perennial herbs; leaves simple, opposite, entire, sessile; flowers usually from the upper leaf axils, on long pedicels; sepals 5–12, united at the base, green; petals 5–12, united at the base; stamens 5–12, attached to the corolla; ovary superior; capsules many-seeded, often enclosed by the dried persistent corolla.

Approximately twenty species are in this genus, all of them in North America.

1. Upper branches opposite; leaves ovate... 1. *S. angularis*
1. Upper branches alternate; leaves lineaer-lanceolate .... 2. *S. campanulata*

1. **Sabatia angularis** (L.) Pursh, Fl. Am. Sept. 137. 1814. Fig. 223.
Chironia angularis L. Sp. Pl. 190. 1753.

Perennial herbs from fibrous roots; stems erect, oppositely branched, to 75 cm tall, 4-angled and usually slightly winged, glabrous or nearly so; leaves opposite, simple, broadly lanceolate to ovate, acute to short-acuminate at the apex, rounded or sometimes subcordate at the sessile base, glabrous, to 3.5 cm long, to 2.5 cm wide, with 3, 5, or 7 veins; flowers several from the upper leaf axils, actinomorphic, on glabrous pedicels to 3 cm long, subtended by smaller lanceolate bracts; sepals 5, united at the base, green, lanceolate, glabrous, to 15 mm long; petals 5, united at the base, pink or rarely white, with a green base, oblanceolate, 1–2 cm long; capsules ovoid, 3–4 mm long, with numerous seeds. July–August.

Wet soil.

IL, IN, KY, MO, OH (FAC+), KS (FAC).

Marsh pink.

This handsome species differs from the similar *S. campanulata* by its opposite rather than alternate branching. The five petals form a green star at the center of the flower. The leaves of *S. angularis* have 3, 5, or 7 veins, while the leaves of *S. campanulata* have only 3 veins.

Fig. 223. *Sabatia angularis*
(Marsh pink). Habit.
Flower (lower left).

2. **Sabatia campanulata** (L.) Torr. Fl. U.S. 1:217. 1824. Fig. 224.
Chironia campanulata L. Sp. Pl. 190. 1753.

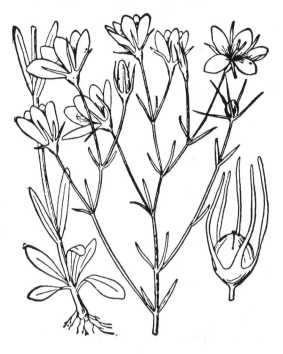

Perennial herb with a woody rootstock; stems upright, branched or unbranched, ridged, glabrous, to 60 cm tall; leaves opposite, ascending to spreading, linear to linear-lanceolate, obtuse to acute at the apex, tapering to the sessile or short-petiolate base, glabrous, sometimes slightly revolute, to 3 cm long, to 7 mm wide; flowers solitary on alternating ascending to spreading branches, each flower on a pedicel to 7 cm long; calyx campanulate, glabrous, 1.5–2.5 mm long, the 5 lobes linear to setaceous, to 15 mm long, up to 2 mm wide; corolla tube cylindrical, white or greenish yellow, 3–5 mm long, the 5 lobes spreading, more or less oblong, obtuse to acute, pink or rose, to 18 mm long, to 7 mm wide, with a patch of yellow at the base; stamens 5; ovary superior; capsules obovoid, glabrous, 3–4 mm long, with several seeds. July–September.

Fig. 224. *Sabatia campanulata* (Slender marsh pink). Habit (center). Lower part of plant (left). Flower without corolla (lower right).

Swampy woods, wet meadows.

IN, KY (FACW).

Slender marsh pink.

The linear-lanceolate leaves and alternate branching in the inflorescence distinguish this species from *S. angularis*.

70. GROSSULARIACEAE—GOOSEBERRY OR CURRANT FAMILY

Shrubs (in our area) or trees; leaves alternate, simple, lobed, without stipules; inflorescence various; flowers perfect, actinomorphic, perigynous to epigynous and with a hypanthium; sepals 4–5, attached to the hypanthium; petals 4–5, often scale-like, attached to the sepals; stamens 4–5; ovary inferior, 1-locular, with numerous ovules; styles 2; fruit a berry.

When considered separate from the Saxifragaceae, this family contains about five genera and 175 species. Only the following has species that sometimes occur in shallow water.

1. **Ribes** L.—Gooseberry; Currant

Shrubs, sometimes with spines; leaves alternate, simple, palmately lobed; flowers in racemes, perfect, with a hypanthium; sepals 4–5, attached to the hypanthium; petals 4–5, shorter than the sepals; stamens 4–5; ovary inferior; fruit a berry with numerous seeds.

Approximately 150 species comprise this genus, which contains both gooseberries and currants.

1. Some part of plant prickly or bristly.
   2. Leaves cuneate at base; fruit glabrous; stamens exserted; upper internodes not bristly ......
   ....3. *R. hirtellum*
   2. Leaves usually truncate at base; fruit bristly; stamens about as long as the calyx; upper internodes bristly ....5. *R. oxyacanthoides*
1. Plants neither prickly nor bristly.
   3. Leaves with resinous glands on lower surface; berries black.
      4. Flowers yellow; racemes drooping .... 1. *R. americanum*
      4. Flowers white; racemes erect or ascending... 4. *R. hudsonianum*
   3. Leaves without resinous glands on lower surface; berries red.
      5. Plants with a skunklike odor; flowers white or rose; fruits bristly; racemes ascending ....
      ....2. *R. glandulosum*
      5. Plants without a skunklike odor; flowers purplish; fruits glabrous; racemes drooping ....
      ....6. *R. triste*

1. **Ribes americanum** Mill. Gard. Dict. ed. 8, 4. 1768. Fig. 225.

Shrubs to 1.5 m tall, without prickles or bristles; branchlets pubescent; leaves alternate, simple, palmately 3- to 5-lobed, suborbicular, doubly serrate, more or less truncate or subcordate at the base, glabrous above, usually somewhat pubescent below with golden yellow resin glands, to 7 cm wide; racemes pendulous, loosely flowered, the branches downy-pubescent; flowers yellow; calyx 8–10 mm long, tubular, with short, obtuse lobes; stamens not exserted; fruit globose, black, glabrous, 6–10 mm in diameter.

April–May.

Moist woods, fens, wet meadows.

IA, IL, IN, KY, MO, OH (FACW). Wild black currant.

The golden yellow resin glands on the lower surface of the leaves are distinctive, although *R. hudsonianum* and *R. oxyacanthoides* are also glandular. *Ribes americanum* has yellow flowers, while *R. hudsonianium* has white flowers. *Ribes oxyacanthoides* has prickly stems.

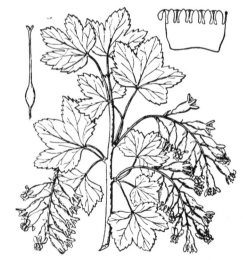

Fig. 225. *Ribes americanum* (Wild black currant). Habit. Inflorescence (lower left). Pistil (upper left). Corolla cut open with stamens (upper right).

2. Ribes glandulosum Grauer, Pl. Min. Cog. 2. 1784. Fig. 226.

Fig. 226. *Ribes glandulosum* (Skunk currant). Habit. Flower (upper right).

Spreading to ascending shrubs without prickles or bristles; leaves alternate, simple, palmately 5- to 7-lobed, doubly serrate, cordate at the base, glabrous except for some pubescence on the veins, without resin glands, with a skunklike odor when crushed, to 7 cm wide; racemes ascending, loosely flowered, the branches glandular-hispid; flowers white or rose; calyx 5–9 mm long, rotate, glandular-hispid; stamens slightly exserted; fruit globose, red, glandular-hispid, 4–7 mm in diameter. May–June.

Swampy woods.

OH (FACW).

Skunk currant.

The common name is derived from the skunklike odor that is emitted when the leaves and berries are crushed.

This species is further distinguished by its lack of spines, its lack of resinous glands on the lower surface of the leaves, its white or rose flowers, its glandular-hispid red berries, and its ascending racemes.

3. Ribes hirtellum Michx. Fl. Bor. Am. 1:111. 1803. Fig. 227.

Erect shrubs to 2 m tall, the lower stems usually prickly, the upper internodes not prickly, the bark shredding; leaves alternate, simple, palmately 3- to 5-lobed, doubly serrate, cuneate to truncate at the base, glabrous or with soft hairs on the veins below, without resinous glands, to 6 cm wide, with softly hairy petioles; flowers in clusters of 1–3, green to purplish; calyx to 10 mm long, narrowly tubular; stamens exserted; fruit globose, black, glabrous, 6–10 mm in diameter. April–July.

Swampy woods, rocky areas.

IA, IL, IN (FACW), OH (FAC).

Low wild gooseberry.

Ribes oxyacanthoides, which is somewhat similar in appearance, has prickles on the upper internodes and stamens that are not exserted.

4. Ribes hudsonianum Richards. Bot. Frankl. Journ., ed. 2, 6. 1823. Fig. 228.

Erect shrubs to 1.5 m tall, the stems without prickles; leaves alternate, simple, palmately 3- to 5-lobed, doubly serrate, cordate at the base, pubescent, with resinous glands on the lower surface, to 8.5 cm wide; racemes ascending to erect, loosely

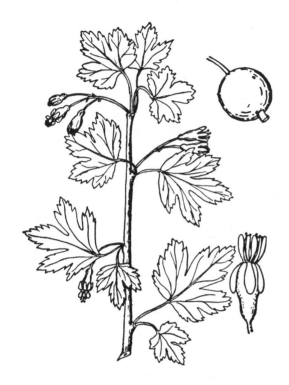

Fig. 227. *Ribes hirtellum* (Low wild gooseberry). Habit. Fruit (upper right). Flower, without petals (lower right).

Fig. 228. *Ribes hudsonianum* (Northern black currant). Habit. Fruit (upper right).

flowered; flowers white; calyx 4–5 mm long, campanulate; stamens included; fruit globose, black, glabrous, to 10 mm in diameter. May–June.

Swampy woods.

IA (OBL).

Northern black currant.

This northern species, which barely enters our range in northeastern Iowa, has glandular-resinous leaves, erect to ascending racemes, and white flowers.

5. **Ribes oxyacanthoides** L. Sp. Pl. 201. 1753. Fig. 229.
Grossularia oxyacanthoides (L.) Mill. Gard. Dict. ed. 8, 4. 1768.

Erect shrubs to 2 m tall, the upper internodes prickly; leaves alternate, simple, palmately 3- to 5-lobed, doubly serrate, truncate at the base, glandular and pubescent on the lower surface, to 3.5 cm wide; flowers in clusters of 1–3, purple; calyx campanulate, glabrous, oblong, obtuse at the tip; stamens as long as the calyx; berries globose, greenish purple, glabrous. May–June.

Wet woods.

NE, OH (not listed by the U.S. Fish and Wildlife Service).

Northern gooseberry.

This species is recognized by its prickly stems, glandular and pubescent lower leaf surfaces, stamens as long as the calyx, and greenish purple fruits.

Fig. 229. *Ribes oxyacanthoides* (Northern gooseberry). Habit. Stem (left). Flower (lower right).

6. **Ribes triste** Pall. Nova Acta Acad. Petrop. 10:378. 1797. Fig. 230.

Decumbent to ascending shrubs, without prickles; leaves alternate, simple, rather shallowly palmately 3- to 5-lobed, doubly serrate, cordate at the base, glabrous to sometimes tomentose on the lower surface, without resinous glands, to 10 cm wide; racemes pendulous, loosely flowered; flowers purplish, on glandular pedicels; calyx rotate; fruit globose, red, bristly, to 6 mm in diameter. June–July.

Bogs, swampy woods.

IL, OH (OBL).

Red currant.

This scraggly shrub is distinguished by its lack of prickles and resinous glands, its purple flowers, its red bristly berries, and its pendulous racemes.

Fig. 230. *Ribes triste* (Red currant). Habit. Flower, face view (upper right).

71. HALORAGIDACEAE—WATER MILFOIL FAMILY

Aquatic monoecious or dioecious herbs; leaves usually finely divided; flowers in the axils of bracts or foliage leaves, unisexual, actinomorphic; sepals 3–4, united below; petals 4 or absent; stamens 3, 4, or 8; ovary superior, with 3 or 4 locules; fruit 3- or 4-parted.

There are eight genera and about one hundred species in this family. Two genera occur in the central Midwest.

1. Leaves crowded, usually whorled; flowers 4–numerous; stamens 4 or 8; fruits 4-lobed. .... 1. *Myriophyllum*
1. Leaves more remote, alternate; flowers 3–numerous; stamens 3; fruit 3-angled .... 2. *Proserpinaca*

1. **Myriophyllum** L.—Water Milfoil

Mostly aquatic herbs; flowers unisexual, axillary in the axils of foliage leaves or bracts, often grouped into emergent spikes; calyx 4-parted; petals 4, or absent; stamens 4 or 8; fruit 4-lobed, at maturity splitting into 4 mericarps.

This genus consists of about twenty to twenty-five species found throughout the world.

1. Leaves all whorled.
   2. Plants dioecious; emergent leaves 2.5–3.5 cm long; petioles 5–7 mm long.......................... .... 1. *M. aquaticum*
   2. Plants monoecious; emergent leaves up to 2 cm long; petioles up to 2 mm long.

3. Bracts not exceeding the flowers; stems white upon drying; stamens 8.
    4. Leaf segments 12 or more per side .... 7. *M. spicatum*
    4. Leaf segments up to 12 per side ....6. *M. sibiricum*
3. Bracts exceeding the flowers; stems not white upon drying; stamens 4 (8 in *M. verticillatum*).
    5. Bracts subtending the staminate flowers deeply cleft; stamens 8 ....
    .... 8. *M. verticillatum*
    5. Bracts subtending the staminate flowers entire or finely toothed; stamens 4.
        6. Bracts linear to lanceolate, up to 1.5 mm long..............................3. *M. hippuroides*
        6. Bracts oblanceolate to elliptic, more than 1.5 mm long..............2. *M. heterophyllum*
1. Leaves alternate or opposite, rarely any of them whorled.
    7. Fruits tuberculate .... 5. *M. pinnatum*
    7. Fruits more or less smooth... 4. *M. humile*

1. **Myriophyllum aquaticum** (Vell.) Verde, Kew Bull. 28:36. 1873. Fig. 231.
Enydria aquatica Vell. Fl. Flum. 57. 1825.
Myriophyllum brasiliense Camb. Fl. Bras. Merid. 2:182. 1829.

Fig. 231. *Myriophyllum aquaticum*
(Parrot's-feather). Habit.

Rooted aquatic herbs; stems stout, some of them submersed, some of them emersed; leaves numerous, all whorled, the emergent ones 2.5–3.5 cm long, on petioles 5–7 mm long, stiff, with 10–26 narrowly linear divisions on each side; flowers in the axils of bracts, unisexual, but only pistillate ones seen in the United States, the plants dioecious, the bracts similar to the leaves in size and shape, the bracteoles filiform and cleft; fruit 1.5–2.0 mm long, 4-sulcate, granular. June–October.

Standing water.

KY, MO (OBL).

Parrot's-feather; water feather

This South American species, commonly grown in aquaria, rarely is found in the wild in the central Midwest. Its stout stems, large stiff leaves, and dioecious nature are distinctive in the genus.

2. **Myriophyllum heterophyllum** Michx. Fl. Bor. Am. 2:91. 1803. Fig. 232.

Submersed aquatic herbs; stem occasionally branched; submersed leaves in whorls of 4–6, 2–5 cm long, capillary-pinnate with 7–10 pairs of divisions; emersed leaves few, pinnatifid; floral bracts whorled, oblanceolate to spatulate, 0.4–2.0 cm long, entire to serrate; flowers axillary, perfect or unisexual; petals 1.5–3.0 mm long, acute; stamens 4; fruit subglobose, 1–2 mm long, deeply 4-sulcate, each section with slightly rounded sides and a dorsal flattened margin with a conspicuous anterior beak. June–September.

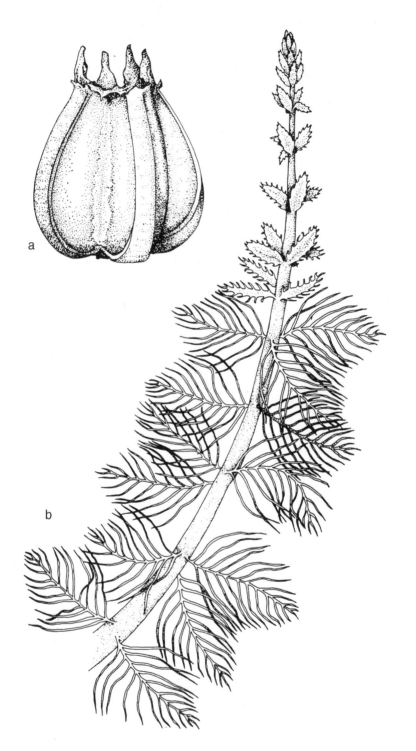

Fig. 232. *Myriophyllum heterophyllum* (Water milfoil).   a. Fruit.   b. Habit.

Ponds, lakes, and still waters of streams.
IA, IL, IN, KS, KY, MO, NE, OH (OBL).
Water milfoil.
This species is characterized chiefly by its floral bracts and fruits. The bracts, which are longer than the flowers, are oblanceolate to spatulate and the margins entire to serrate. The fruits are distinctly 4-angled and each section has a flattened, smooth dorsal ridge.

3. **Myriophyllum hippuroides** Nutt. ex Torr. & Gray, Fl. N. Am. 1:530. 1840.
Fig. 233.
Myriophyllum verticillatum L. var. *cheneyi* Fassett, Rhodora 41:524. 1939.

Submerged herbs; stems sparsely branched; submerged leaves verticillate or subverticillate, to 4.5 cm long, capillary-pinnate with 5–10 pairs of divisions; emersed leaves deeply toothed; floral bracts whorled, linear to lanceolate, 0.5–2.0 cm long, entire or shallowly toothed; flowers axillary, perfect or unisexual; petals 1.5–3.0 mm long; stamens 4; fruit ovoid, 1.6–2.0 mm long, deeply 4-sulcate, each section somewhat papillose with a dorsal ridge with slightly undulate margins and an anterior beak. June–September.
Ponds, lakes, water in slow moving streams.
IL (OBL).
Mare's-tail water milfoil.
This species is distinguished by its entire or slightly toothed bracts and its 4-angled fruits that have smooth to undulate ridges. It is similar to *M. verticillatum*, but the following differences are noteworthy:
The submerged leaves of *M. hippuroides* are whorled and scattered, while the leaves of *M. verticillatum* are only whorled. The floral bracts of *M. hippuroides* are entire or with 2–4 pairs of shallow teeth, while the floral bracts of *M. verticillatum* are deeply toothed. The flower of *M. hippuroides* has four stamens; the flower of *M. verticillatum* has eight stamens. The fruits of *M. hippuroides* are ovoid and strongly 4-sulcate with each section flattened and covered with minute papillae; on the dorsal edge of each segment is a smooth ridge with occasional undulations along the margins. The fruits of *M. verticillatum* are ovoid, plump, and not deeply divided; the individual sections are rounded, smooth, and without a dorsal ridge.

4. **Myriophyllum humile** (Raf.) Morong, Bull. Torrey Club 18:242. 1891.
Fig. 234.
Burshia humilis Raf. Rep. 5:361. 1808.

Highly variable submersed, floating, or terrestrial perennial herbs, rooting at the nodes if terrestrial; stems to 8 cm long if terrestrial, to 1 m long if submersed or floating; leaves of terrestrial plants alternate, linear, entire and up to 1 cm long and up to 4 mm wide, or divided into 2–4 subopposite leaflets; leaves of submersed and floating plants finely divided, subverticillate to scattered, to 3.5 cm long, to 3 cm wide, with up to 8 pairs of capillary divisions; inflorescence with bracts similar to the leaves; flowers perfect, subtended by minute bracteoles about 0.5 mm long, the plants monoecious; calyx 4-parted; petals 4, 0.8–1.0 mm long; stamens 4; fruit 4-sulcate, ovoid, 1.0–1.4 mm long, shiny, more or less smooth. June–October.

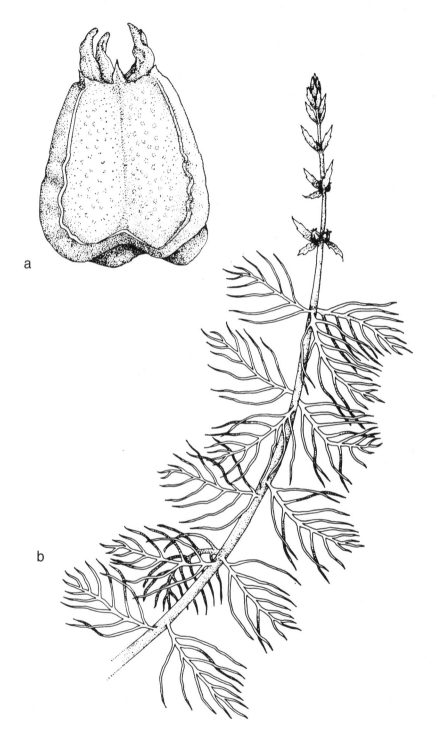

Fig. 233. *Myriophyllum hippuroides* (Mare's-tail water milfoil).   a. Fruit.   b. Habit.

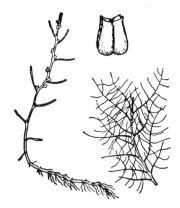

Standing water; muddy shores of ponds. IL, OH (OBL).

Low water milfoil.

The other water milfoil with most of its leaves alternate or opposite is *M. pinnatum*, but that species has tuberculate fruits.

5. Myriophyllum pinnatum (Walt.) BSP. Prel. Cat. N. Y. 16. 1888. Fig. 235.

Potamogeton pinnatum Walt. Fl. Carol. 90. 1788.

Myriophyllum scabratum Michx. Fl. Bor. Am. 2:190. 1803.

Fig. 234. *Myriophyllum humile* (Low water milfoil). Habit (left). Lower leaves (lower right). Fruit (upper right).

Either submerged or terrestrial herbs; stems sparsely branched; submerged leaves verticillate, subverticillate, or scattered, 1–3 cm long, capillary-pinnate with about 5 pairs of divisions; emersed leaves developing occasionally and resembling bracts; floral bracts whorled, linear, 0.5–2.0 cm long, conspicuously toothed with 3–5 pairs of teeth; flowers axillary, perfect or unisexual; petals 1.5–2.0 mm long, rounded above; stamens 4; fruit ovoid, 1.3–2.0 mm long, deeply 4-sulcate, each section with a dorsal, tuberculate, slightly concave ridge and anterior beak. June–October.

Rooted in muddy shores or in shallow waters of ponds and lakes.

IA, IL, IN, KS, KY, MO, NE, OH (OBL).

Water milfoil.

Several characteristics distinguish this species: the submerged pinnatifid leaves are both whorled and scattered along the stem; the floral bracts, which are borne on an emersed spike, are linear, irregularly toothed, and much longer than the flower; each lobe of the 4-sectioned fruit has a conspicuous dorsal tuberculate ridge terminated at the apex by a beak.

6. Myriophyllum sibiricum Komarov, Feddes Rep. Spec. Nov. Regni Veg. 13:168. 1914. Fig. 236.

Myriophyllum exalbescens Fern. Rhodora 21:120. 1919.

Submersed herbs; green turions often present in the autumn; stems simple or sometimes branched, not thickened beneath the inflorescence; submersed leaves in whorls of 3–4, 1.2–3.0 cm long, capillary-pinnate with 5–12 pairs of divisions; emersed leaves, if present, similar to the submersed; floral bracts whorled, spatulate to obovate, up to 3 mm long, entire to denticulate; flowers axillary, perfect or unisexual; petals 2.5 mm long, obovate, pink in the staminate flowers; stamens 8; fruit ovoid, 2.3–3.0 mm long, 4-sulcate, the sections rounded, smooth to slightly tuberculate and obscurely beaked. June–October.

Pools, quiet streams.

IA, IL, IN, KS, MO, NE, OH (OBL).

Water milfoil.

This species is similar to the European *M. spicatum*, differing by its fewer leaf divisions and its more or less smooth fruits.

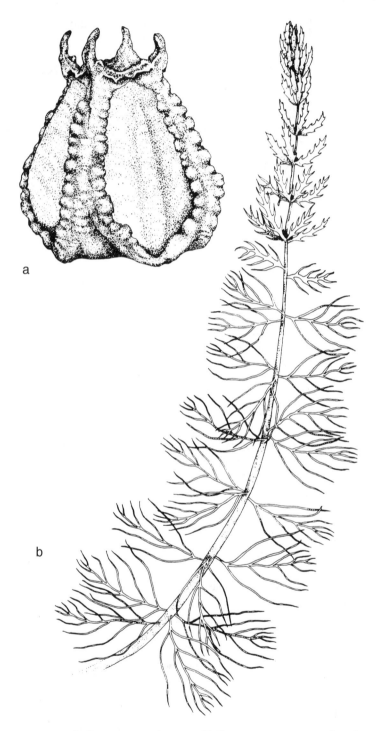

Fig. 235. *Myriophyllum pinnatum* (Water milfoil).   a. Fruit.   b. Habit.

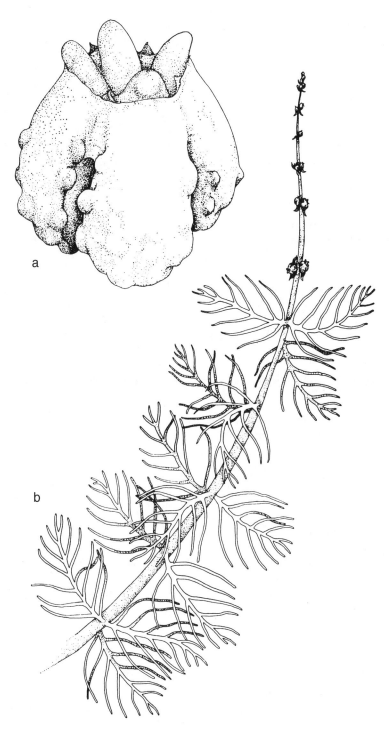

Fig. 236. *Myriophyllum sibiricum* (Water milfoil).   a. Fruit.   b. Habit.

7. Myriophyllum spicatum L. Sp. Pl. 992. 1753. Fig. 237.

Submersed herbs; turions absent; stems simple or sometimes branched, thickened beneath the inflorescence; submersed leaves in whorls of 3–4, 1.5–3.5 cm long, capillary-pinnate with 12 or more pairs of divisions; emersed leaves, if present, similar to the submersed; floral bracts whorled, the lower toothed or entire, exceeding the flowers, the upper bracts ovate, entire, equaling or shorter than the flowers; flowers axillary, perfect or unisexual, 2.5 mm long, reddish in the perfect and staminate flowers; stamens 8; fruit 4-sulcate, subglobose, 2–3 mm long, tuberculate.

Lakes and ponds.

IA, IL, IN, KY, MO, NE, OH (OBL).

European water milfoil.

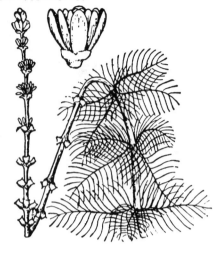

Fig. 237. *Myriophyllum spicatum* (European water milfoil). Habit. Flowering cluster (above).

This European species is very similar to the native *M. sibiricum* but differs by its more numerous leaf divisions and its larger, tuberculate fruits.

8. Myriophyllum verticillatum L. Sp. Pl. 992. 1753. Fig. 238.

Submerged herbs; stems branched occasionally; leaves in whorls of 4–5, 0.8–4.5 cm long, capillary-pinnate with 9–13 division pairs; emersed leaves, if developed, similar to the submersed leaves; floral bracts whorled, linear, up to 1.5 cm long, deeply pinnatifid with 8–15 pairs of segments; flowers axillary, perfect or unisexual; petals 2.5 mm long, spatulate; stamens 8; fruit ovoid, 2.0–2.5 mm long, 4-sulcate, each section rounded with a shallow concavity on the dorsal surface and a conspicuous beak at the summit. June–October.

Shallow water of ponds, streams, and lakes.

IA, IL, IN, NE, OH (OBL).

Common water milfoil.

This species may be distinguished readily on the basis of leaves, bracts, and fruits. The capillary-pinnate leaves are borne only in whorls; the bracts are deeply pinnatifid and exceed the flower; the fruit is 4-angled with each section smooth and rounded with a slight vertical concavity on the dorsal margin.

2. **Proserpinaca** L.—Mermaid-weed

Mostly aquatic perennials; stems with short spines; leaves alternate, the submersed ones pinnatifid, the emersed ones pinnatifid or serrate; flowers perfect, small, borne in the axils of the emersed leaves, usually solitary or occasionally with 2–5 flowers in a group; sepals 3; petals 0; stamens 3; ovary superior, 3-locular, with 3 styles; fruit indehiscent, dry, triangular, 3-seeded, bony.

There are two species in this genus.

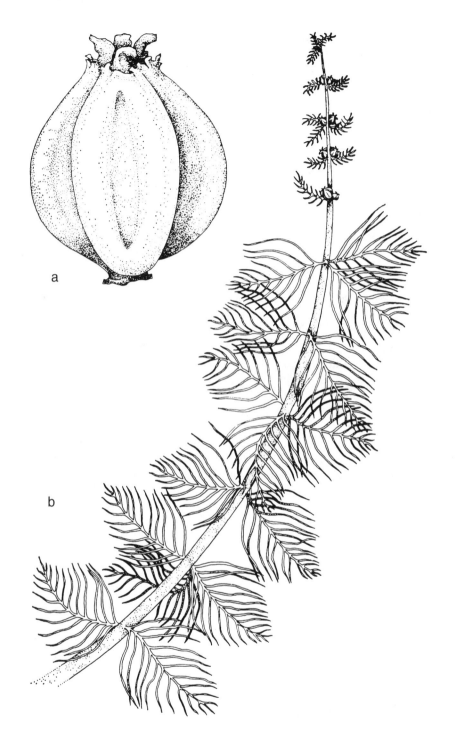

a

b

Fig. 238. *Myriophyllum verticillatum* (Common water milfoil).   a. Fruit.   b. Habit.

1. **Proserpinaca palustris** L. Sp. Pl. 88. 1753. Fig. 239.

Proserpinaca palustris L. var. *amblyogona* Fern. Rhodora 11:120. 1907.

Proserpinaca palustris L. var. *crebra* Fern. & Grisc. Rhodora 37:177–178. 1935.

Submersed or terrestrial perennial herbs; stems prostrate, the branches erect; submersed leaves, if present, alternate, 1–7 cm long, capillary-pinnate to shallowly incised; emersed leaves lanceolate to oblanceolate, 1.5–8.5 cm long, serrate; flowers axillary, solitary or in groups of 2–5, perfect; calyx 3-sided, trilobed; petals 0; stamens 3; fruits pyramidal-ovoid, 3-angled, 3–6 mm long, 2–6 mm wide, sometimes wing-angled, the sides flat, concave, or convex, smooth to tuberculate. June–October.

Shallow water and shores of ponds and lakes; roadside ditches.

IA, IL, IN, KY, MO, OH (OBL).

Mermaid-weed.

Variation occurs in the fruits of this species. Typical var. *palustris* has fruits 4–6 mm wide, wing-angled, and with concave sides. Var. *crebra* has fruits 2–4 mm wide, acutely angled but not winged, and with flat sides. Var. *amblyogona* has fruits 2–4 mm wide, obtusely angled but not winged, and with convex sides.

This species differs from *P. pectinata*, which does not occur in the central Midwest, by its merely serrate terrestrial leaves.

72. HAMAMELIDACEAE—WITCH HAZEL FAMILY

Trees or shrubs; leaves alternate, simple, stipulate; flowers perfect or unisexual, borne in heads or spikes; calyx 4-parted or absent; petals 4, free, or absent; stamens 4 or 8; ovary subinferior to inferior; fruit a capsule or a head of capsules.

This family consists of twenty genera and about one hundred species, found mostly in temperate regions of the world.

1. Liquidambar L.—Sweet Gum

Trees; monoecious; winter buds scaly; leaves alternate, simple, palmately lobed, toothed, with deciduous stipules; flowers unisexual, borne in racemes, bracteate; staminate flowers without sepals or petals and with numerous stamens; pistillate flowers with a reduced calyx, no petals, 4 aborted stamens, and a partly inferior ovary of 2 united carpels; fruit a spherical head of hardened capsules with curved styles, each capsule usually with 1 seed.

Liquidambar is a genus of two Asian species and one North American species.

1. Liquidambar styraciflua L. Sp. Pl. 999. 1753. Fig. 240.

Trees to 35 m tall, with a trunk up to 1.5 m in diameter, the branches forming a pyramidal crown; bark gray, becoming scaly; winter buds ovoid, up to 6 mm long, with many orange-brown scales; twigs reddish brown at first, sometimes developing corky wings, often in an irregular pattern; leaves alternate, simple, palmately 5- or 7-lobed, serrate, truncate or more or less subcordate at the base, green on both sides, glabrous on the upper surface, glabrous except for tufts of hairs in the axils of the veins below, up to 15 cm wide, with slender, usually glabrous petioles up to 15 cm long; staminate flowers in globose heads up to 6 mm in diameter, with several heads in a terminal raceme, the flowers subtended by hairy bracts; pistillate flow-

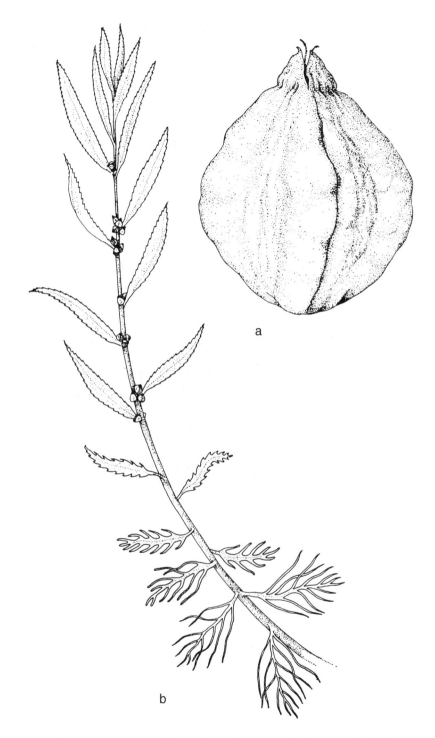

Fig. 239. *Proserpinaca palustris* (Mermaid weed).   a. Fruit.   b. Habit.

ers in a solitary, globose head up to 12 mm in diameter, borne on long pendulous peduncles; fruit a globose head of ripened carpels, the head up to 3 cm in diameter; seeds light brown, winged, up to 1 cm long. April–May.

Wet woods, mesic woods.

IL, IN, MO (FACW), KY, OH (FAC).

Sweet gum.

The star-shaped leaves and spiny-looking globose fruits are distinctive for this species. Some specimens produce heavy, corky wings on the branchlets, while other specimens are devoid of the corky wings.

Fig. 240. *Liquidambar styraciflua* (Sweet gum).

a. Branch with leaves.
b. Leaf.
c. Twig.
d. Fruiting head.
e. Seed.
f. Twig with corky wings.
g. Flowering branch.
h. Flower.

73. HIPPURIDACEAE—MARE'S-TAIL FAMILY

Only the following genus comprises this family.

1. Hippuris L.—Mare's-tail

Only the following species comprises this genus.

1. **Hippuris vulgaris** L Sp. Pl. 4. 1753. Fig. 241.

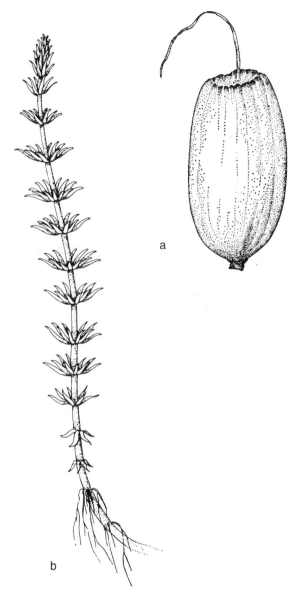

Submersed herbs; stems erect, unbranched; submersed leaves verticillate, linear to filiform, often reduced to scales; emersed leaves in whorls of 6–12, linear to oblong, 0.4–3.5 cm long, entire; flowers axillary, perfect or unisexual; sepals and petals none; hypanthium enclosing ovary in the perfect flowers; stamen 1; style 1; fruit indehiscent, ellipsoid, nutlike, 1-seeded. June–October.

Open water.

IA, IL, IN, NE (OBL).

Mare's-tail.

Hippuris vulgaris is characterized by having its entire, linear leaves arranged in whorls of 6–12 on unbranched, erect stems. The tiny flowers are axillary in the leaves of the middle and upper portion of the plant. The ellipsoid fruit is smooth.

Fig. 241. *Hippuris vulgaris* (Mare's-tail). a. Fruit. b. Habit.

74. HYDROPHYLLACEAE—WATERLEAF FAMILY

Annual or perennial herbs; leaves alternate, simple or compound, without stipules; flowers perfect, actinomorphic, in cymes, racemes, or panicles, blue or purple or white; sepals 5, united below; petals 5, united below; stamens 5, usually attached to the corolla; ovary superior, 1- to 2-locular; styles 2 or 2-cleft; fruit a capsule.

This family consists of about twenty genera and 250 species, with many of the species occurring in the western United States.

Only the following genus has species that occur in shallow water in the central Midwest.

1. Hydrolea L.- False Fiddleleaf

Perennial herbs; leaves alternate, simple, entire, often with spines in the axils of the leaves; flowers in cymes or panicles, blue; sepals 5, united below; petals 5, united below; stamens 5; ovary superior, 2-locular; styles 2; fruit a capsule.

Hydrolea consists of eleven species, most of them in the tropics.

1. Leaves and stems pubescent; leaves ovate; calyx hirsute; flowers in terminal panicles...........
....*1. H. ovata*
1. Leaves and stems glabrous; leaves lanceolate; calyx glabrous or minutely pubescent; flowers in axillary cymes ....*2. H. uniflora*

1. Hydrolea ovata Nutt. Trans. Am. Phil. Soc. II, 5:196. 1833–37. Fig. 242.

Perennial herbs with slender rhizomes; stems erect, to 1 m tall, softly pubescent; leaves alternate, simple, ovate to lance-ovate, acute at the apex, rounded at the base, pubescent on both surfaces, to 6 cm long, to 4 cm wide, usually with short axillary spines, short-petiolate; flowers several in terminal panicles; sepals 5-parted, the lobes linear-lanceolate to lanceolate, glandular-hirsute, 6–9 mm long; corolla blue, 5-parted, 10–17 mm long; stamens 5, nearly as long as the corolla; ovary superior; capsule globose, hispid, 4–5 mm in diameter. July–August.

Wet woods, swamps, wet roadside ditches, often in standing water.

KY, MO (OBL).

Ovate-leaved false fiddleleaf.

This species is distinguished by its axillary spines, ovate leaves, hirsute calyx, and corollas 6–9 mm long.

2. Hydrolea uniflora Raf. Autik. Bot. 24. 1840. Fig. 243.

Perennial herbs; stems creeping, often rooting at the nodes, glabrous or nearly so, to 75 cm long; leaves alternate, simple, lanceolate, acuminate at the apex, tapering to the base, glabrous or minutely pubescent on both surfaces, to 10 cm long to 3.5 cm wide, usually with axillary spines, short-petiolate; flowers 1–10 in dense, axillary cymes; sepals 5-parted, the lobes lanceolate to ovate, glabrous, to 8 mm long; corolla blue, 5-parted, 7–10 mm long; stamens 5, shorter than the corolla; ovary superior; capsule globose, glabrous, 4–5 mm in diameter. June–September.

Swampy woods, wet roadside ditches, often in shallow water.

IL, IN, KY, MO (OBL).

Few-flowered false fiddleleaf.

This species is recognized by its axillary spines, lanceolate leaves, glabrous calyx, and corolla 7–10 mm long.

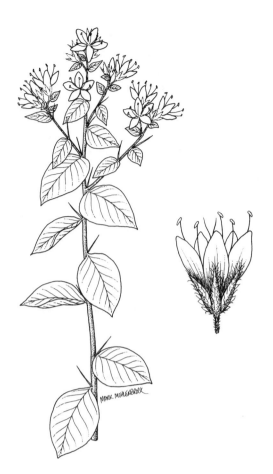

Fig. 242. *Hydrolea ovata*
(Ovate-leaved false fiddleleaf).
Habit (left). Flower (right).

Fig. 243. *Hydrolea uniflora*
(Few-flowered false fiddleleaf).
Habit (center). Flower (lower left).
Flower opened out (lower right).

75. HYPERICACEAE—ST. JOHN'S-WORT FAMILY

Herbs or shrubs; leaves opposite, simple, entire, often glandular-dotted, without stipules; inflorescence terminal or axillary, solitary or more commonly in cymes or corymbs; flowers perfect, actinomorphic; sepals 4–5, free; petals 4–5, free; stamens 5–numerous, often in fascicles, the anthers dehiscing longitudinally; ovary superior, 1- to 5-locular, with numerous ovules; fruit a capsule, usually terminated by the persistent styles in the form of a beak.

This family consists of fifty genera and about twelve hundred species, many of them in the tropics. Two genera have aquatic representatives in the central Midwest.

This family is sometimes called the Clusiaceae, sometimes the Guttiferae.

1. Flowers yellow; stamens in multiples of 5's... 1. *Hypericum*
1. Flowers pink or purple; stamens 9..2. *Triadenum*

1. **Hypericum** L.—St. John's-wort

Herbs or shrubs with usually erect stems; leaves opposite, simple, entire, often glandular-punctate; flowers mostly cymose, occasionally solitary in the axils of the leaves, perfect; sepals 5, free, essentially equal; petals 5, generally longer than the sepals, free, yellow or orange, convolute in bud; stamens 5–numerous, sometimes in fascicles; ovary superior, 1- to 5-locular; styles 3–5; fruit a 1- to 5-celled capsule with the styles persisting, often as a beak.

When considered separate from genera such as *Ascyrum* and *Triadenum*, *Hypericum* consists of 375 species, mostly in north temperate regions.

The number of styles is usually three or five. Sometimes when there are three styles, they appear united as a single beak.

Several species of *Hypericum* in the central Midwest are upland species.

1. Erect shrubs to 2 m tall.
  2. Styles 4 or 5...8. *H. lobocarpum*
  2. Styles 3.
    3. Leaves with axillary fascicles; capsules lanceoloid, to 3 mm across.......4. *H. densiflorum*
    3. Leaves without axillary fascicles; capsules ovoid, 4.5–6.0 mm across ..11. *H. nudiflorum*
1. Herbs or, if woody, less than 1 m tall.
  4. Petals 8 mm long or longer; stamens 20 or more.
    5. Styles 5 (rarely 4).
      6. Flowers 4–7 mm across; capsules 2–3 cm long; some of the leaves partly clasping; bark neither white nor papery.. 12. *H. pyramidatum*
      6. Flowers 1.5–3.0 cm across; capsules up to 1 cm long; leaves not clasping; stems with white papery bark.. 7. *H. kalmianum*
    5. Styles 3, often united to appear as a single beak.
      7. Stems arising from a creeping, stoloniferous base, entirely herbaceous ....
      ....1. *H. adpressum*
      7. Plants without a creeping, stoloniferous base, although underground rhizomes may be present; stems often woody near base.........................5. *H. denticulatum*
  4. Petals at most only 6 mm long; stamens 5–12 (rarely 20).
    8. Bracts foliaceous, resembling the foliage leaves ....2. *H. boreale*
    8. Bracts linear-setaceous, much reduced from the foliage leaves.
      9. Leaves ovate to orbicular.

10. Capsules 3.0–3.5 mm long; plants much branched, not virgate ....
.... 10. *H. mutilum*
10. Capsules 4 mm long or longer; plants sparingly branched, virgate ....
.... 6. *H. gymnanthum*
9. Leaves linear to lanceolate.
11. Leaves linear, 1- to 3-nerved; some or all the sepals less than 4 mm long ....
.... 3. *H. canadense*
11. Leaves lanceolate, 5- to 7-nerved; some or all the sepals 4 mm long or longer........
.... 9. *H. majus*

1. **Hypericum adpressum** Bart. Comp. Fl. Phil. 2:15. 1818. Fig. 244.

Perennial herbs from creeping stolons; stems ascending to erect, sparingly branched, glabrous, to 70 cm tall; leaves opposite, simple, narrowly lanceolate to oblong, acute to obtuse at the apex, tapering to the sessile base, to 4 cm long, to 1 cm wide, punctate above, glabrous, with numerous axillary fascicles of smaller leaves; flowers numerous, to 2 cm across, in terminal cymes, the slender, glabrous pedicels up to 6 mm long; sepals 5, lanceolate to ovate, acute, punctate, glabrous, up to 6 mm long; petals 5, yellow, free, up to 12 mm long; stamens numerous, distinct, long-persistent after flowering; styles 3, more or less united into a beak; capsules ovoid, to 6 mm long, incompletely 3-celled; seeds numerous. July–August.
Wet ground.
IL, IN, KY, MO (OBL).
Creeping St. John's-wort.
This species is rather easily distinguished by its creeping stoloniferous base, differing from *H. ellipticum* by its long-persistent stamens.

2. **Hypericum boreale** (Britt.) Bickn. Bull. Torrey Club 22:213. 1894. Fig. 245.
Hypericum canadense L. var. *boreale* Britt. Bull. Torrey Club 18:365. 1891.

Perennial herbs; stems decumbent, ascending, only slightly 4-angled, sparsely branched, glabrous, to 40 cm tall; leaves simple, opposite, elliptic to oblong, obtuse, more or less rounded at the sessile base, to 2 cm long, to 8 mm wide, punctate, glabrous, 3- to 5-nerved; flowers few to several, to 5 mm across, in cymes subtended by foliaceous bracts; sepals 5, free, elliptic, obtuse to subacute, to 4 mm long, epunctate, glabrous; petals 5, free, yellow, to 5 mm long; stamens up to 12, free; styles 3, free; capsules ellipsoid to oblongoid, to 5 mm long, 1-celled, usually purplish; seeds numerous, pale brown, appearing minutely wrinkled. July–September.
Marshes, sandy soil.
IA, IL, IN, OH (OBL).
Northern St. John's-wort.
Of the species of *Hypericum* with small flowers, this is the only one with foliaceous bracts that are the same size as the foliage leaves.

3. **Hypericum canadense** L. Sp. Pl. 785. 1753. Fig. 246.

Slender perennial herbs with short, leafy offshoots; stems erect, 4-angled, freely branched, glabrous, mostly epunctate, to 75 cm tall; leaves simple, opposite, linear to linear-oblanceolate, acute to obtuse, narrowed or rounded to the nearly sessile or

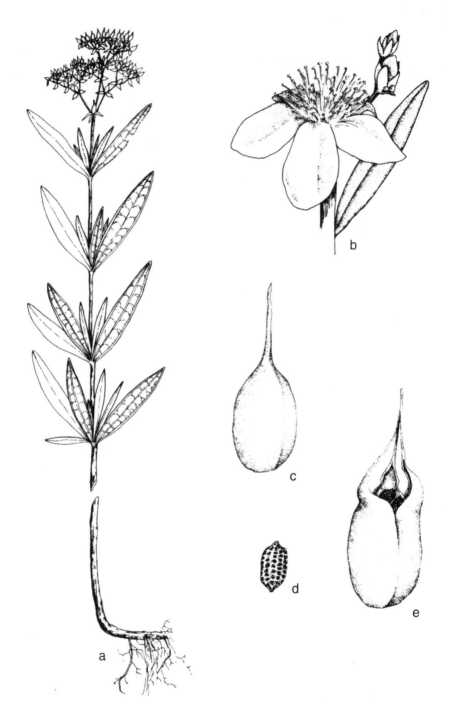

Fig. 244. *Hypericum adpressum*
(Creeping St. John's-wort).

a. Habit.
b. Flower.
c. Capsule.

d. Seed.
e. Dehiscing capsule.

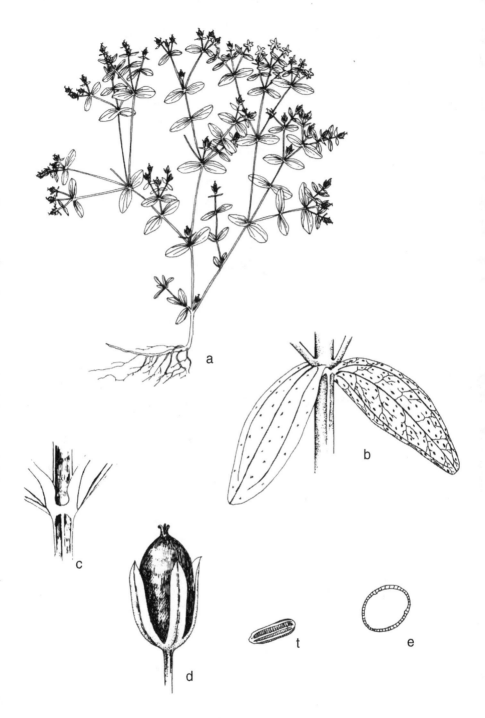

Fig. 245. *Hypericum boreale*
(Northern St. John's-wort).

a. Habit.
b. Leaves.
c. Node.

d. Capsule.
e. Cross-section of capsule.
f. Seed.

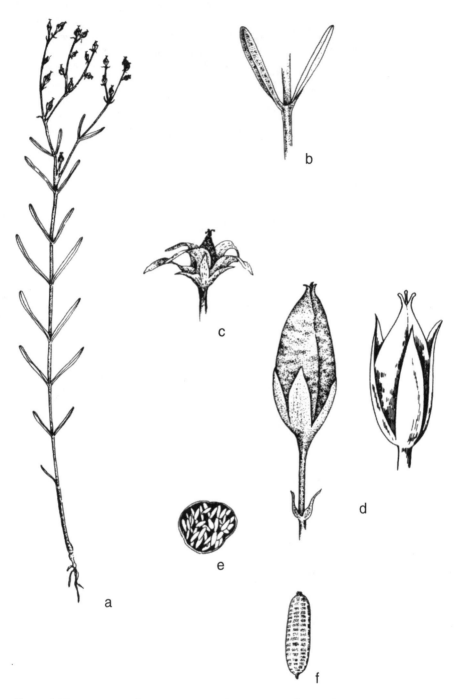

Fig. 246. *Hypericum canadense*
(Canada St. John's-wort).

a. Habit.
b. Node.
c. Flower.

d. Capsules.
e. Cross-section of capsule.
f. Seed.

rarely subclasping base, to 3.5 cm long, to 4.5 (–6.0) mm wide, punctate, glabrous, 1- or 3-nerved; flowers several, to 5 mm broad, in cymes subtended by setaceous bracts; sepals 5, linear-lanceolate, acute, to 4.5 mm long, glabrous, essentially punctate; petals 5, free, yellow, to 5 mm long; stamens 5 or 10; styles 3, free; capsules conical, usually reddish, to 6 mm long, 1-celled; seeds numerous, pale brown, smooth or nearly so. July–September.

Wet meadows, marshes, banks of ponds and lakes, occasionally in standing water.

IA, IL, IN, KY, OH (FACW).

Canada St. John's-wort.

This species has the narrowest leaves of any of the small-flowered species of *Hypericum*. The linear leaves have only 1 or 3 veins.

4. Hypericum densiflorum Pursh, Fl. Am. Sept. 376. 1814. Fig. 247.

Hypericum prolificum L. var. *densiflorum* (Pursh) Gray, Man. ed. 3, 84. 1867.

Shrubs with brownish bark; stems erect, much branched, glabrous, to 2 m tall; leaves simple, opposite, narrowly lanceolate, obtuse to subacute at the apex, tapering to the sessile base, to 4 cm long, to 7 mm wide, punctate, glabrous, 1-nerved; flowers numerous, 1.0–1.5 cm across, in crowded cymes, the moderately stout, glabrous pedicel to 8 mm long; sepals 5, free, elliptic to lanceolate, more or less punctate, glabrous, to 5 mm long; petals 5, free, yellow, up to 15 mm long; stamens numerous, free; styles 3; capsules ovoid, to 6.5 mm long, 3-celled; seeds numerous, black. June–August.

Swampy areas.

KY (FAC+).

Densely-flowered St. John's-wort.

This species is very similar to *H. lobocarpum*, differing by its slightly smaller flowers, 3 styles, and 3-celled capsules.

5. Hypericum denticulatum Walt. Fl. Carol. 190. 1788. Fig. 248.

Hypericum virgatum Lam. Encycl. 4:158. 1797.

Hypericum denticulatum Walt. var. *recognitum* Fern. & Schub. Rhodora 50:208. 1948.

Slender perennial herbs; stems erect, virgate, 4-angled, glabrous, to 60 cm tall; leaves opposite, simple, spreading to ascending, elliptic to narrowly ovate, obtuse to acute at the apex, rounded or tapering to the sessile base, to 2.5 cm long, to 1.5 cm wide, punctate or epunctate, glabrous, 1-nerved; flowers few to several, to about 1 cm across, scattered and racemose along the branches, subtended by small, linear bracts; sepals 5, lanceolate to narrowly ovate, acute, glabrous, punctate or epunctate, to 6 mm long; petals 5, free, copper-yellow, 8–10 mm long; stamens numerous, free; styles 3, free; capsules ovoid, to 5 mm long, 1-celled; seeds numerous, black. June–August.

Moist woods and gravelly hills; seeps.

IL, KY, OH (FACW–).

Coppery St. John's-wort.

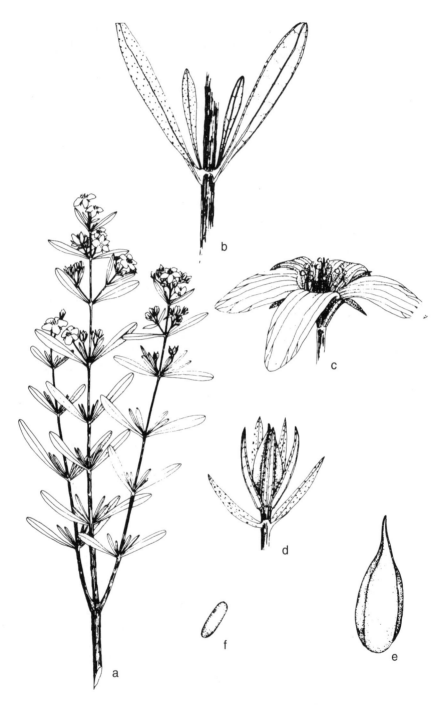

Fig. 247. *Hypericum densiflorum*
(Densely-flowered St. John's-wort).

a. Habit.
b. Node.
c. Flower.

d. Capsule with sepals.
e. Capsule.
f. Seed.

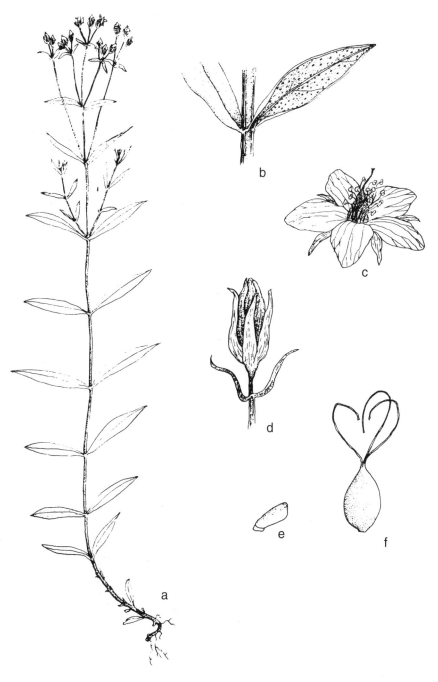

Fig. 248. *Hypericum denticulatum*
(Coppery St. John's-wort).

a. Habit.
b. Node.
c. Flower.

d. Opening bud.
e. Seed.
f. Capsule.

This species is recognized at once by its virgate manner of growth. Its free styles also separate it from the sometimes similar-appearing *H. sphaerocarpum*, which is not considered to be a wetland species.

6. **Hypericum gymnanthum** Engelm. & Gray, Boston Journ. Nat. Hist. 5:212. 1847. Fig. 249.

Slender annual herbs; stems erect, 4-angled, sparsely branched or unbranched, glabrous, sometimes punctate, to 80 cm tall; leaves simple, opposite, ovate, acute

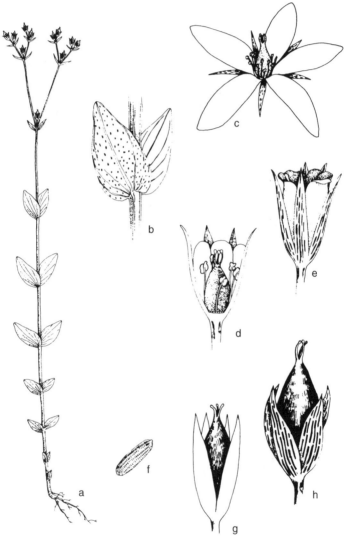

Fig. 249. *Hypericum gymnanthum*
(Virgate St. John's-wort).

a. Habit.
b. Node.
c. Flower.
d. Longitudinal view of unexpanded flower.

e. Unexpanded flower.
f. Seed.
g, h. Capsules.

to obtuse, cordate-clasping, to 2 cm long, to nearly 1 cm wide, punctate, glabrous, 3- to 5- (7-) nerved; flowers numerous, to 5 mm across, in cymes subtended by setaceous bracts; sepals 5, narrowly lanceolate, acute, to 5 mm long, glabrous, essentially epunctate; petals 5, free, yellow, to 3 mm long; stamens up to 12, free; styles 3, free; capsules ovoid, 4 mm long or longer, 1-celled; seeds numerous, pale brown, smooth or nearly so. July–September.

Wet soil, infrequent in standing water.

IL, IN, MO, OH (OBL), KS (FACW+).

Virgate St. John 's-wort.

Although this species has leaves somewhat reminiscent of those of *H. mutilum*, the petals of *H. gymnanthum* are at least 3.5 mm long. The leaves of *H. gymnanthum* are sometimes subcordate and clasping at the base.

7. **Hypericum kalmianum** L. Sp. Pl. 783. 1753. Fig. 250.

Small shrubs with papery, often whitish, bark; stems erect, much branched, glabrous, to 1.2 m tall; leaves simple, opposite, linear-oblong to oblanceolate, revolute, obtuse to subacute, tapering to the sessile or very short-petiolate base, to 4 (−5) cm long, up to 7.5 mm wide, punctate above, 1-nerved, glabrous, usually glaucous beneath, often with axillary clusters of small leaves; flowers 1–10, up to 3.5 cm across, in terminal corymbs, the stout, glabrous pedicels 4 mm long or much longer; some or all the sepals 5, foliaceous, oblong, acute, glabrous, sometimes punctate, up to 15 mm long; petals 5, free, yellow, to 30 mm long; stamens numerous, free; styles usually 5, rarely 4 or 6, united below into a single beak; capsules ovoid, up to 1 cm long (excluding the beak), completely (4-) 5- (6-) celled; seeds numerous, black, smooth. June–August.

Bogs, fens.

IL, IN (FACW), OH (FAC).

Kalm's St. John's-wort.

This shrubby species has flowers that may have a width of 35 mm. It has fewer and somewhat larger flowers than the similar *H. lobocarpum*.

8. **Hypericum lobocarpum** Gattinger ex Coulter, Bot. Gaz. 11:275. 1886. Fig. 251. *Hypericum densiflorum* Pursh var. *lobocarpum* (Gattinger) Svenson, Rhodora 42:11. 1940.

Shrubs with brownish bark; stems erect, much branched, glabrous, to 1.5 m tall; leaves simple, opposite, narrowly lanceolate, obtuse, tapering to the sessile base, to 4 cm long, to 7 mm wide, punctate above, glabrous, 1-nerved; flowers numerous, to 1.5 (−1.8) cm across, in crowded cymes, the moderately stout, glabrous pedicels 2–9 mm long; sepals 5, elliptic to lanceolate, more or less punctate, glabrous, to 5 mm long; petals 5, free, yellow, up to 15 mm long; stamens numerous, free; styles (4−) 5, more or less united into a single beak; capsules ovoid, to 6.5 mm long, completely (4-) 5-celled; seeds numerous, black. June–August.

Wet woods; sometimes in dry, acidic soils.

IL, KY, MO (FAC+), The U.S. Fish and Wildlife Service considers this species to be the same as *H. densiflorum*.

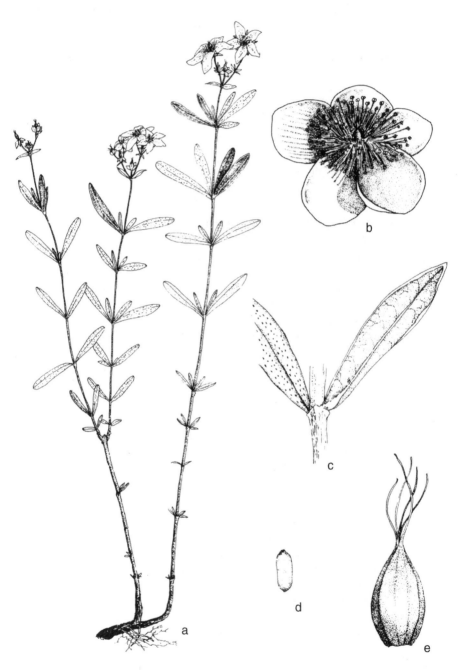

Fig. 250. *Hypericum kalmianum*
(Kalm's St. John's-wort).

a. Habit.
b. Flower.
c. Node.

d. Seed.
e. Capsule.

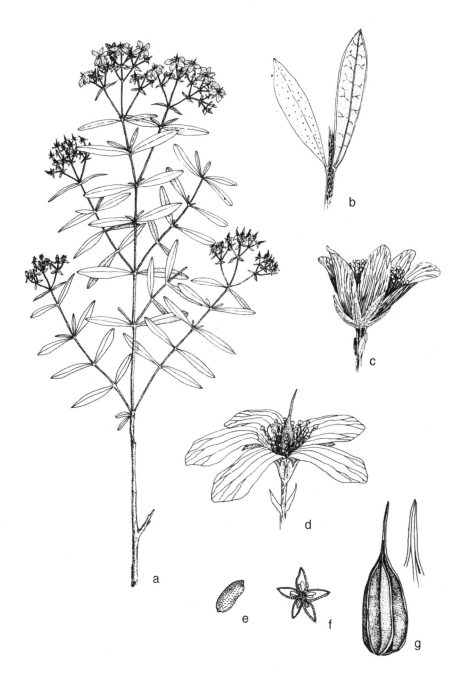

Fig. 251. *Hypericum lobocarpum*
(Lobe-fruited St. John's-wort).

a. Habit.
b. Node.
c, d. Flower.

e. Seed.
f. Cross-section of capsule.
g. Capsule.

Lobe-fruited St. John's-wort.

This species is often considered to be a variety of *H. densiflorum*. However, the number of styles and number of cells in the capsule in *H. lobocarpum* are (4–) 5, while there are only 3 styles and 3 cells in the capsule of *H. densiflorum*.

9. **Hypericum majus** (Gray) Britt. Mem. Torrey Club 5:225. 1894. Fig. 252.
Hypericum canadense L. var. *majus* Gray, Man. ed. 5, 86. 1867.

Slender perennial herbs with short, leafy offshoots; stems erect, 4-angled, sparsely branched, glabrous, epunctate, to 80 cm tall; leaves simple, opposite, lanceolate to oblong-lanceolate, subacute to acute, tapering or more or less rounded at the sessile or somewhat clasping base, to nearly 5 cm long, to 1.5 cm wide, punctate, glabrous, (3-) 5- to 7-nerved; flowers mostly numerous, to 8 mm across, in cymes subtended by setaceous bracts; sepals 5, lanceolate, acute to acuminate, to 7 mm long, glabrous, essentially epunctate; petals 5, free, yellow, to 5 (–7) mm long; stamens 5 or 10, free; styles 3, free; capsules ellipsoid, conical, to 10 mm long, 1-celled; seeds numerous, pale brown, striatulate. July–September.

Moist soil, marshes, mud flats.

IA, IL, IN, KS, MO, NE, OH (FACW).

Small-flowered St. John's-wort.

Hypericum majus differs from other small-flowered species by its 5- to 7-nerved leaves.

10. **Hypericum mutilum** L. Sp. Pl. 787. 1753. Fig. 253.
Hypericum parviflorum Willd. Sp. Pl. 3:1456. 1802.
Hypericum mutilum L. var. *parviflorum* (Willd.) Fern. Rhodora 41:549. 1939.

Slender annual herbs; stems ascending to erect, 4-angled, freely branched, glabrous, to 75 cm tall (usually shorter); leaves simple, opposite, oblong to ovate, obtuse to less commonly acute, more or less rounded at the sessile or clasping base, to 2.2 cm long, to 1.2 cm wide, punctate, glabrous, 5-nerved; flowers numerous, to 5 mm across, in cymes subtended by setaceous bracts; sepals 5, linear to linear-lanceolate, acute, 3–6 mm long, glabrous, essentially epunctate; petals 5, free, yellow, to 5 mm long; stamens up to 12, free; styles 3, free; capsules ovoid to ellipsoid, to 3.5 mm long, 1-celled; seeds numerous, pale brown, smooth or nearly so. June–September.

Moist soil, marshes, mud flats.

IA, IL, IN, KY, MO, OH (FACW), KS, NE (FACW+).

Dwarf St. John's-wort.

The only other small-flowered wetland *Hypericum* with ovate laves is *H. gymnanthum*, but this latter species has larger capsules and is sparingly branched and virgate.

11. **Hypericum nudiflorum** Michx. Fl. Bor. Am. 2:78. 1803. Fig 254.

Shrub to 2 m tall, much branched; stems glabrous, brown, 4-angled; leaves simple, opposite, elliptic to oblong-lanceolate, entire, acute to obtuse at the apex, tapering to the base, glabrous, punctate, to 6 cm long, to 2.5 cm wide, sessile with the midvein decurrent on the stem; flowers in cymes, up to 1.5 (–1.8) cm across, with

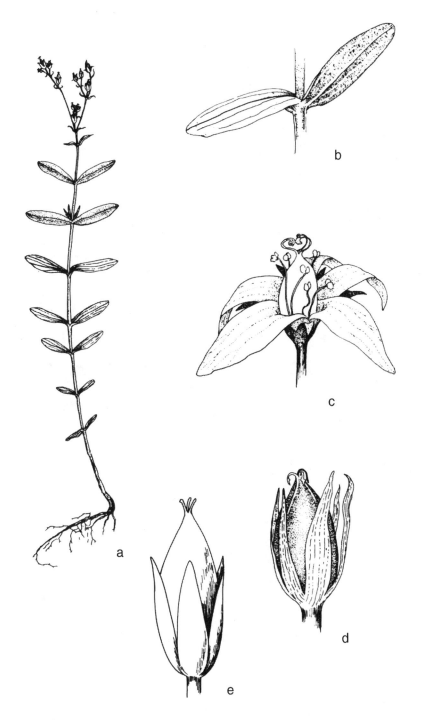

Fig. 252. *Hypericum majus*
(Small-flowered St. John's-wort).

a. Habit.
b. Node.

c. Flower.
d, e. Capsules.

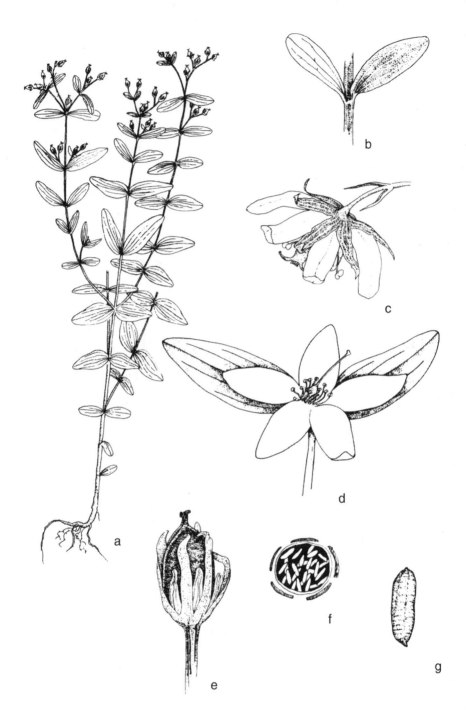

Fig. 253. *Hypericum mutilum*
(Dwarf St. John's-wort).

a. Habit.
b. Node.
c, d. Flower.

e. Flower without petals.
f. Cross-section of capsule.
g. Seed.

subulate bracts 2–3 mm long, on short glabrous pedicels; sepals 5, free, oblong, 2–4 mm long; petals 5, free, copper-yellow, 6–10 mm long; stamens numerous; ovary superior; styles 3 but appearing as 1; capsules ovoid, 4–9 mm long, 1-locular, glabrous; seeds linear-oblongoid, 1.5–2.0 mm long, dark brown, reticulate. June–July.

Marshes, wet woods.

KY (OBL), but not found recently.

Southern shrubby St. John's-wort.

This is a shrubby species that may reach a height of two meters. There are no axillary fascicles of leaves.

Fig. 254. *Hypericum nudiflorum* (Southern shrubby St. John's-wort). Habit (right). Capsule (center). Bark (left).

12. **Hypericum pyramidatum** Ait. Hort. Kew. 3:103. 1789. Fig. 255.

Stout herbaceous perennial herbs; stems erect, branched, glabrous, to 2 m tall, the branchlets more or less angular; leaves opposite, simple, ovate-oblong to broadly lanceolate, acute or occasionally obtuse at the apex, sessile and sometimes slightly clasping at the base, the largest to 10 cm long, to 3.5 cm wide, punctate above, glabrous; flowers relatively few, to 7 cm across, in terminal cymes; sepals 5, ovate-lanceolate, acute, glabrous, epunctate, to 1.2 cm long, to 8 mm wide; petals 5, free, bright yellow, to 3 cm long; stamens numerous, united into 5 fascicles; styles 5, free above, united below; capsules ovoid, to 3 cm long, beaked but with the persistent styles spreading at the tip, 3-celled but often appearing more because of the turned-in placentae; seeds numerous, black, glabrous. July–August.

Banks of rivers and streams.

IA, IL, IN, MO (FAC+), OH (FAC), KS, NE (not listed).

Giant St. John's-wort.

The flowers of this handsome species may measure up to 7 cm across. The capsule is 2–3 cm high and is terminated by the persistent style which has free and spreading tips.

Fig. 255. *Hypericum pyramidatum*
(Giant St. John's-wort).

a. Habit.
b. Flower.

c. Stamen.
d. Capsule.

2. Triadenum Raf.—Pink St. John's-wort

Herbs with ascending or erect stems; leaves opposite, entire, usually punctate; flowers cymose, the cymes terminal and axillary, perfect; sepals 5, free, equal, petals 5, generally longer than the sepals, pinkish or flesh-colored, imbricate in the bud; stamens usually 9, in fascicles of 3 each, with 3 large orange glands alternating with each fascicle of stamens; ovary superior; styles 3; fruit a 3-celled capsule.

Some botanists place the species of *Triadenum* in *Hypericum*, but the color of the flowers and the number of stamens justify its segregation as a distinct genus. Ten species in North America and Asia comprise this genus.

1. Leaves apparently without punctations .... 2. *T. tubulosum*
1. Leaves conspicuous punctate, at least on the lower surface.
   2. Leaves petiolate .... 4. *T. walteri*
   2. Leaves sessile.
      3. Sepals obtuse at the apex, up to 5 mm long; styles up to 1.5 mm long .... 1. *T. fraseri*
      3. Sepals acute at the apex, 5–8 mm long; styles 2–3 mm long .... 3. *T. virginicum*

1. **Triadenum fraseri** (Spach) Gl. Phytologia 2:289. 1947. Fig. 256.

Elodea fraseri Spach, Ann. Sci. Nat. II, 5:168. 1836.

Hypericum virginicum L. var. *fraseri* (Spach) Fern. Rhodora 38:434. 1936.

Perennial herbs from strongly stoloniferous bases; stems erect, sparingly branched, glabrous, epunctate, often reddish throughout, to 75 cm tall; leaves simple, opposite, elliptic to oblong-ovate, obtuse and sometimes emarginate at the apex, rounded or subcordate at the often slightly clasping base, subcoriaceous, glabrous, with translucent punctations on the lower surface, paler beneath, to 5 cm long, to 2.5 cm wide; flowers several in cymes in the upper axils, the cymes on peduncles 5–15 mm long; sepals 5, elliptic to oblong, obtuse, glabrous, epunctate, 3–5 mm long; petals 5, free, flesh-colored, 5–8 mm long; styles 0.5–1.5 mm long; capsules 7–12 mm long, ovoid, obtuse to subacute; seeds numerous, dark, brown, honeycombed. July–September.

Bogs, wooded swamps.

IA, IL, IN, KS, OH (OBL).

Fraser's pink St. John's-wort.

This species is distinguished from the similar *T. virginicum* by its shorter, obtuse sepals and its very short styles.

2. **Triadenum tubulosum** (Walt.) Gl. Phytologia 2:289. 1947. Fig. 257.

Hypericum tubulosum Walt. Fl. Car. 191. 1788.

Triadenum longifolium Small, Bull. Torrey Club 25:40. 1898.

Perennial herbs with strongly stoloniferous bases; stems erect to ascending, sometimes branched from near the base, glabrous, epunctate, stramineous except for the sometimes reddish base, to nearly 1 m tall; leaves simple, opposite, oblong to oblong-lanceolate, obtuse and mucronulate at the apex, truncate, rounded, or rarely subcordate at the sessile base, thin, glabrous, apparently epunctate, pale beneath, to 15 cm long, to 5 cm wide; flowers few in cymes in the uppermost axils, the cymes borne on peduncles up to 1 cm long; sepals 5, narrowly oblong, acute,

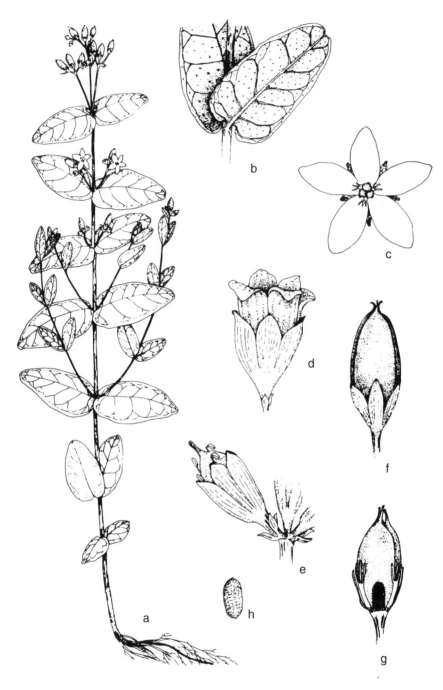

Fig. 256. *Triadenum fraseri*
(Fraser's pink St. John's-wort).

a. Habit.
b. Node.
c. Flower.

d. Unexpanded flower.
e, f, g. Capsules.
h. Seed.

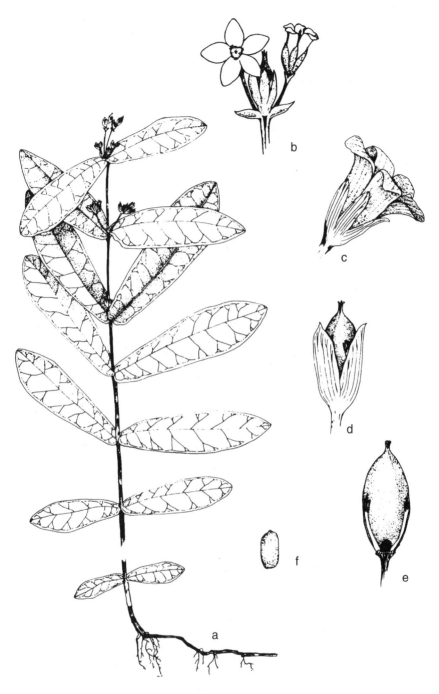

Fig. 257. *Triadenum tubulosum*
(Marsh St. John's-wort).

a. Habit.
b. Flowers.
c. Unexpanded flower.

d, e. Capsules.
f. Seed.

glabrous, epunctate, 4–6 mm long; petals 5, free, pinkish, 5–7 mm long; styles 0.8–1.2 mm long; capsules 9–12 mm long, acute; seeds numerous, brown, honey-combed. August–September.

Swampy woods, sometimes in depressions with standing water.

IL, IN, KS, KY, MO, OH (OBL).

Marsh St. John's-wort.

This species often forms dense colonies because of its stoloniferous habit. The character that distinguishes this species from other members of the genus is the absence of punctations on the leaves. The leaves rarely clasp at the base, which further distinguishes *T. tubulosum* from *T. fraseri* and *T. virginicum*.

3. **Triadenum virginicum** (L.) Raf. Fl. Tell. 3:79. 1836. Fig. 258.
Hypericum virginicum L. Sp. Pl. ed. 2, 1104. 1763.

Perennial herbs from stoloniferous bases; stems erect, branched, glabrous, epunctate, often reddish throughout, to nearly 1 m tall; leaves simple, opposite, elliptic to oblong-ovate, obtuse and sometimes emarginate at the apex, rounded at the sessile or seldomly clasping base, membranaceous to subcoriaceous, glabrous, with translucent punctations on the lower surface, sometimes paler beneath, to 5 cm long, to 2 cm wide; flowers several in cymes in the upper axils, the cymes on peduncles to 1.7 cm long; sepals 5, lanceolate, acute, glabrous, epunctate, 5–8 mm long; petals 5, free, flesh-colored, 7–10 mm long; styles 2–3 mm long; capsules 8–12 mm long, narrowly ovoid, acute; seeds numerous, brown, rather inconspicuously honeycombed. August–September.

Bogs.

IL, IN, OH (OBL).

Marsh St. John's-wort.

Triadenum virginicum has larger sepals and petals and longer styles than the similar *T. fraseri*. In addition, the sepals are decidedly pointed in *T. virginicum*.

4. **Triadenum walteri** (Gmel.) Gl. Phytologia 2:289. 1947. Fig. 259.
Hypericum petiolatum Walt. Fl. Carol. 191. 1788, non L. (1763).
Hypericum walteri Gmel. Syst. Nat. 2:1159. 1791.
Elodea petiolata (Walt.) Pursh, Fl. Am. Sept. 2:379. 1814.
Elodes petiolata (Walt.) Gray, Man. ed. 5, 86. 1867.
Triadenum petiolatum (Walt.) Britt. in Britt. & Brown, Illustr. Fl. 2:437. 1897.
Hypericum tubulosum Walt. var. *walteri* (Gmel.) Lott. Journ. Arn. Arb. 19:279. 1938.

Perennial herbs from stoloniferous bases; stems erect, often branched, glabrous, epunctate, usually reddish throughout, to nearly 1 m tall; leaves simple, oppo-site, narrowly oblong to oblong-lanceolate, obtuse and sometimes mucronulate at the apex, tapering to a petiole, thin, glabrous, with translucent punctations on the lower surface, green or only slightly paler beneath, to 12 cm long, to 3 cm wide; petioles slender, to 1.5 cm long; flowers several in cymes in the upper axils, the cymes borne on peduncles up to 5 mm long; sepals 5, narrowly oblong, more or less obtuse, glabrous, epunctate, 3–5 mm long; petals 5, free, pinkish, 5–7 mm long; styles 0.8–1.2 mm long; capsules 7–12 mm long, subacute; seeds numerous, brown, subprominently honeycombed. July–September.

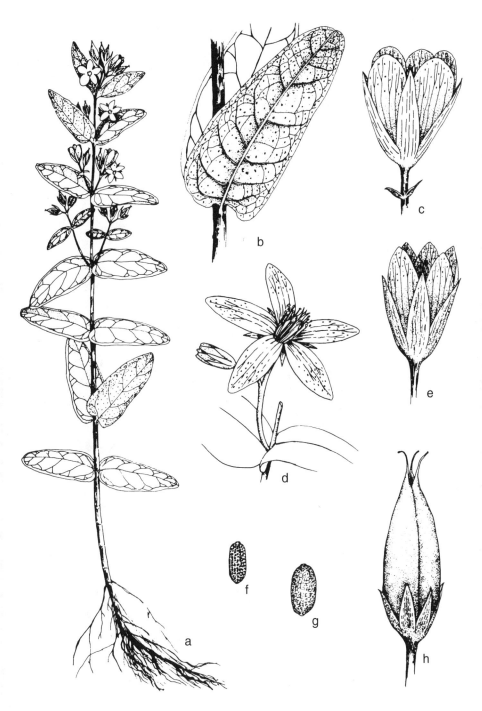

Fig. 258. *Triadenum virginicum*
(Marsh St. John's-wort).

a. Habit.
b. Node.
c, e. Unexpanded flowers.

d. Flower.
f, g. Seeds.
h. Capsule.

335

Wooded swamps, sometimes in depressions with standing water.
IL, IN, KS, KY, MO, OH (OBL).
Petiolate marsh St. John's-wort.

This species is readily distinguished from all other species of this genus by its petiolate leaves.

Triadenum walteri is commonly found in cypress swamps, often growing on fallen logs.

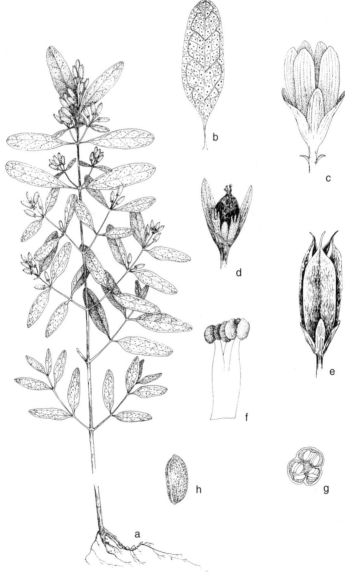

Fig. 259. Triadenum walteri a. Habit. e. Capsule.
(Petiolate marsh St. John's-wort). b. Leaf. f. Cluster of stamens.
c. Unexpanded flower. g. Cross-section of capsule.
d. Developing capsule. h. Seed.

76. JUGLANDACEAE—WALNUT FAMILY

Trees or shrubs, often aromatic; leaves alternate, pinnately compound, without stipules; flowers unisexual, the staminate in elongate, pendulous catkins, often from the twigs of the previous year, the pistillate in short catkins at the apex of the current season's twigs, or solitary and terminal on the twigs; staminate flowers solitary in the axils of bracts, with 4 sepals attached to a pair of bracteoles, or the sepals absent, with 2-many stamens; pistillate flowers subtended by a cluster of united bracteoles, with a 1- to 4-parted calyx, an inferior ovary, 2 free styles, and a solitary ovule; fruit a nut surrounded by a husk.

This family consists of seven genera and approximately sixty species. Only the following genus has a species that sometimes occurs in standing water.

1. Carya Nutt.—Hickory

Trees; leaves alternate, pinnately compound; staminate flowers in narrow, elongate catkins in clusters of 3, with a 2- to 3-lobed calyx subtended by bracts and 3–10 stamens; pistillate flowers solitary or sometimes in spikes of 2–10 flowers at the apex of branches, without a calyx but with a 4-lobed involucre of bracts, the ovary inferior; fruit a nut surrounded by a more or less dehiscent husk.

Carya consists of about eighteen species found in North America and Asia. Only the following species may occur in standing water.

1. **Carya aquatica** (Michx. f.) Nutt. Gen. N. Am. Pl. 2:222. 1818. Fig. 260.
Juglans aquatica Michx. f. Hist. Arb. For. Am. Sept. 1:182. 1810.

Tree to 25 m tall, with a trunk diameter up to 0.7 m and a narrow crown; bark reddish brown, furrowed, becoming somewhat scaly at maturity; twigs slender, reddish brown or gray, smooth or occasionally hairy; leaf scars alternate, 3-lobed, scarcely elevated,

Fig. 260. *Carya aquatica* (Water hickory). Habit (left). Fruits (lower right).

with several bundle traces; buds pointed, reddish brown with yellow scales, usually pubescent, up to 5 mm long; leaves alternate, pinnately compound, with 7–17 leaflets, the leaflets lanceolate, more or less falcate, acute at the apex, tapering to the slightly asymmetrical base, up to 10 cm long, up to 4.5 cm wide, finely serrate, dark green and glabrous or nearly so on the upper surface, brownish and glabrous or somewhat pubescent on the lower surface; staminate and pistillate flowers borne separately but on the same tree, appearing when the leaves are partly unfolded, the staminate several in slender, pendulous catkins up to 4.5 cm long, the pistillate fewer, in shorter spikes; perianth absent; fruit short-ellipsoid, usually tapering to either end, flattened, up to 3.5 cm long and two-thirds as broad, the husk 4-winged, dark brown but with yellow scales, thin, splitting only about halfway to the base, the nut flattened, 4-angled, reddish brown, the shell thin, the seed bitter. March–April.

Swamps, swampy woods.

IL, IN, KY, MO, OH (OBL).

Water hickory; bitter pecan.

The flattened fruits and nuts of this species are distinctive.

77. LAMIACEAE—MINT FAMILY

Herbs (in our area) or low shrubs, sometimes aromatic; stems 4-angled; leaves opposite, simple; flowers variously arranged; calyx 5-parted, tubular below, the teeth or lobes equal or unequal; corolla (4-) 5-lobed, united below, usually bilabiate; stamens 2 or 4, attached to the corolla tube; ovary 4-lobed, with a bifid style; fruit a cluster of four 1-seeded nutlets.

This family consists of approximately two hundred genera and thirty-two hundred species found in most parts of the world. Many genera and species in the central Midwest are upland.

1. Lobes of corolla almost equal, not bilabiate.
  2. Calyx lobes equal .... 2. *Isanthus*
  2. Calyx lobes unequal.
    3. Flowers white, fertile stamens 2 .... 3. *Lycopus*
    3. Flowers blue; fertile stamens 4 .... 4. *Mentha*
1. Lobes of corolla unequal, bilabiate, or the upper lip absent.
    4. Upper lip of corolla seemingly absent; leaves gray on the lower surface........9. *Teucrium*
    4. Upper lip of corolla present; leaves green on the lower surface.
      5. Calyx crested on the upper side... 7. *Scutellaria*
      5. Calyx without a crest on the upper side.
        6. Leaves entire or merely with 1–4 low teeth on each margin.
          7. Flowers in interrupted spikes .... 8. *Stachys*
          7. Flowers in axillary or terminal cymes or corymbs.
            8. Inflorescence in dense terminal heads arranged in cymes or corymbs............
            .... 6. *Pycnanthemum*
            8. Inflorescence in the axils of the leaves ....1. *Calamintha*
        6. Leaves serrate, dentate, crenate, or pinnatifid.
          9. Flowers crowded in heads arranged in cymes or corymbs.... 6. *Pycnanthemum*
          9. Flowers in axillary whorls or terminal racemes or panicles.
            10. Flowers much longer than the subtending bracts .... 5. *Physostegia*
            10. Flowers not longer than or only a little longer than the subtending bracts
            .... 8. *Stachys*

1. **Calamintha** Moench—Calamint

Herbs (in the central Midwest) or low shrubs; stems 4-angled; leaves opposite, simple; flowers usually in axillary clusters; calyx slightly bilabiate, tubular, 5-lobed; corolla bilabiate, 4 lobed, longer than the calyx; stamens 4, at least the longest pair usually exserted; ovary deeply 4-lobed, with a bifid style; nutlets 4, glabrous.

This genus consists of about six species.

1. **Calamintha arkansana** (Nutt.) Shinners, Sida 1:72. 1962. Fig. 261.
Cunila glabella Michx. Fl. Bor. Am. 1:13. 1803, misapplied.
Hedeoma arkansana Nutt. Trans. Am. Phil. Soc. 5:186. 1834.
Calamintha glabella (Michx.) Benth. in DC. Prodr. 12:230. 1848.
Clinopodium glabellum (Michx.) Kuntze, Rev. Gen. Pl. 515. 1891.
Satureja arkansana (Nutt.) Briq. Nat. Pflanzen. IV, 3a:302. 1896.

Fig. 261. *Calamintha arkansana* (Slender calamint). Habit (center). Flower (upper left). Calyx (next to upper left). Lower leaves (lower right).

Aromatic perennial herbs; stems creeping or erect, slender, usually much branched, glabrous or sometimes pubescent at the nodes, to 40 cm long; cauline leaves simple, opposite, linear to narrowly elliptic, acute at the apex, tapering to the sessile or very short-petiolate base, entire or occasionally with 1–4 small teeth per side, to 2.5 cm long, to 5 (–8) mm wide, glabrous, punctate, strongly aromatic when crushed; leaves of creeping stems elliptic to narrowly ovate; flowers few to several in the axils of the leaves, on filiform pedicels up to 15 mm long, subtended at the base by a pair of linear bracts; calyx 5-toothed, the teeth much shorter than the tube, glabrous, punctate, usually with about 13 strong nerves, 4.0–4.5 mm long; corolla bilabiate, purple, 7–10 mm long, pubescent on the outer surface; stamens 4, the longest pair just barely exserted; nutlets 4, ovoid, 0.8–1.0 mm long, brown, more or less pitted. May–July.

Springy areas, wet meadows, calcareous fens, in gravelly streams; also in depressions on cliffs.

IL, MO (FACW), OH (FACU), as *Satureja arkansana*.

Slender calamint.

This species at one time or another has been placed in *Calamintha, Satureja, Clinopodium,* and *Cunila.*

Calamintha arkansana is the only wetland mint in the central Midwest with essentially entire leaves and pedicellate flowers in the axils of the leaves. Narrow-leaved species of *Pycnanthemum* with entire leaves have sessile flowers in terminal cymes.

2. **Isanthus** Michx.—False Pennyroyal

Annual herb; stems 4-sided; leaves opposite, simple, entire; flowers axillary, bracteate, zygomorphic, perfect; sepals 5, regular, united below; corolla 5-lobed; stamens 4; ovary superior, 4-lobed; nutlets 4.

Only the following species is in this genus.

1. **Isanthus brachiatus** (L.) BSP. Prel. Cat. N.Y. 44. 1888. Fig. 262. *Trichostema brachiatum* L. Sp. Pl. 598. 1753.

Annual herbs with fibrous roots; stems slender, branched, glandular-puberulent, 4-sided, to 40 cm tall; leaves opposite, simple, elliptic to lanceolate, acute at the apex, tapering to the sessile base, entire or occasionally toothed, puberulent, 3-nerved; flowers 1–3 in the axils, 2–3 mm across, on pedicels to 1 cm long; calyx campanulate, with 5 equal lobes; corolla tubular, 5-lobed, blue, 1–2 mm long; stamens 4; ovary superior, 4-lobed; nutlets 4, puberulent, 2.5–3.0 mm long. August–September.

Dry habitats.

IA, IL, IN, KS, KY, MO, NE, OH (UPL).

False pennyroyal.

Fig. 262. *Isanthus brachiatus* (False pennyroyal). Habit (center). Calyx (lower left). Flower (lower right).

This slender species has a regular calyx and axillary flowers.

It is nearly always found in dry habitats, but in Iowa it occurs in wet situations.

3. **Lycopus** L.—Water Horehound

Perennial herbs; stems erect, 4-angled; leaves opposite, simple, toothed or lobed, without stipules; flowers in axillary clusters, calyx 4- to 5-lobed, campanulate; corolla 4- to 5-lobed, the lobes nearly equal, united below into a short tube; fertile stamens 2, sometimes with 2 minute staminodia; ovary superior, 4-lobed; fruit a cluster of 4 nutlets.

About fifteen species, all in north temperate regions of the world, comprise this genus. The flowers of the species are all very similar in appearance, differing mostly in the characteristics of the calyx.

1. Some or all the leaves deeply pinnatifid .... 1. *L. americanus*
1. Leaves toothed, never deeply pinnatifid.
  2. Leaves petiolate.
    3. Leaves and stems green, the leaves often with long hairs on the lower surface; calyx up to 1 mm long; plants without tubers .... 7. *L. virginicus*
    3. Leaves and stems often suffused with purple, the leaves glabrous or merely puberulent on the lower surface; calyx 1.3–2.0 mm long; plants usually with tubers .... 5. *L. rubellus*
  2. Leaves sessile or nearly so.
    4. Nutlets tuberculate at the summit.
      5. Uppermost leaves less than 1 cm wide; leaves tapering to base .... 3. *L. angustifolius*
      5. Uppermost leaves more than 1 cm wide; leaves rounded at base ....2. *L. amplectens*
    4. Nutlets not tuberculate at the summit.
      6. Stems glabrous; tubers present; calyx teeth to 1 mm long ....6. *L. uniflorus*
      6. Stems pubescent; tubers absent; calyx teeth 2–3 mm long .... 4. *L. asper*

1. **Lycopus americanus** Muhl. ex Bart. Fl. Phil. Prodr. 15. 1815. Fig. 263.
Lycopus americanus Muhl. var. *scabrifolius* Fern. Rhodora 47:180–181. 1945.

Perennial herbs with stolons but without tubers; stems erect, glabrous, usually with a tuft of short hairs at each node, to 1 m long; leaves opposite, simple, petiolate, lanceolate to linear-lanceolate, acuminate at the apex, tapering to the base, serrate or deeply pinnatifid, glabrous or less commonly scabrous, to 8 cm long, to 3.5 cm wide; flowers crowded in axils of the leaves, subtended by lanceolate bracts 1.4–1.5 mm long; calyx lobes 4, 1.5 mm long, subulate at the tip; corolla lobes 4, white, about as long as the calyx; stamens 2; ovary 4-lobed, superior; nutlets 4, concave at the summit, each nutlet obovoid, 1.0–1.5 mm long. July–September.

Wet ground, occasionally in standing water.

IA, IL, IN, KS, KY, MO, NE, OH (OBL).

Lobe-leaved water horehound.

This is the only species of *Lycopus* in which some of the leaves are deeply pinnatifid. During winter, remains of the spherical fruiting clusters persist on the dead, leafless stems. Plants with scabrous leaves may be called var. *scabrifolius*.

2. **Lycopus amplectens** Raf. Autik. Bot. 38. 1840. Fig. 264.

Perennial herbs with stolons and tubers; stems erect, glabrous, usually with a tuft of short white hairs at each node, to 60 cm tall; leaves opposite, simple, lanceolate to oblong, acute at the apex, more or less rounded at the sessile base, serrate, glabrous, to 6 cm long, to 3 cm wide; flowers crowded in axils of the leaves, subtended by minute bracts about 0.5 mm long; calyx lobes 4, 1.5–2.0 mm long, acute at the tip; corolla lobes 4, white, 2–3 mm long, longer than the calyx; fertile stamens 2; ovary 4-lobed, superior; nutlets 4, each nutlet truncate and tuberculate at the summit, obovoid, 1.2–1.5 mm long. July–September.

Wet ground, occasionally in standing water.

IL, IN (OBL).

Coastal plain water horehound.

This primarily coastal plain species is distinguished by its glabrous leaves and its nutlets tuberculate at their summit. The similar appearing *L. rubellus* has petiolate leaves.

Fig. 263. *Lycopus americanus* (Lobe-leaved water horehound). Habit. Nutlet (lower left).

Fig. 264. *Lycopus amplectens* (Coastal plain water horehound). Habit (left).
Nutlet (upper right).

3. **Lycopus angustifolius** Ell. Sketch. Bot. S. Car. 26. 1816. Fig. 265.
Lycopus rubellus Moench. var. *angustifolius* H. E. Ahles, Journ. Elisha Mitchell Sci. Sci. 80. 173. 1964.

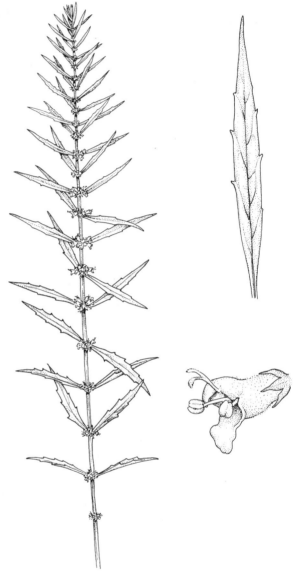

Perennial herbs with stolons and tubers; stems densely pubescent, branched or unbranched, usually with a tuft of white hairs at each node, to 80 cm tall; leaves simple, opposite, linear to lanceolate, to 12 cm long, to 3 cm wide, the uppermost leaves much narrower, less than 1 cm wide, with feltlike pubescence on both surfaces, serrate, sessile; flowers up to 15 per leaf axil, subtended by bracts 1–2 mm long; calyx 5-toothed, 2.3–3.0 mm long, the teeth narrowly triangular; corolla 4-lobed, white, 3–4 mm long; stamens slightly exserted; nutlets tuberculate at the summit, angular, 1.0–1.2 mm long. July–September.

Wet soil, rarely in standing water.

MO (OBL). The U.S. Fish and Wildlife Service does not distinguish this plant from *L. rubellus*.

Narrow-leaved water horehound.

This is primarily a coastal plain species

Fig. 265. *Lycopus angustifolius* (Narrow-leaved water horehound). Branch with leaves and flowers (left). Leaf (upper right). Flower (lower right).

which barely enters the range covered by this book. Its narrow upper leaves with feltlike pubescence on both surfaces are distinctive.

4. Lycopus asper Greene, Pittonia 3:329. 1898. Fig. 266.

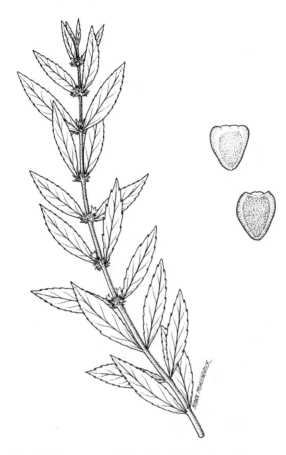

Perennial herbs with stolons and tubers; stems 4-angled, erect, pubescent, usually with tufts of white hairs at the nodes, branched or unbranched, to 1.2 m tall; leaves opposite, simple, oblanceolate to oblong, acute to acuminate at the apex, more or less rounded to tapering at the sessile base, serrate, glabrous or somewhat pubescent, scabrous, punctate, to 8 cm long, to 3 cm wide; flowers crowded in axillary clusters, subtended by lanceolate, ciliate bracts 2.5–3.5 mm long, much longer than the flowers; calyx 5-toothed, tubular below, the teeth narrowly triangular, glandular-pubescent; corolla 4-lobed, white to pinkish, 4–7 mm long; stamens 2, not exserted; ovary 4-lobed, superior; nutlets 4, truncate at the summit, angular, obovoid, smooth, 1.7–2.0 mm long. June–September.

Fig. 266. *Lycopus asper* (Rough water horehound). Habit (left). Nutlets (right).

Marshes, wet meadows, disturbed fens, around springs, occasionally in standing water.

IA, IL, IN, KS, MO, NE, OH (OBL).

Rough water horehound.

This is the only species of *Lycopus* with strongly scabrous leaves, although a variety of *L. americanus* has slightly scabrous leaves.

5. Lycopus rubellus Moench, Meth. Suppl. 146. 1802. Fig. 267.

Perennial herbs with stolons and usually with tubers; stems 4-angled, branched or unbranched, glabrous or puberulent, usually with tufts of short hairs at the nodes, to 90 cm tall; leaves opposite, simple, lanceolate to lance-ovate, acute to acuminate at the apex, tapering or somewhat rounded at the base, glabrous or pu-

berulent below, shallowly serrate, often flushed with red, with a winged petiole, to
12 cm long, to 3.5 cm wide; flowers 5–20 in the axils of leaves, the bracts linear, 1–2
mm long, pubescent; calyx 4- or 5-toothed, 2.0–2.5 mm long, the teeth narrowly
triangular; corolla 5-lobed, white, glandular-pubescent, 2.5–5.0 mm long; stamens
2, slightly exserted; ovary 4-lobed, superior; nutlets 4, tuberculate at the summit,
angular, 1.0–1.4 mm long. August–September.

Ditches, along shores of lakes and ponds, fens.

IL, IN, KS, KY, MO, OH (OBL).

Reddish water horehound.

The leaves, which are often reddish tinged, are petiolate and often winged, and
the calyx teeth are narrowly triangular.

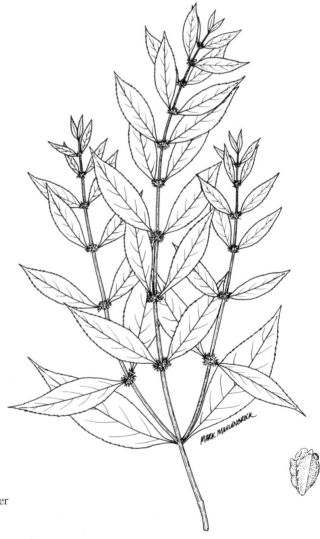

Fig. 267. *Lycopus
rubellus* (Reddish water
horehound). Habit.
Nutlet (lower right).

6. **Lycopus uniflorus** Michx. Fl. Bor. Am. 1:14. 1803. Fig. 268.

Perennial herbs with stolons and tubers; stems 4-angled, erect, branched or un-branched, glabrous or nearly so, usually with tufts of short hairs at the nodes, up to 60 cm long; leaves opposite, simple, elliptic to lanceolate to ovate, acute at the apex, sessile or with a very short, winged petiole, to 10 cm long, to 4 cm wide, glabrous or sometimes sparsely pubescent on the lower surface, punctate; flowers 2–12 in the leaf axils, the bracts subulate, up to 1 mm long, ciliate; calyx 4- or 5-toothed, up to 1 mm long, the teeth broadly triangular; corolla 5-lobed, white, glandular-pubescent, 2.5–3.5 mm long; stamens 2, exserted; ovary 4-lobed, superior; nutlets with or without tubercles at the summit, angular, 1.0–1.2 mm long. July–September.

Bogs, along streams, around lakes and ponds.

IA, IL, IN, KS, KY, MO, NE, OH (OBL).

Few-flowered water horehound.

This species is distinguished by its nearly glabrous stems and its calyx only up to 1 mm long.

Fig. 268. *Lycopus uniflorus* (Few-flowered water horehound). Habit. Calyx (right).

7. Lycopus virginicus L. Sp. Pl. 21. 1753. Fig. 269.

Perennial herbs with stolons but usually without tubers; stems 4-angled, erect, branched or unbranched, glabrous or puberulent, usually with tufts of short hairs at the nodes, to 75 cm tall; leaves opposite, simple, ovate to ovate-lanceolate, acuminate at the apex, tapering to the base, sharply serrate, glabrous or sometimes puberulent, to 9 cm long, to 4.5 cm wide, at least the lower leaves petiolate; flowers several in crowded axillary clusters, subtended by short, oblong bracts; calyx 4- or 5-toothed, the teeth up to 1 mm long, narrowly ovate to ovate, glabrous; corolla 4- or 5-lobed, white, about twice as long as the calyx; stamens 2, included; ovary 4-lobed, superior; nutlets 4, as long as the calyx. July–October.

Wet soil, occasionally in standing water.

IA, IL, IN, KS, KY, MO, NE, OH (OBL).

Virginia water horehound.

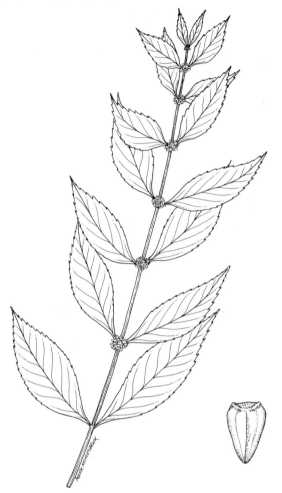

Fig. 269. *Lycopus virginicus* (Virginia water horehound). Habit. Nutlet (lower right).

This species is similar to *L. uniflorus*, both of which have calyx teeth less than 1 mm long. *Lycopus virginicus* differs by the absence of tubers, the mostly puberulent stems, and the included stamens.

4. **Mentha** L.—Mint

Perennial herbs with rhizomes or stolons; stems 4-angled; leaves opposite, simple, serrate; flowers in axillary or terminal spikes or heads, blue to lavender; calyx 4- or 5-lobed, actinomorphic or bilabiate; corolla 4-lobed, tubular below, usually bilabiate or nearly actinomorphic; stamens 4, exserted; ovary 4-lobed, superior; nutlets 4.

Most of the two hundred species in this genus are native to Europe, Asia, and Australia. Only the following adventive species may be found in shallow water in the central Midwest.

1. Leaves and stems usually pubescent; calyx tube 2–3 mm long ....1. *M. arvensis*
1. Leaves and stems glabrous, or with pubescence on the veins of the lower surface of the leaves; calyx tube 3–4 mm long.
  2. Leaves petiolate.. 2. *M. piperita*
  2. Leaves sessile... 3. *M. spicata*

1. **Mentha arvensis** L. Sp. Pl. 577. 1753. Fig. 270.
Mentha glabrata Vahl, Symb. Bot. 3:75. 1806.
Mentha arvensis L. f. *glabrata* (Benth.) S. R. Stewart, Rhodora 46:333. 1944.
Mentha arvensis L. var. *villosa* (Benth.) S. R. Stewart, Rhodora 46:333. 1944.
Mentha arvensis L. f. *glabra* (Benth.) S. R. Stewart, Rhodora 46:333. 1944.

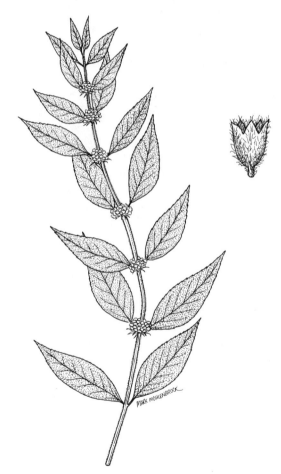

Perennial herbs with a strong, mint odor; stems erect, 4-sided, branched or unbranched, to 75 cm long, the angles retrorsely pubescent; leaves opposite, simple, lanceolate to ovate, acute at the apex, rounded or tapering to the base, serrate, glabrous or pubescent, to 10 cm long, to 6.5 cm wide; flowers crowded in axillary glomerules, subtended by bracts; bracts lanceolate to ovate, similar to the cauline leaves but smaller, much longer than the flowers; calyx tubular, 5-lobed, pubescent, 2–3 mm long; corolla pink to purple, bilabiate, 4–7 mm long; stamens 4, included; fruit a cluster of 4 nutlets, 1.5–1.7 mm long, smooth. July–September.

Wet soil, wet meadows, disturbed fens and marshes.

IA, IL, IN, KS, KY, MO, NE, OH (FACW).

Field mint.

Fig. 270. *Mentha arvensis* (Field mint). Habit. Calyx (upper right).

The strong mint odor is often smelled before the plant is actually seen in the field.

There is considerable variation in leaf shape and pubescence of stems. Typical var. *arvensis* has ovate leaves more or less rounded at the base and stems that are pubescent on both the angles and sides. If the pubescence is only on the angles, the plants may be called f. *glabra*. Plants with lanceolate leaves that taper to the base may be called var. *villosa*. If the leaves are glabrous or nearly so, the plants may be called f. *glabrata*.

2. **Mentha piperita** L. Sp. Pl. 576. 1753. Fig. 271.

Strongly aromatic perennial herbs from slender rhizomes; stems erect, 4-angled, usually branched, glabrous or nearly so, to 1 m tall; leaves opposite, simple, lanceolate to lance-ovate, acute at the apex, tapering or sometimes rounded at the base, serrate, glabrous or nearly so above, usually pubescence on the veins beneath, to 4.5 cm wide, the petioles up to 15 mm long, glabrous or nearly so; flowers crowded in terminal or sometimes interrupted spikes up to 8 cm long, up to 1.2 cm thick, subtended by lanceolate, acuminate bracts, the lowermost sometimes foliaceous; calyx 5-toothed, tubular below, 3–4 mm long, the teeth usually with spreading hairs and shorter than the glabrous tube; corolla 4-lobed, tubular below, pale blue to nearly white, 4–6 mm long; stamens 4, slightly exserted; nutlets 4, ovoid, usually slightly roughened. July–September.

Wet soil, marshes, disturbed fens.

IA, IL, IN, MO (OBL), KS, KY, NE, OH (FACW+).

Peppermint.

This native of Europe occasionally grows as an escape in wet situations. The oil that is extracted from the leaves is used commercially.

Mentha piperita differs from *M. arvensis* by its longer calyx tube and by its nearly glabrous leaves and stems.

3. **Mentha spicata** L. Sp. Pl. 576. 1753. Fig. 272.

Perennial herbs from rhizomes; stems erect, 4-sided, branched or unbranched, glabrous or pubescent, to 75 cm tall; leaves opposite, simple, lanceolate to ovate, acute at the apex, tapering to the base, serrate, glabrous on the upper surface, pubescent on the veins on the lower surface, with the odor of spearmint, to 8 cm long, to 4 cm wide, the petioles to 1.5 cm long; flowers in terminal and axillary spikes, bracteate, the bracts glabrous or pubescent, subulate, glandular; sepals 5, 3–4 mm long, campanulate below, glabrous, the lobes subulate, hirsute; petals 5, united below, 2-lipped, lavender; stamens 4, included; ovary superior, 4-lobed; nutlets 4. July–September.

Wet meadows, fens, around ponds; native to Europe.

IA, IL, IN, KY, MO, OH (FACW+), KS, NE (FACW).

Spearmint.

This species, an escape from cultivation, has the strong odor of spearmint. It also differs from *M. spicata* by its petiolate leaves.

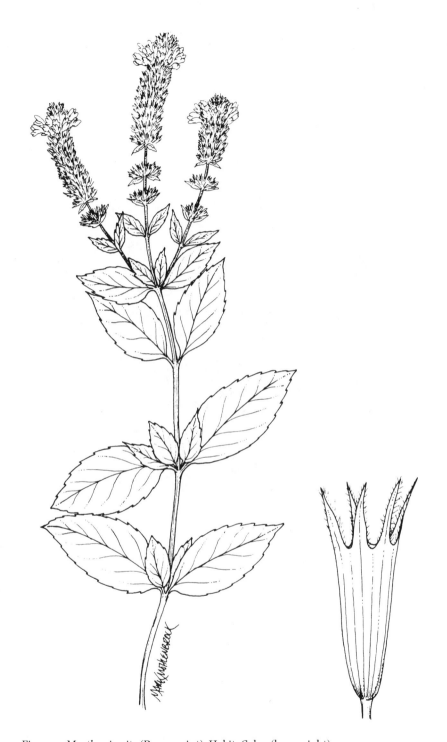

Fig. 271. *Mentha piperita* (Peppermint). Habit. Calyx (lower right).

Fig. 272. *Mentha spicata* (Spearmint). Habit. Flower (lower left).

5. **Physostegia** Benth.—False Dragonhead; Obedience Plant

Perennial herbs; stems 4-angled, erect; leaves opposite, simple; flowers in spikes or spikelike racemes, usually large and showy, subtended by bracts; calyx equally 5-parted, tubular below, usually 10-nerved; corolla bilabiate, much longer than the calyx; stamens 4, usually included; ovary 4-lobed; nutlets 4, usually triangular, smooth.

This genus consists of fifteen species, all in North America. Some of them may occur in shallow water.

1. At least some of the leaves clasping the stem.
    2. Leaves sharply serrate; corolla 2.5–3.5 cm long .... 1. *P. angustifolia*
    2. Leaves entire or undulate-dentate; corolla 1.0–1.5 cm long.
        3. Leaves linear to narrowly lanceolate, up to 1 cm wide .... 2. *P. intermedia*
        3. Leaves lanceolate to narrowly ovate, up to 2 cm wide .... 3. *P. parviflora*
1. None of the leaves clasping the stem.
    4. Leaves linear to narrowly lanceolate, stiff, up to 1 cm wide; spikes very slender, greatly interrupted ....1. *P angustifolia*
    4. Leaves lanceolate to ovate, not stiff, most of them at least 1.5 cm wide; spikes not interrupted .... 4. *P. virginiana*

1. **Physostegia angustifolia** Fern. Rhodora 45:462–463. 1943. Fig. 273.

Fig. 273. *Physostegia angustifolia* (Narrow-leaved false dragonhead). Habit.

Perennial herbs with a thickened rootstock; stems stiff, erect, to 1.5 m tall, glabrous; leaves opposite, simple, firm, appressed to the stem or ascending, linear to lanceolate, acute at the apex, or the lower ones obtuse, clasping at the base, to 12 cm long, to 1.5 (–2.0) cm wide, glabrous or nearly so, appressed-serrate, strongly reduced in size in the upper part of the stem; spikes to 30 cm long, the flowers subtended by lanceolate bracts up to 2/3 as long as the calyx; calyx 3–10 mm long, lengthening during fruiting, without stipitate glands; corolla lavender or whitish with purple spots, (15–) 20–30 mm long; nutlets ovoid, 2–3 mm long, brown. May–July.

Wet ditches, sloughs, along streams.

IL, MO (FAC), KS (FACW).

Narrow-leaved false dragonhead.

This species has the clasping leaves of *P. parviflora* but the large flowers of *P. virginiana*.

2. **Physostegia intermedia** (Nutt.) Gray, Proc. Am. Acad. 8:371. 1872. Fig. 274. *Dracocephalum intermedium* Nutt. Trans. Am. Phil. Soc. 5:187. 1833.

Perennial herbs with rhizomes; stems 4-angled, slender, erect, sometimes spongy at base if submersed, mostly unbranched, to 1.5 m tall, glabrous; leaves opposite, simple, linear-lanceolate to narrowly oblong, acute to acuminate at the apex, narrowed to the sessile and often clasping base, the lowermost leaves on petioles, entire or undulate-dentate, not stiff, 2–3 cm long, to 1 cm wide; flowers in terminal, very slender spikes, often remote, subtended by bracts about as long as the calyx; calyx tubular, with 5 equal teeth, 4–5 mm long, puberulent but not glandular; corolla bilabiate, lavender, 10–15 mm long; stamens 4; nutlets 4, triangular, brown, 2.0–2.5 mm long and nearly as broad. May–July.

Wet meadows, wet prairies, marshes.

KY, MO (FACW–).

Slender false dragonhead.

This species is distinguished by its narrow leaves that are entire or undulate-dentate with some of them clasping the stem.

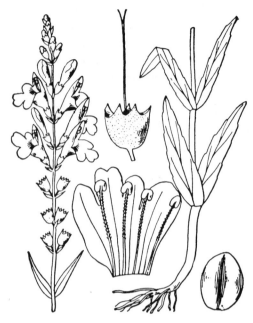

Fig. 274. *Physostegia intermedia* (Slender false dragonhead). Upper part of plant (left). Calyx (above center). Corolla cut open (below center). Lower leaves (next to right). Seed (lower right).

3. **Physostegia parviflora** Nutt. ex Benth. in DC. Prodr. 12:434. 1848. Fig. 275.

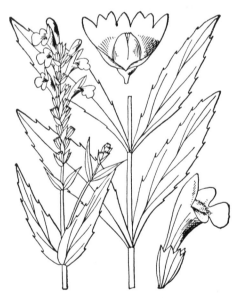

Fig. 275. *Physostegia parviflora* (Small-flowered false dragonhead). Upper part of plant (left). Calyx and fruit (above center). Leaves (below center). Flower (lower right).

Perennial herbs with a thickened rootstock or short rhizomes; stems 4-angled, to 75 cm tall, glabrous; leaves opposite, simple, spreading to ascending, lanceolate to narrowly ovate, obtuse to acute, sessile and some of them clasping at the base, to 10 cm long, to 2 cm wide, glabrous or nearly so, remotely toothed, the leaves on the upper part of the stem scarcely reduced in size from the lower leaves; spikes to 10 cm long, the flowers subtended by narrowly ovate bracts up to 1/2 as long as the calyx; calyx 4–7 mm long, with stipitate glands; corolla bilabiate, purple, 9–12 (–15) mm long; nutlets 4, ovoid, 2.0–3.5 mm long, brown. May–July.

Wet meadows, along streams.

IA, IL (FACW–), NE (NI).

Small-flowered false dragonhead.

This species has the smallest flowers of any *Physostegia* in the central Midwest. The flowers tend to be a darker purple than those of the other species.

4. **Physostegia virginiana** (L.) Benth. Lab. Gen. & Sp. 504. 1834. Fig. 276. *Dracocephalum virginianum* L. Sp. Pl. 594. 1753.

Perennial herbs from slender rhizomes; stems 4-angled, up to 1.5 m tall, glabrous; leaves opposite, simple, spreading to ascending, narrowly elliptic to lanceolate to oblanceolate, acute to acuminate at the apex, cuneate at the sessile but non-clasping base, to 15 cm long, to 4 cm wide, serrate, glabrous or nearly so, the uppermost sometimes strongly reduced; spikes up to 20 cm long, the narrowly ovate bracts about half as long as the calyx; calyx 5–12 mm long, without stipitate glands; corolla bilabiate, pink to rose to white with purple spots, 15–35 mm long; nutlets 4, ovoid, 2.5–3.5 mm long, brown. July–October.

Marshes, bogs, wet prairies, low woods.

IA, IL, IN, KS, MO, NE (FACW), KY, OH (FACU)

Obedience plant.

This species has the largest flowers and the largest leaves of any in the genus. The common name is derived from the flower that retains its position when moved.

Fig. 276. *Physostegia virginiana* (Obedience plant). Habit. Calyx (upper right).

6. **Pycnanthemum** Michx.—Mountain Mint

Perennial herbs; stems 4-angled; leaves opposite, simple, sometimes aromatic; flowers small, usually white with purple dots, borne in terminal and sometimes axillary cymes or heads; calyx with 5 equal or unequal lobes; corolla bilabiate, 4-lobed; stamens 4; ovary deeply 4-lobed; nutlets 4 in a cluster.

Approximately twenty species comprise this genus, all in North America.

Several species occur in uplands, with only the following sometimes found in shallow water.

1. Leaves not more than 3 times longer than broad; calyx with uniform teeth up to 1 mm long; bracts white-hairy .... 1. *P. muticum*
1. Leaves at least 3 times longer than broad; calyx bilabiate, with some of the teeth 2–3 mm long; bracts not white-hairy.
    2. Stems glabrous; larger leaves up to 3.0 (–5.5) mm wide, with 1 or 3 veins, not aromatic ....
    ....2. *P. tenuifolia*
    2. Stems pubescent; larger leaves (5–) 6–10 mm wide, with 5 or 7 veins, aromatic ....
    .... 3. *P. virginiana*

1. **Pycnanthemum muticum** (Michx.) Pers. Syn. Pl. 2:128. 1807. Fig. 277.
Brachystemon muticum Michx. Fl. Bor. Am. 1:6. 1803.
Koellia muticum (Michx.) Britt. Mem. Torrey Club 4:145. 1894.

Perennial herbs; stems 4-angled, rather stout, erect, to 1.2 m tall, short-pubescent to pilose; leaves opposite, simple, lance-ovate to ovate, acute to acuminate at the apex, rounded at the sessile or short-petiolate base, punctate, serrate, glabrous on the upper surface, glabrous or pubescent on the veins on the lower surface, to 6 cm long, 1.5–4.0 cm wide; flowers crowded in heads 8–20 mm across, forming a terminal corymb or cyme, the heads and cymes subtended by white-hairy ovate bracts; calyx equally 5-toothed, 3 mm long, pubescent in the upper half, the teeth up to 1 mm long; corolla bilabiate, 3.5–4.5 mm long, white with purple dots, pubescent on both surfaces; stamens 4; nutlets 4, oblongoid, brown, 1.0–1.2 mm long. July–September.

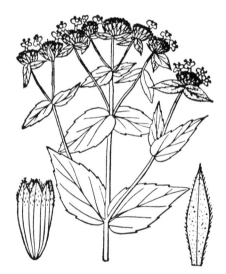

Fig. 277. *Pycnanthemum muticum* (Short-toothed mountain mint). Habit (center). Calyx (lower left). Bract (lower right).

Bogs, wet meadows, wet woods; also in drier habitats.

IL, MO (FAC), KY, OH (FACW).

Short-toothed mountain mint.

This is the only wetland *Pycnanthemum* in the central Midwest that has ovate leaves and white-hairy bracts.

2. Pycnanthemum tenuifolium Schrad. Hort. Gotting 10. 1824. Fig. 278.
Origanum flexuosum Walt. Fl. Carol. 165. 1788, misapplied.
Pycnanthemum flexuosum (Walt.) BSP. Prel. Cat. U. S. 42. 1888.

Fig. 278. *Pycnanthemum tenuifolium* (Narrow-leaved mountain mint). Habit. Flower (lower left).

Perennial herbs with slender rhizomes; stems 4-angled, slender, stiff, erect, to 85 cm tall, usually unbranched, glabrous; leaves opposite, simple, scarcely aromatic, linear to linear-lanceolate, acute at the apex, tapering to the sessile base, entire, 1- or 3-nerved, glabrous, punctate, to 4 cm long, 1.5–3.0 (–4.5) mm wide, often with short leafy branches in the axils; flowers crowded in heads 5–10 mm across, forming a terminal corymb subtended by appressed green bracts; calyx more or less unequally 5-toothed, 4–5 mm long, glabrous or with short whitish hairs, the teeth 2–3 mm long; corolla bilabiate, 3–4 mm long, white with purple dots, white-pubescent on the outer surface; stamens 4; nutlets 4, oblongoid, nearly black, 0.8–1.0 mm long. July–September.

Bogs, wet meadows, wet prairies, often in dry habitats.
IA, IL, IN, KS, MO, NE (FAC), KY, OH (FACW).
Narrow-leaved mountain mint.

This species is very similar to narrow-leaved forms of *P. virginianum*, differing by having only 1 or 3 veins per leaf. The leaves of *P. tenuifolium* are only slightly it at all aromatic.

3. **Pycnanthemum virginianum** (L.) Durand & Jackson ex Robins. & Fern. in Gray, Man. Bot., ed. 7, 703. 1908. Fig. 279.

Satureja virginiana L. Sp. Pl. 567. 1753.

Koellia virginiana (L.) MacM. Met. Minn. 452. 1852.

Perennial herbs with slender rhizomes; stems 4-angled, rather stout, rather stiff, pubescent on the angles, to 1 m tall; leaves opposite, simple, strongly aromatic, linear-lanceolate to lanceolate, acute to acuminate at the apex, tapering or rounded at the sessile or short-petiolate base, entire, 5- or 7-nerved, glabrous above, usually short-pubescent below, punctate, to 4 (–5) cm long, to 6 (–10) mm wide, often with short leafy branches in the axils, strongly aromatic; flowers crowded in heads 5–12 mm across, forming a terminal corymb subtended by stiff, appressed bracts; calyx more or less unequally 5-toothed, 3.5–4.5 mm long, short-pubescent and often glandular, the teeth 2–3 mm long; corolla bilabiate, 3–4 mm long, white with purple dots, pubescent on the outer surface; stamens 4; nutlets 4, triangular-oblongoid, brown, 0.8–1.0 mm long. July–September.

Calcareous fens, wet meadows, marshes.

IA, IL, IN, MO (FACW+), KS, NE (NI), KY, OH (FAC).

Virginia mountain mint.

This species differs from *P. tenuifolium* by its strongly aromatic leaves that have 5 or 7 veins.

Fig. 279. *Pycnanthemum virginianum* (Virginia mountain mint). Habit (center). Calyx (lower left). Flower cut open (lower right).

7. **Scutellaria** L.—Skullcap

Perennial herbs; stems 4-angled; leaves opposite, simple; flowers in racemes or solitary in the axils of the leaves; calyx bilabiate, the upper lip bearing a conspicuous tubercle; corolla bilabiate, tubular below, usually blue; stamens 4, concealed by the upper lip of the corolla; ovary superior, 4-lobed, nutlets 4, on a gynophore.

There are about 225 species in the genus, most of them in temperate regions of the world. The crest on the upper lip of the calyx is distinctive. Three species may be found in shallow standing water, particularly in marshes and fens, in the central Midwest.

1. Flower solitary in the leaf axils.. 1. *S. galericulata*
1. Flowers several in racemes.
   2. Flowers 20–25 mm long; leaves obtuse at the apex, entire .... 2. *S. integrifolia*
   2. Flowers 5–7 mm long; leaves acute at the apex, sharply serrate .... 3. *S. lateriflora*

1. **Scutellaria galericulata** L. Sp. Pl. 2:599. 1753. Fig. 280.
Scutellaria epilobiifolia Lam. Bull. Bot. Gen. 300. 1832.

Perennial herbs with slender rhizomes; stems 4-angled, erect, to 60 cm tall, with recurved hairs; leaves opposite, simple, lanceolate to lance-ovate, acute to acuminate at the apex, rounded or subcordate at the base, crenate or serrate, glabrous to puberulent, at least on the lower surface, to 4.5 cm long, to 2 cm wide, on petioles up to 4 mm long; flower solitary in the axils of the leaves on pedicels 1–3 mm long; calyx bilabiate, 3–5 mm long, short-pubescent; corolla bilabiate, blue, to 2.2 cm long; stamens 4; ovary 4-lobed; nutlets 4 in a cluster. June–August.

Wet habitats, particularly marshes and fens.

IA, IL, IN, KY, MO, OH (OBL), KS, NE (FACW+).

Large skullcap.

This is the only wetland *Scutellaria* with large flowers borne singly in the axils of the leaves. If our plants differ from the European *S. galericulata*, they should be called *S. epilobiifolia*.

2. **Scutellaria integrifolia** L. Sp. Pl. 59. 1753. Fig. 281.

Perennial herbs with slender rhizomes; stems 4-angled, erect, rather slender, branched or unbranched, densely and minutely pubescent, to 75 cm tall; leaves opposite, simple, oblong to ovate, obtuse at the apex, tapering or somewhat rounded at the base, entire, or the lowermost sometimes crenate, densely and minutely pubescent, punctate, 1.5–3.5 cm long, 1.0–2.5 cm wide, sessile or on pubescent petioles; flowers in terminal racemes or panicles, on pubescent pedicels 3–4 mm long; calyx bilabiate, 2.5–3.0 mm long, pubescent, usually with stipitate glands , becoming conspicuously veiny and up to 10 mm long in fruit; corolla bilabiate, blue or sometimes pinkish, 20–25 mm long; stamens 4, not exserted; nutlets 4, ovoid to orbicular, 1.0–1.2 mm in diameter, dark brown or black, tuberculate. May–August.

Bogs, wet meadows, wet woods.

KY, OH (FACW), MO (FACW–).

Hyssop-leaved skullcap.

This is the only wetland *Scutellaria* in the central Midwest with obtuse leaves and downy pubescent stems.

Fig. 280. *Scutellaria galericulata* (Large skullcap). Habit. Flower (lower left).

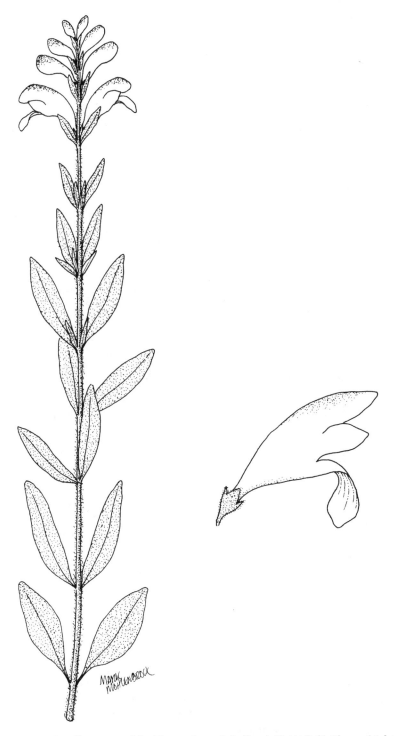

Fig. 281. *Scutellaria integrifolia* (Hyssop-leaved skullcap). Habit (left). Flower (right).

3. **Scutellaria lateriflora** L. Sp. Pl. 598. 1753. Fig. 282.

Perennial herbs with slender rhizomes and stolons; stems 4-angled, erect, usually branched, glabrous or somewhat pubescent on the angles, to 1.2 m tall; leaves opposite, simple, lance-ovate to ovate, acute to acuminate at the apex, rounded or truncate at the base, sharply serrate, to 10 cm long, to 5 cm wide, glabrous or pubescent on the veins and margins; flowers often in 1-sided terminal and axillary racemes, on glabrous pedicels to 2 mm long; calyx bilabiate, 1.8–2.2 mm long, pubescent, sometimes with glandular hairs, becoming as much as 4 mm long in fruit; corolla bilabiate, blue or rarely white, 5–7 mm long; stamens 4, not exserted; nutlets 4, suborbicular, about 1 mm in diameter, tan, pitted. July–September.

Calcareous fens, swampy woods, wet meadows.

IA, IL, IN, MO (FACW), KS, KY, NE, OH (FACW+).

Mad-dog skullcap.

This is the only wetland *Scutellaria* in the central Midwest with small flowers only 5–7 mm long.

8. **Stachys** L.—Hedge Nettle

Perennial herbs (in our area) with thickened rhizomes; stems 4-angled, glabrous or bristly hairy; leaves opposite, simple; flowers purplish, in whorls arranged in spikes or racemes; calyx 5-toothed, tubular below, the teeth and the tube about equal in length; corolla bilabiate, 4- or 5-cleft; stamens 4, usually not exserted; ovary superior, deeply 4-lobed, the style bifid; nutlets 4 in a cluster, obovoid to oblongoid.

Fig. 282. *Scutellaria lateriflora* (Mad-dog skullcap). Habit (right). Flower (lower left).

There are approximately 175 species in this genus, found in all continents except Australia and Antarctica.

Stachys differs from other members of the Lamiaceae with bilabiate corollas and four stamens by its nearly equal calyx teeth and by the corolla tube not longer than the calyx.

1. Sides of the stems pubescent; lower leaf surfaces pubescent throughout; calyx pubescent.
  2. Calyx pilose with gland-tipped hairs; leaves sessile..4. *S. palustris*
  2. Calyx hirsute and pilose, eglandular; leaves sessile or petiolate.
    3. Leaves petiolate, the petioles at least 10 mm long ....4. *S. palustris*
    3. Leaves sessile or nearly so..5. *S. pilosa*
1. Sides of the stems glabrous or sometimes merely setose; lower leaf surfaces glabrous, or pubescent only on the veins; calyx glabrous, or occasionally with a few setae, or sometimes puberulent in *S. aspera*.
  4. Leaves obtuse to subacute, firm, entire or with low teeth; nutlets less than 2 mm long.
    5. Stems glabrous or nearly so; leaves linear to narrowly oblong, entire or nearly so; calyx glabrous or with setae on the angles..3. *S. hyssopifolia*
    5. Stems retrorsely hispid; leaves oblong, with low teeth; calyx puberulent, sometimes with setae on the angles .... 1. *S. aspera*
  4. Leaves acuminate, not firm, sharply serrate; nutlets 2 mm long or longer.
    6. Calyx and leaves glabrous; some or all of the petioles 10 mm long or longer ....
    .... 6. *S. tenuifolia*
    6. Calyx usually with a few setae on the angles; leaves sessile or nearly so, hispid.............
    ....2. *S. hispida*

1. **Stachys aspera** Michx. Fl. Bor. Am. 2:5. 1803. Fig. 283.
Stachys palustris L. var. *aspera* (Michx.) Gray, Man. ed. 2, 317. 1856.
Stachys asperifolia Michx. var. *ambigua* Gray Syn. Fl. 2, 1:387. 1878.
Stachys ambigua (Gray) Britt. Mem. Torrey Club 5:285. 1894.

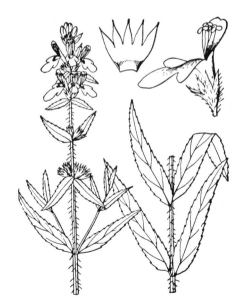

Perennial herbs with thick, whitish rhizomes; stems 4-angled, slender, branched or unbranched, retrorsely hispid on the angles, to 75 cm tall; leaves opposite, simple, narrowly oblong, obtuse to subacute at the apex, rounded or tapering to

Fig. 283. *Stachys aspera* (Hairy hyssop-leaved hedge nettle). Upper part of plant (left). Leaves (lower right). Calyx (upper center). Flower (upper right).

the base, glabrous or sparsely hairy on the upper surface, serrate, hirsute on the lower surface, to 7 cm long, to 1.5 cm wide, the petioles up to 8 mm long; flowers in interrupted spikes, arranged in whorls, each flower subtended by broadly lanceolate to ovate bracts; calyx with 5 more or less equal lobes and tubular below, the teeth and tube each about 3–4 mm long, the teeth glabrous, the tube pubescent and sometimes setose on the angles; corolla bilabiate, pink with purple speckles, 10–15 mm long; stamens 4; nutlets in clusters of 4, ellipsoid to nearly globose, black, 1.5–2.2 mm long. June–September.

Marshes, wet prairies, swamps.

IA, IL, IN, MO (FACW+), OH (FACW).

Hairy hyssop-leaved hedge nettle.

This species is distinguished by its glabrous calyx teeth, stems that are pubescent only on the angles, and petioles 8 mm long or less.

2. **Stachys hispida** Pursh, Fl. Am. Sept. 407. 1814. Fig. 284.
Stachys tenuifolia Willd. var. *hispida* (Pursh) Fern. Rhodora 45:468. 1943.
Stachys tenuifolia Willd. var. *platyphylla* Fern. Rhodora 45:468–469. 1943.

Perennial herbs with thickened rhizomes; stems 4-angled, branched or unbranched, to 1 m tall, glabrous or bristly pubescent on the angles; leaves opposite, simple, narrowly lanceolate to narrowly ovate, acuminate at the apex, rounded or tapering to a sessile or short-petiolate base, serrate, usually hispid on the veins beneath, to 15 cm long, to 6 cm wide; flowers in whorls arranged in terminal, interrupted racemes, with 6–8 flowers per whorl, each flower subtended by ciliate bracts; calyx with 5 nearly equal teeth, tubular below, the glabrous teeth 2–3 mm long, the tube 3–4 mm long, usually with a few setae on the angles; corolla bilabiate, pink with purple spots, 10–12 mm long; stamens 4; nutlets in clusters of 4, obovoid, dark brown, rugose, 2.0–2.8 mm long. June–September.

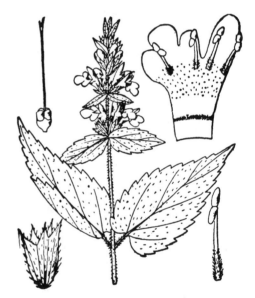

Fig. 284. *Stachys hispida* (Hispid hedge nettle). Habit (center). Calyx (lower left). Ovary (upper left). Corolla cut open (upper right). Stamen (lower right).

Swampy woods, wet meadows, marshes.

IA, IL, IN, MO (FACW+), KY, OH (OBL), KS, NE (not listed by U.S. Fish and Wildlife Service).

Hispid hedge nettle.

This species is distinguished by the sides of the stem glabrous and the angles setose, its acuminate, serrate leaves, its calyces with setae on the angles, its often sessile leaves, its ciliate bracts, and its nutlets at least 2 mm long.

Plants with narrowly lanceolate leaves to 4 cm wide are var. *hispida*. Plants with narrowly ovate leaves to 6 cm wide are var. *platyphylla*.

3. **Stachys hyssopifolia** Michx. Fl. Bor. Am. 2:4. 1803. Fig. 285.

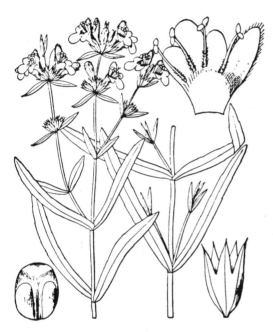

Perennial herbs with thick, whitish rhizomes; stems 4-angled, slender, branched or unbranched, glabrous, to 75 cm tall; leaves opposite, simple, firm, linear to narrowly oblong, obtuse to subacute at the apex, rounded or tapering to the base, usually entire, pale green, glabrous, except sometimes with a few hairs at the base, to 7 cm long, up to 1 cm wide, sessile or very short-petiolate; flowers in interrupted spikes, arranged in whorls, each flower subtended by linear bracts; calyx with 5 more or less equal teeth and tubular below, the teeth and tube each about 3–4 mm long, glabrous, or with setae on the angles; corolla bilabiate, pink with purple speckles, 10–15 mm long; stamens 4; nutlets in clusters of 4, ellipsoid to nearly glabrous, black, 1.5–2.0 mm long. June–October.

Fig. 285. *Stachys hyssopifolia* (Smooth hyssop-leaved hedge nettle). Habit (next to left). Leaves (next to right). Seed (lower left). Calyx (lower right). Flower cut open (upper right).

Bogs, shores of lakes and ponds.

IL, IN, MO (FACW+).

Smooth hyssop-leaved hedge nettle.

This mostly coastal plain species is distinguished by its firm, pale green, narrow leaves and its calyx teeth about as long as the calyx tube. Its usually linear leaves and its glabrous stems are also distinctive.

4. **Stachys palustris** L. Sp. Pl. 580. 1753. Fig. 286.

Perennial herbs from slender rhizomes and tubers; stems 4-angled, erect, usually unbranched, pubescent on the sides and angles, to 1 m tall; leaves opposite, simple, lanceolate, acute at the apex, tapering to the base, serrate, pubescent on

Fig. 286. *Stachys palustris* (Woundwort). Habit (left). Section of stem (right).

both surfaces, to 2 cm wide, sessile or on petioles at least 10 mm long; flowers sessile in the axils of the leaves or in whorls in terminal spikes, usually 6 flowers per whorl; calyx with 5 nearly equal teeth, tubular below, glandular-pubescent and hirsute or sometime eglandular, the teeth slightly shorter than the tube, the tube 3.5–5.0 mm long; corolla bilabiate, to 2 cm long, purple; stamens 4, included; nutlets 4, ovoid. July–September.

Wet meadows, wet prairies, swampy woods, floodplain forests.

IL, OH (OBL).

Woundwort.

This species differs from other members of the genus except *S. pilosa* by the sides of the stems pubescent. It differs from the sessile-leaved *S. pilosa* by its eglandular calyx. Plants with leaves sessile or nearly so and with glandular hairs on the calyx are var. *palustris*. Plants with leaves on petioles at least 10 mm long and with eglandular hairs on the calyx are var. *phaneropoda.*

Stachys palustris is a native of Europe.

5. **Stachys pilosa** Nutt. Journ. Acad. Nat. Sci. Phila. 7:48. 1834. Fig. 287.
Stachys palustris L. var. *homotricha* Fern. Rhodora 10:85. 1908.
Stachys palustris L. var. *nipigonensis* Jennings, Journ. Wash. Acad. Sci. 10:459–460. 1920.
Stachys palustris L. var. *pilosa* (Nutt.) Fern. Rhodora 45:475. 1943.
Stachys palustris L. var. *homotricha* Fern. Rhodora 10:85. 1908.

Perennial herbs from slender rhizomes and tubers; stems 4-angled, erect, usually unbranched, pubescent on the sides and angles, to 1 m tall; leaves opposite, simple, narrowly lanceolate, oblong to oval, obtuse to acute to acuminate at the apex, tapering or somewhat rounded at the base, serrate, pubescent on both surfaces, 1.5–5.0 cm wide, sessile or nearly so; flowers sessile in the axils of the leaves or in whorls in terminal spikes, usually 6 flowers per whorl; calyx with 5 nearly equal teeth, tubular below, hirsute, eglandular, the teeth slightly shorter than the tube, the tube 3.5–5.0 mm long; corolla bilabiate, to 2 cm long, pink or pale purple; stamens 4, included; nutlets 4, ovoid. July–September.

Wet meadows, wet prairies, swampy woods, floodplain woods.

IA, IL, IN, KS, MO, NE, OH (OBL), as *S. palustris.*

Marsh hedge nettle.

Plants with their largest leaves oblong to oval with a subacute or obtuse apex are var. *pilosa.* Those plants with lanceolate, acuminate leaves and with dense long hairs on the sides of the stems are var. *homotricha.* Plants with lanceolate, acuminate leaves and the angles of the stems retrorsely long-hirsute are var. *nipigonensis.*

6. **Stachys tenuifolia** Willd. Sp. Pl. 3:100. 1801. Fig. 288.

Perennial herbs with thickened rhizomes; stems 4-angled, branched or unbranched, to 1 m tall, glabrous or bristly pubescent on the angles; leaves opposite, simple, lanceolate to ovate, acuminate at the apex, rounded or tapering to the petiolate base, serrate, glabrous on both surfaces, to 10 cm long, to 6 cm wide, the petioles 1–2 cm long; flowers in whorls arranged in terminal, interrupted racemes,

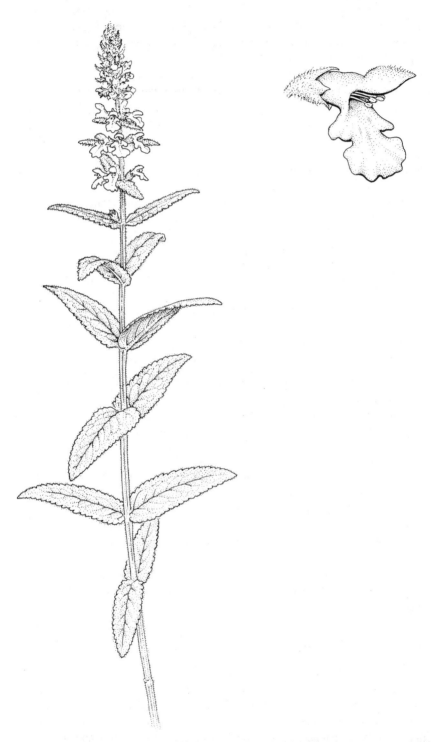

Fig. 287. *Stachys pilosa* (Marsh hedge nettle). Habit (left). Flower (right).

with 6–8 flowers per whorl, each flower subtended by usually eciliate bracts; calyx with 5 nearly equal teeth, tubular below, the glabrous teeth 2–3 mm long, about half as long as the tube, glabrous or bristly-pubescent; corolla bilabiate, pink with purple spots, 10–12 mm long; stamens 4; nutlets in clusters of 4, obovoid, dark brown, rugose, 2.0–2.8 mm long. June–September.

Marshes, swamps, fens, wet meadows, wet woods.

IA, IL, IN, MO (OBL), KS, NE (FACW), KY, OH (FACW+).

Stalked hedge nettle.

This species is distinguished from other wetland species of *Stachys* in the central Midwest except *S. palustris* var. *phaneropoda* by its long-petiolate leaves. It differs from this latter plant by its calyx teeth that are about half as long as the calyx tube. In *S. palustris* var. *phaneropoda*, the calyx teeth and the calyx tube are about equal in length.

9. **Teucrium** L.—Germander; Wood Sage

Annual or perennial herbs; stems 4-angled; leaves opposite, simple, usually toothed; flowers in racemes (in our area), subtended by well developed bracts; calyx campanulate, 5-parted, the lobes equal or unequal; corolla 5-lobed, all the lobes forming a lower lip, the upper lip absent, tubular below; stamens 4; ovary superior, 4-lobed; nutlets 4.

Approximately one hundred species comprise the genus, all in temperate parts of the world. This is the only genus of mints in North America in which the upper lip of the corolla is completely missing.

Only the following plant occurs in wet areas in the central Midwest.

1. **Teucrium canadense** L. var. **virginianum** (L.) Farw. Pap. Mich. Acad. Sci. 1:97. 1923. Fig. 289.
Teucrium virginianum L. Sp. Pl. 564. 1753.
Teucrium occidentale Gray, Syn. Fl. 2:349. 1878.
Teucrium boreale Bickn. Bull. Torrey Club 28:171. 1901.
Teucrium canadense L. var. *occidentale* (Gray) McClintock & Epling, Brittonia 5:499. 1945.
Teucrium canadense L. var. *boreale* (Bickn.) Shinners, Sida 1:183. 1963.

Perennial herbs with slender rhizomes; stems 4-angled, erect, to 1 m tall, softly gray-pubescent; leaves opposite, simple, ovate-lanceolate, acute to acuminate at the apex, more or less rounded at the base, crenate to dentate, to 10 cm long, to 6 cm wide, green above, gray below, with soft gray appressed or spreading pubescence on the lower surface, on petioles up to 1.5 cm long; racemes mostly terminal, to 15 cm long, the flowers subtended by linear bracts, with at least the lower bracts longer than the flowers; calyx 5-lobed, 2-lipped, 7–10 mm long, with or without glandular hairs; corolla 5-lobed on the lower lip, with the upper lip absent, pink to purple, 12–18 mm long; stamens 4; ovary shallowly 4-lobed; nutlets 4 in a cluster. June–September.

Most wetland habitats.

IA, IL, IN, KS, KY, MO, NE, OH (FACW–).

Wood sage; American germander.

The soft pubescence which gives the lower surface of the leaves a distinctive gray appearance is diagnostic.

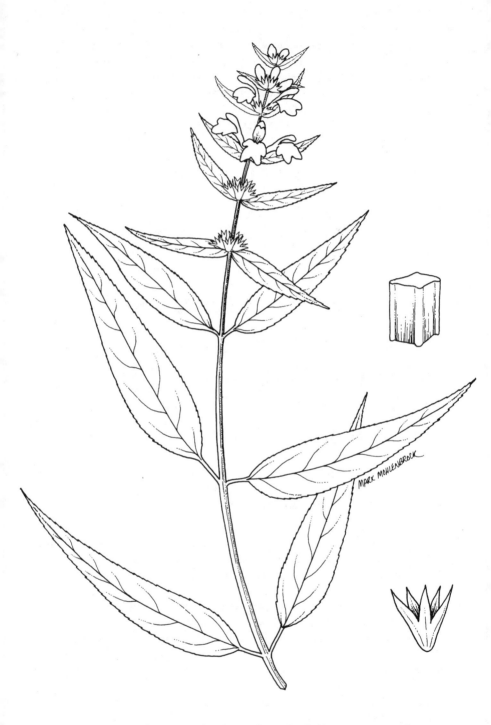

Fig. 288. *Stachys tenuifolia* (Stalked hedge nettle). Habit (left). Cross-section of stem (upper right). Calyx (lower right).

Fig. 289. *Teucrium canadense* (Wood sage). Habit. Calyx (lower left).

78. LAURACEAE—LAUREL FAMILY

Trees or shrubs, usually aromatic; leaves simple, alternate, without stipules; inflorescence various; flowers perfect or unisexual, actinomorphic, often aromatic; sepals 4–6, united below or sometimes free; petals 0; stamens usually 9 or 12, sometimes some of them reduced to staminodia, attached to the calyx; ovary superior, 1- to 2-locular, with 1 ovule in each locule; fruit a drupe or berry.

This family consists of about forty genera and approximately two thousand species, mostly in the tropics. Only the following genus is represented in shallow water in the central Midwest.

1. **Lindera** Thunb.—Spicebush

Aromatic shrubs (in our area); leaves alternate, simple, entire, deciduous (in our area); flowers unisexual, the plants dioecious; sepals 6, free; petals 0; stamens 9; pistillate flowers with 12–18 staminodia; ovary superior, 2-locular; fruit a drupe.

About one hundred species comprise this genus, most of them in Asia.

1. Leaves tapering to the base; petioles 6–12 mm long .... 1. *L. benzoin*
1. Leaves rounded or subcordate at the base; petioles 2–6 mm long .... 2. *L. melissaefolium*

1. **Lindera benzoin** (L.) Blume, Mus. Bot. Ludg. 1:324. 1851. Fig. 290.
Laurus aestivalis L. Sp. Pl. 370. 1753.
Laurus benzoin L. Sp. Pl. 371. 1753.
Benzoin aestivale (L.) Nees, Syst. Laurin. 495. 1836.
Benzoin benzoin (L.) Coulter, Mem. Torrey Club 5:164. 1894.
Benzoin aestivale (L.) Nees var. *pubescens* Palmer & Steyerm. Ann. Mo. Bot. Gard. 22:545. 1939.
Lindera benzoin (L.) Blume var. *pubescens* (Palmer & Steyerm.) Rehder, Journ. Arn. Arb. 20:412. 1939.

Shrubs to 5 m tall, much branched; bark smooth, brown; branchlets dark brown, glabrous or sometimes pubescent; leaves alternate, simple, elliptic to

Fig. 290. *Lindera benzoin* (Spicebush). Habit in fruit (left). Flower (right).

oblong-obovate, acute to short-acuminate at the apex, cuneate at the base, entire, glabrous or less commonly pubescent beneath, pale on the lower surface, to 12 cm long, up to half as wide, the petioles 6–12 mm long, glabrous or pubescent; flowers in nearly sessile clusters from nodes of the previous season; flowers yellow, aromatic, up to 3 mm across, appearing before the leaves; sepals 6, free, about 1.5 mm long, linear to linear-lanceolate; petals 0; stamens 9, as long as or a little longer than the sepals; drupes red, ellipsoid, up to 1 cm long, up to 6 mm in diameter. March–April.

Rich woods, wet ground, rarely in standing water.

IA, IL, IN, KS, KY, MO, NE, OH (FACW–).

Spicebush.

Although this is a common shrub of wetlands, it is not often found in standing water. On rare occasions, specimens may be found with hairy leaves, twigs, and petioles. These closely resemble the rare *L. melissaefolium* but differ by their cuneate leaves and longer petioles. The pubescent specimens may be called var. *pubescens.*

All parts of the plant are aromatic, particularly the leaves when crushed. The red berries ripen in August and September.

2. **Lindera melissaefolium** (Walt.) Blume, Mus. Bot. Ludg. 1:324. 1857. Fig. 291.
Laurus melissaefolia Walt. Fl. Carol. 134. 1788.
Benzoin melissaefolium (Walt.) Nees, Syst. 494. 1836.

Shrubs to 3 m tall; twigs gray or brownish, densely pubescent; leaves alternate, simple, entire, ovate-lanceolate, elliptic, or oblong, acute to acuminate at the apex, rounded or subcordate at the base, densely pubescent on both surfaces, to 12 cm long, to 3 cm wide, on pubescent petioles 2–6 mm long; flowers in clusters, yellow, aromatic, up to 3 mm across, appearing before the leaves, on pedicels about equaling the calyx; sepals 6, free, linear to linear-lanceolate, about 1.5 mm long; petals 0; stamens 9; drupes red, ellipsoid, 5–10 mm long. February–March.

Standing water in shallow ponds.

MO (OBL).

Pondberry.

This species, whose range is primarily in the southeastern United States, reaches our range only in southeastern Missouri where it grows in a shallow pond. Because of its rarity throughout its range, it is listed as Federally Threatened by the U.S. Fish and Wildlife Service.

Hairy forms of *Lindera benzoin* resemble this species, but the leaves of *L. benzoin* taper to the base and have longer petioles.

79. LEITNERIACEAE—CORKWOOD FAMILY

Dioecious shrubs or small trees; leaves alternate, simple, without stipules; flowers unisexual, axillary, opening before the leaves expand; staminate flowers many in catkins; pistillate flowers few; calyx 0; corolla 0; stamens 3–12; ovary superior, 1-locular, with 1 ovule; fruit a drupe.

Only the following genus comprises the family.

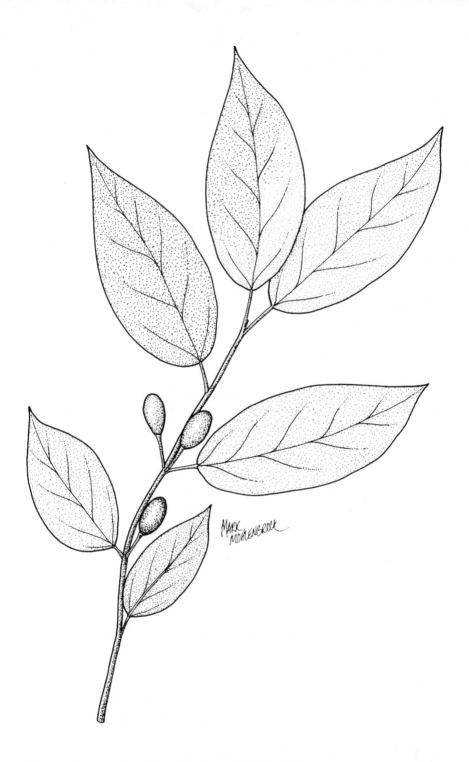

Fig. 291. *Lindera melissaefolium* (Pondberry). Habit in fruit.

1. **Leitneria** Chapm.—Corkwood

Only the following species is in the genus.

1. **Leitneria floridana** Chapm. Fl. South. U. S. 427–428. 1860. Fig. 292.

Shrubs or small trees to 7 m tall, with extensive subterranean offshoots; bark red-brown, smooth, with conspicuous light-colored lenticels; leaves alternate, simple, entire, oblong to elliptic to obovate, acute at the apex, tapering to the base, glabrous on the upper surface, silky-hairy, at least on the veins, on the lower surface, often silky-hairy throughout when young, to 17 cm long, to 5 cm wide, on slender petioles up to 4 cm long; flowers borne before the leaves appear from wood of the previous season, the male and female flowers in catkins, usually on different plants; staminate catkins erect to ascending, becoming pendulous, brown, to 5 cm long, to 1.5 cm wide, without a perianth, with (5–) 10–12 stamens, subtended by small, ovate scales; pistillate catkins erect, reddish, to 2 cm long, to 1 cm wide, without a perianth, with a solitary flower in the axils of 2 bractlets and with about 4 additional bractlike segments and a superior ovary; drupes erect, ellipsoid, brown, to 1.5 cm long. March–April.

Swampy woods, wet ditches.

MO (OBL).

Corkwood.

This species, primarily of the southeastern United States, grows in a few wet ditches and swampy woods in the bootheel of Missouri. Its wood is very light, being lighter than cork.

The smooth twigs with conspicuous lenticels and the leaves which somewhat resemble the leaves of pawpaw are distinctive.

80. LENTIBULARICEAE—BLADDERWORT FAMILY

Aquatic or wetland herbs; stems rooting or free floating, simple or much dissected; inflorescence scapose, racemose, perfect, zygomorphic, with or without subtending bracts; calyx 2- to 5-cleft, united below; corolla usually bilabiate, usually with a spur at back of flower; stamens 2, attached to the base of the corolla; ovary superior, 1-locular; fruit a capsule, with numerous seeds.

About five genera and two hundred species comprise this family, although *Utricularia* may be divided into several genera based on characters of the bracts and the fruits. The family is found worldwide. Only the following genus occurs in the central Midwest.

1. **Utricularia** L. Bladderwort

Aquatic or sometimes creeping herbs; stems unbranched or highly divided into capillary or filiform segments, often bearing bladders which buoy the plant to the water surface, or sometimes with whorls of inflated floats; flowers 1–several on leafless scapes, with or without bracts; calyx 2-lipped; corolla bilabiate, zygomorphic; stamens 2; ovary superior, 1-locular; fruit a capsule with many seeds.

This genus has no roots. The finely divided structures apparently are the stems, rather than the leaves. Most of the plants bear the bladders that are specialized structures designed to trap minute organisms, usually crustaceans. The plant is

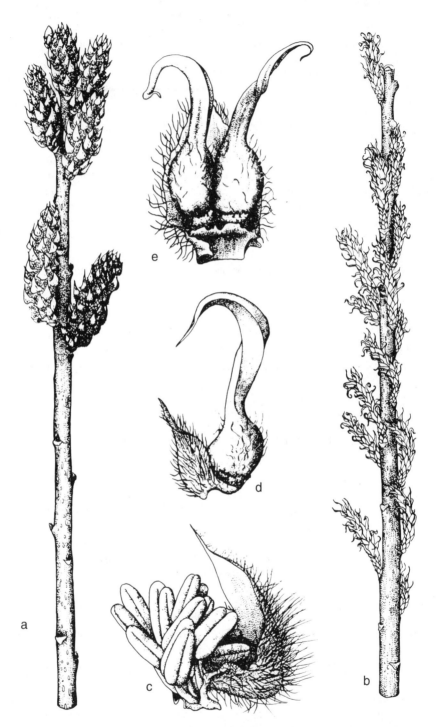

Fig. 292. *Leitneria*   a. Twig with staminate catkins.   c. Staminate flower with bract.
floridana (Corkwood).   b. Twig with pistillate catkins.   d, e. Pistillate flowers with bract.

able to break down the bodies of the trapped organisms and utilize the nitrogenous material derived from them.

Some species of *Utricularia* are free floating in the water. In other aquatic species, the lower part of the plants is buried in the mucky bottoms of the ponds or streams. In terrestrial species, most of the stems are buried just beneath the surface of wet sand or mud.

Plants that are usually terrestrial and have stems that creep in mud or wet sand are *U. cornuta, U. resupinata,* and *U. subulata.* Plants that are aquatic but usually have stems that creep in mud at the base of the plants include *U. gibba, U. fibrosa, U. minor, U. intermedia,* and *U. ochroleuca.* Plants that are aquatic and have whorls of inflated floats are *U. inflata* and *U. radiata.* Plants with purple rather than yellow flowers are *U. purpurea* and *U. resupinata.* Plants with the divisions of the stems flat instead of terete are *U. minor, U. intermedia,* and *U. ochroleuca.*

1. Plants usually terrestrial, the stems creeping in mud or wet sand.
  2. Flowers purple; bract one, tubular.. 11. *U. resupinata*
  2. Flowers yellow; bracts 2 or more, flat.
    3. Calyx enclosing the fruit; a pair of bractlets as well as bracts at base of pedicels ....
    .... 1. *U. cornuta*
    3. Calyx not enclosing the fruit; bractlets absent .... 12. *U. subulata*
1. Plants aquatic, either free floating or creeping in mud at base of plant in water.
  4. Branches in whorls or opposite; lobes of lower lip of corolla saccate; flowers purple..........
  .... 9. *U. purpurea*
  4. Branches alternate or absent; no corolla lobes saccate; flowers yellow.
    5. Inflated floats in a whorl near base of flowering stalk .... 10. *U. radiata*
    5. Inflated floats absent.
      6. Divisions of stem flat.
        7. Apex of stem segments entire; bladders borne on the stem ....7. *U. minor*
        7. Apex of stem segments serrulate; bladders borne on separate branches.
          8. Branches bearing bladders without lateral divisions; spur of flower as long as
          lower lip of corolla ....5. *U. intermedia*
          8. Branches bearing bladders with small dissected divisions; spur of flower about
          half as long as lower lip of corolla ....8. *U. ochroleuca*
      6. Divisions of stem terete.
        9. Stems free floating.
          10. Flowers 2–5 per scape, without bracts; cleistogamous flowers present..............
          ....3. *U. geminiscapa*
          10. Flowers 6 or more per scape, with bracts; cleistogamous flowers absent............
          ....6. *U. macrorhiza*
        9. Stems creeping near bottom of plant in shallow water.
          11. Spur of flower stout, much shorter than the lower lip of the corolla; mature stems
          with 2 segments...4. *U.gibba*
          11. Spur of flower slender, nearly as long as the lower lip of the corolla; mature stems
          with 3 segments ....2. *U. fibrosa*

1. **Utricularia cornuta** Michx. Fl. Bor. Am. 1:12. 1803. Fig. 293.
Stomoisia cornuta (Michx.) Raf. Fl. Tellur. 4:108. 1838.

Mostly terrestrial plants, the stems usually creeping just beneath the surface of mud or wet sand, much divided into filiform segments; bladders numerous, falcate, about 1 mm wide; scape filiform, wiry, (1–) to 3–5 (–8) flowered, to 35 cm tall;

flowers on ascending pedicels 1–2 mm long, subtended by one oblong, flat, acute to acuminate bract and two bractlets; calyx bilabiate, green; corolla yellow, bilabiate, 15–25 mm long, the spur long-pointed, to 14 mm long, positioned downward; capsules globose, beaked, 2.5–3.5 mm in diameter, with several seeds; seeds about 0.1 mm long, reticulate. July–August.

Sandy borders of ponds, sometimes under water, particularly after a heavy rain.

IL, IN, OH (OBL).

Horned bladderwort.

This mostly terrestrial species is distinguished by its long, slender spur of the corolla. The fruiting capsule is beaked.

Fig. 293. *Utricularia cornuta* (Horned bladderwort). Habit. Capsule (above center). bladder (far left). Stamens (center right). Bracts (lower right).

2. **Utricularia fibrosa** Walt. Fl. Carol. 64. 1788. Fig. 294.

Free floating aquatic herbs or creeping in mud in pond at bottom of plant; stems divided into usually three filiform segments; bladders small, numerous; scapes very slender, wiry, to 35 cm long, (1–) 2- to 6-flowered; flowers on slender ascending pedicels to 15 mm long, subtended by a 3-lobed bract; calyx bilabiate, green; corolla yellow, bilabiate, 8–10 mm long, the emarginate stout spur nearly as long as the lower lip of the corolla; capsules globose, 3–4 mm in diameter, with several seeds. June–August.

Ponds.

KY (OBL).

Bladderwort.

This coastal plain species is distinguished by its few, small flowers and its 3-lobed bracts. Some botanists call this plant *U. striata* LeConte.

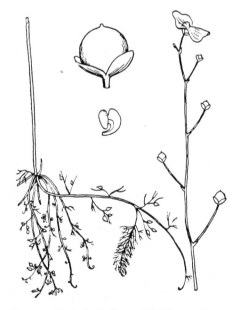

Fig. 294. *Utricularia fibrosa* (Bladderwort). Habit. Capsule (above). Bladder (center). Flowering branch (far right).

3. **Utricularia geminiscapa** Benj. Linnaea 20:305. 1847. Fig. 295.

Fig. 295. *Utricularia geminiscapa* (Hidden-fruited bladderwort). Habit.

Free floating aquatic herbs; stems sparsely branched into alternate, filiform, terete segments, with few to many bladders; scapes slender, to 12 cm tall, 2- to 5-flowered; flowers on strongly ascending pedicels up to 6 mm long, without bracts; calyx more or less bilabiate, greenish; corolla yellow, bilabiate, 6–8 mm long, the obtuse spur a little shorter than the lower lip of the corolla; cleistogamous flowers present, on short pedicels; capsules globose, with numerous seeds. July–August.

Shallow water.

IA, IN, OH (OBL).

Hidden-fruited bladderwort.

This rather delicate bladderwort differs by its terete branches, free floating habit, bractless flowers only 2–5 per scape, and the presence of some cleistogamous flowers.

4. **Utricularia gibba** L. Sp. Pl. 18. 1753. Fig. 296.

Stems creeping in mud at bottom of pond, the branches alternate, divided into usually two segments, the segments capillary to filiform, bearing a few bladders; scapes to 8 cm tall, rather slender, bearing 1–3 flowers; flowers on slender, ascending pedicels up to 8 mm long, subtended by one small bract, or the bract absent; calyx lobes more or less equal, greenish, to 2.5 mm long; corolla bilabiate, yellow, 6–12 mm long, the obtuse, slender spur much shorter than the lower lip of the corolla; stamens 5; capsules globose, 2.0–3.5 mm in diameter, with several seeds. June–September.

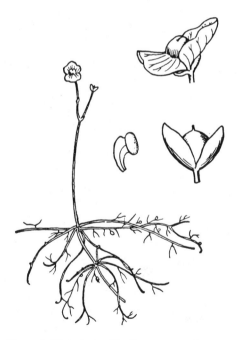

Fig. 296. *Utricularia gibba* (Long-spurred bladderwort). Habit (left). Flower (upper right). Spur of flower (center). Fruit with persistent calyx (center right).

Shallow water.

IA, IL, IN, KS, KY, MO, OH (OBL).

Long-spurred bladderwort.

This bladderwort has stems that creep on mud at the bottom of ponds. It differs from the similar *U. fibrosa* by its shorter, more slender spur of the corolla and its leaves divided into two rather than three segments.

5. **Utricularia intermedia** Hayne in Schrad. Journ. Bot. 1800:18. 1801. Fig. 297.

Aquatic herbs with submersed stems buried in soil at bottom of pond; stems divided into flat, serrulate segments; bladders borne on separate, undivided branches, not subtended by leafy bracts; scapes slender, to 25 cm tall, (1–) 2- to 4-flowered; flowers on ascending pedicels 10–15 mm long, subtended by 1–several bracts; calyx bilabiate, green; corolla bilabiate, yellow, 10–15 mm long, the spur about as long as the lower lip of the corolla; capsules up to 3 mm in diameter, with numerous seeds. June–August.

Shallow water.

IA, IL, IN, OH (OBL).

Bladderwort.

This species is distinguished by its much divided flat, minutely serrulate linear stems with filiform segments. It differs from *U. minor* which has entire divisions of the stems and its very short spur of the corolla.

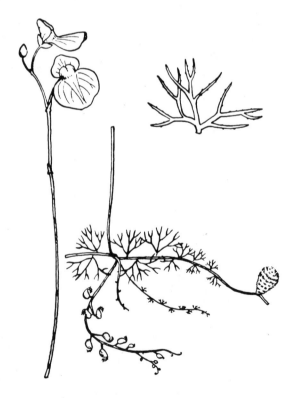

Fig. 297. *Utricularia intermedia* (Bladderwort). Habit with leaves (center). Flowering branch (left). Part of branch (upper right).

6. Utricularia macrorhiza LeConte, Ann. Lyc. N. Y. 1:73. 1824. Fig. 298.

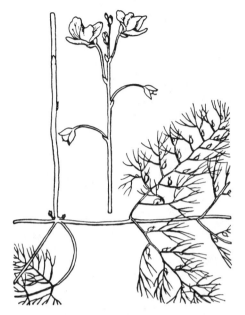

Free floating aquatic herbs; stems much divided into filiform segments; bladders numerous; scapes stout, to 30 cm long, 4- to 20-flowered; flowers on ascending to recurved pedicels up to 20 mm long, subtended by 1–5 bracts 4–6 mm long; calyx bilabiate, green; corolla bilabiate, yellow, 10–20 mm long, the spur about 2/3 as long as the lower lip of the corolla; capsules globose, 3–4 mm in diameter, with numerous seeds. May–August.

Shallow water.

IA, IL, IN, KS, KY, MO, NE, OH (OBL).

Common bladderwort.

This species has often been called *U. vulgaris*, but that is a European species that differs slightly from ours.

Fig. 298. *Utricularia macrorhiza* (Common bladderwort). Habit. Flowering branch (center).

Utricularia macrorhiza differs from other free floating species of bladderworts with filiform branches by having 6 or more flowers per scape.

7. Utricularia minor L. Sp. Pl. 18. 1753. Fig. 299.

Aquatic herbs with submersed stems buried in soil at bottom of pond; stems divided into flat, entire, linear segments; bladders borne on separate, undivided branches, often few in numbers; scapes slender, to 15 cm long, 2- to 6-flowered; flowers on recurved pedicels to 15 mm long, subtended by 2–5 minute bracts; calyx bilabiate, green; corolla bilabiate, yellow, 5–10 mm long, the spur much shorter than the lower lip of the corolla; capsules up to 1.5 mm in diameter, with numerous seeds. June–September.

Shallow water.

IA, IL, IN, OH (OBL).

Small-flowered bladderwort.

This species differs from the similar *U. intermedia* by its entire stem segments and its smaller flowers with a very short spur.

8. Utricularia ochroleuca R. Hartman, Bot. Not. 30. 1857. Fig. 300.

Aquatic herbs with submersed stems buried in soil at bottom of pond; stems divided into flat, serrulate, linear segments; bladders borne on separate, divided branches subtended by leafy bracts; scapes very slender, to 15 cm tall, 1- to 4-flowered; flowers on ascending pedicels up to 12 mm long, subtended by 1–several

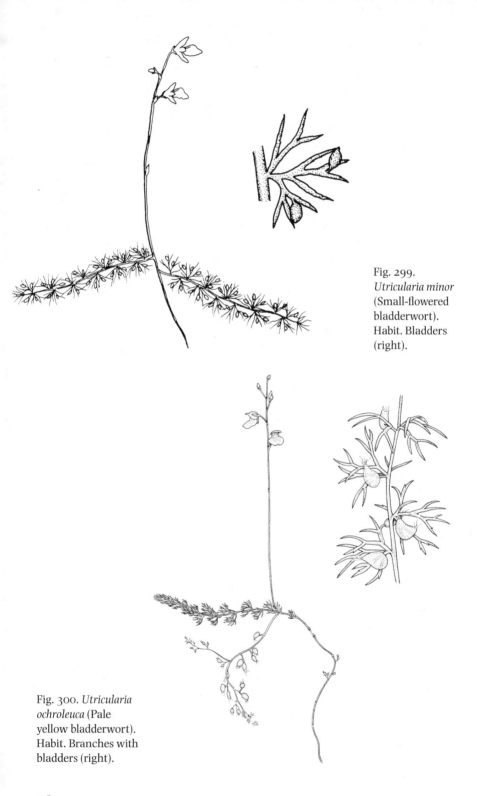

Fig. 299.
Utricularia minor
(Small-flowered
bladderwort).
Habit. Bladders
(right).

Fig. 300. *Utricularia
ochroleuca* (Pale
yellow bladderwort).
Habit. Branches with
bladders (right).

bracts; calyx bilabiate, green; corolla bilabiate, pale yellow, 10–15 mm long, the spur about half as long as the lower lip of the corolla; capsule up to 3 mm in diameter, with numerous seeds. June–August.

Shallow water.

IL, OH (OBL).

Pale yellow bladderwort.

The distinguishing features of this rare species are the flat, serrulate divisions of the stem and the spur of the corolla that is about half as long as the lower lip of the corolla.

9. **Utricularia purpurea** Walt. Fl. Carol 64. 1788. Fig. 301.
Vesculinia purpurea (Walt.) Raf. Fl. Tellur. 4:109. 1838.

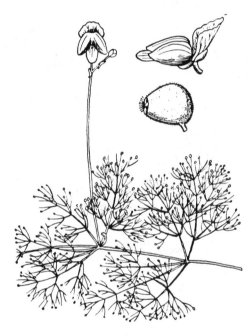

Free floating aquatic herbs; stems submerged, whorled, much divided; bladders ovoid, broader than high; scapes emergent, slender, sometimes more or less spongy, to 15 cm tall, bearing (1–) 2–5 flowers; flowers on recurved to ascending pedicels 3–10 mm long, subtended by 1–2 bilobed, flat bracts; calyx bilabiate, green; corolla bilabiate, purple, 10–12 mm long, the spur obtuse at tip, 2–5 mm long, with a yellow spot; capsules globose, 3–4 mm in diameter, with several seeds; seeds orbicular, with numerous subacute tubercles. July–September.

Ponds, ditches, streams.

IL, IN (OBL).

Purple-flowered bladderwort.

This free floating aquatic is distinguished by its purple flowers with short, obtuse spurs and its tuberculate seeds.

Fig. 301. *Utricularia purpurea* (Purple-flowered bladderwort). Habit (left). Flower (upper right). Bladder (below flower).

10. **Utricularia radiata** Small, Fl. S. E. U. S. 1090. 1903. Fig. 302.
Utricularia inflata Walt. var. *minor* Chapm. Fl. S. U. S. 282. 1860.

Free floating aquatic herbs; submersed stems much divided, linear-filiform; stems at surface of water in whorls of 4–7, inflated, much divided into filiform segments near the tip; bladders wider than high, with slender hairs at one side; scapes slender, to 6 cm tall above the whorl of inflated stems, (1–) 3–4 (–6) flowered; flowers on ascending pedicels up to 20 mm long, subtended by flat, lobed or unlobed bracts, obtuse to subacute, broader than long, 2–3 mm long; calyx bilabiate, green; corolla bilabiate, yellow, 10–15 mm long, the obtuse spur shorter than the lower lip of the corolla; capsules globose, 3–5 mm in diameter, with several seeds; seeds covered with curved, flattened tubercles. May–August.

Ponds, lakes, swamps. IN (OBL).

Small inflated bladderwort.

This species is similar to *U. inflata* because of its inflated floats but differs by its fewer flowers per scape and by its floats that are much branched only near their tips.

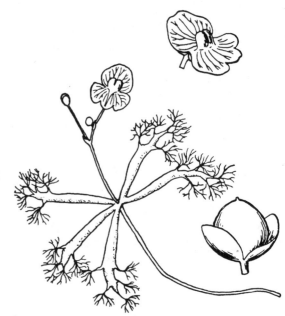

Fig. 302. *Utricularia radiata* (Small inflated bladderwort). Habit. Flower (upper right). Capsule with persistent sepals (lower right).

11. **Utricularia resupinata** B. D. Greene in Bigel. Fl. Bost. ed. 3, 10. 1840. Fig. 303.

Usually terrestrial herbs, with stems slender, rooted in muck, sometimes with emergent, acicular, segmented branchlets to 3 cm long; bladders, when present, ovoid, borne on both the emergent and submerged stems; scapes slender, wiry, to 10 cm tall, bearing one flower; flower on an erect pedicel 10–15 mm long, subtended by a single, tubular bract and no bractlets; calyx bilabiate, green; corolla

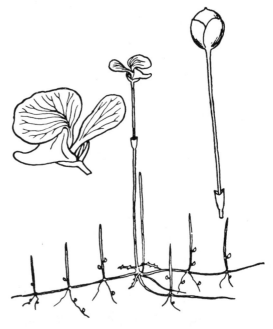

Fig. 303. *Utricularia resupinata* (One-flowered bladderwort). Habit. Flower (upper left).

bilabiate, purple, 10–12 mm long, the spur obtuse at tip, 4–5 mm long, positioned horizontally; capsules globose, 2.0–3.5 mm in diameter, with numerous seeds, the seeds obpyramidal, angular, strongly reticulate. July–August.

Ponds, lakes, ditches.

IN (OBL).

One-flowered bladderwort.

This is the only species of *Utricularia* in the central Midwest with a single-flowered scape with the flower subtended by a single tubular bract and no bractlets. The stems are little branched, and the emergent ones are acicular. The flowers are purple.

This species is primarily found in the coastal plain.

12. **Utricularia subulata** L. Sp. Pl. 18. 1753. Fig. 304.
Setiscapella subulata (L.) Barnhart, Fl. Minn. 170. 1913.

Terrestrial plants, the stems usually buried just beneath the surface of wet sand or mud, filiform, undivided; bladders broadly ovoid, broader than tall, reticulate; scapes filiform, wiry, to 20 cm tall, with (1–) 2–4 (–8) flowers; flowers on ascending pedicels 5–20 mm long, subtended by peltate bracts 1–2 mm long; cleistogamous flowers sometimes present; calyx bilabiate, green; corolla bilabiate, yellow, to 12 mm long, the lobes broader than long, the spur subacute, about as long as the lower lip of the corolla; capsules 2–3 mm long, with several seeds; seeds obovoid, 0.2 mm long, strongly reticulate. May–September.

Wet sandy areas, sometimes in depressions that fill temporarily with water.

IN, MO (OBL).

Slender bladderwort.

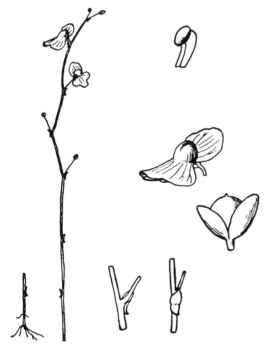

This primarily terrestrial species has filiform stems and scapes. The yellow flowers are subtended by peltate bracts. Sometimes a few or all the flowers may be highly reduced and cleistogamous.

Fig. 304. *Utricularia subulata* (Slender bladderwort). Habit. Bare scape (lower left). Spur (upper right). Flower (right center). Capsule with persistent sepals (lower right). Bladders (center, below).

81. LIMNANTHACEAE—MEADOWFOAM FAMILY

Annual herbs; leaves alternate, pinnately compound or dissected, without stipules; flowers solitary, actinomorphic, perfect; sepals 3–5, free; petals 3–5, free; stamens 3–5 or 6–10; ovary superior, with 2–5 carpels; fruit a mericarp.

Two genera and eleven species, all in North America, comprise this family.

1. **Floerkea** Willd.—False Mermaid

Annual herbs; leaves alternate, pinnately compound or deeply dissected; flower solitary from the leaf axils; sepals 3, free; petals 3, free; stamens 3 or 6; ovary superior with 2–3 carpels; fruit a mericarp.

Only the following species is in the genus.

1. **Floerkea proserpinacoides** Willd. Neue Schrift. Ges. Nat. Fr. 3:448. 1801. Fig. 305.

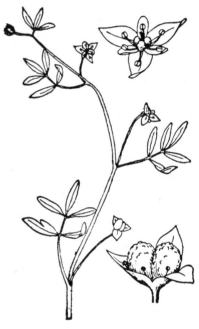

Annual herbs with fibrous roots; stems sprawling to ascending, weak, glabrous, to 30 cm long; leaves alternate, pinnately divided into 3–7 leaflets or deeply pinnately lobed, the leaflets linear, to linear-oblong, obtuse to acute at the apex, tapering to the base, to 2 cm long, to 7 mm wide; flower solitary in the axils of the leaves, to 4 mm across, on pedicels up to 1 cm long; sepals 3, free; petals 3, free, white; stamens 3 or 6; ovary superior, bilobed; fruit a bilobed mericarp, globose, 2.0–2.5 mm long, warty. April–May.

Marshes, mesic woods.

IA, IL, IN, MO (FAC+), KY, OH (FAC). False mermaid.

This delicate wildflower is distinguished by its 3-parted white flower, its pinnately divided leaves, and its bilobed fruits.

Fig. 305. *Floerkea proserpinacoides* (False mermaid). Habit. Fruits (lower right). Flower (upper right).

82. LINACEAE—FLAX FAMILY

Herbs; leaves mostly alternate, simple, entire, sessile; flowers in cymes, racemes, or panicles, actinomorphic, perfect; sepals (4–) 5, free; petals (4–) 5, free; stamens 5; ovary superior; capsules with numerous seeds.

This family contains six genera and 220 species found in many parts of the world.

1. **Linum** L.—Flax

Herbs; leaves alternate or opposite, simple, entire, sessile; flowers in cymes, racemes, or panicles, perfect, actinomorphic; sepals 5, free; petals 5, free; stamens 5; ovary superior; capsule 10-locular; seeds numerous.

Nearly two hundred species are in this genus.

1. Leaves stiff; inner sepals with conspicuous glandular cilia.................................. 1. *L. medium*
1. Leaves thin; inner sepals eglandular or nearly so.. 2. *L. striatum*

1. **Linum medium** (Planch.) Britt. in Britt. & Brown. Ill. Fl. 2:349. 1897. Fig. 306.
Linum virginianum L. var. *medium* Planch. Lond. J. Bot. 7:480. 1848.
Linum virginianum L. var. *texanum* Planch. London J. Bot. 7:481. 1848.
Linum medium (Planch.) Britt. var. *texanum* (Planch) Fern. Rhodora 37:428. 1935.
Cathartolinum medium (Planch.) Small, N. Am. Fl. 25:72. 1907.

Fig. 306. *Linum medium* (Stiff yellow flax).
Upper part of plant (left). Capsule (center).
Lower part of plant (right).

Perennial herbs; stems erect, branched, glabrous, blue-green, terete, to nearly 1 m tall; leaves simple, the lowest sometimes opposite, alternate above, stiff, linear to linear-lanceolate to elliptic-obovate, obtuse to subacute at the apex, tapering to the sessile base, strongly ascending, to 2.5 cm long, to 6 mm wide; inflorescence branches stiffly spreading, with several flowers; flowers 5–10 mm across, on pedicels to 10 mm long; sepals 5, free, lanceolate to lance-ovate, acute glandular along the margin, to 5 mm long; petals 5, free, yellow, 5–8 mm long; stamens 5; ovary superior; capsules depressed-globose, glabrous, 1.5–2.5 mm in diameter, splitting into 5 or 10 valves, with numerous seeds. June–August.

Dry woods, but also calcareous pond shores around Lake Michigan, and in excavated sandy areas where water stands.

IA, IL, IN, KY, MO, OH (FACU), KS (FAC).

Stiff yellow flax.

This species is distinguished from *L. striatum* by its glandular-marginal sepals and its leaves that are mostly all alternate.

Two very different varieties, which should perhaps be treated as separate species, occur in our area. Typical var. *medium*, found in Ohio, has elliptic-obovate leaves, sepals equaling or shorter than the capsules, and capsules splitting into 10 valves at maturity. Var. *texanum*, known from the remainder of our area except Nebraska, has linear to linear-lanceolate leaves, sepals longer than the capsules, and capsules splitting into 5 valves at maturity.

This is a species of dry habitats except around Lake Michigan where it grows in calcareous pond margins and in sandy areas that have been excavated.

2. Linum striatum Walt. Fl. Carol. 118. 1788. Fig. 307.

Linum diffusum Wood, Bot. & Flor. 66. 1870.

Cathartolinum striatum (Walt.) Small, N. Am. Fl. 25:71. 1907.

Perennial herbs; stems decumbent to ascending, branched, glabrous, green, striate-angular, to 1 m tall; leaves simple, opposite on the lower half of the stem, alternate above, thin, elliptic to obovate, subacute to acute at the apex, tapering to the sessile base, strongly ascending, to 3.5 cm long, to 1 cm wide; inflorescence branches spreading, with several flowers; flowers 4–10 mm across, on pedicels to 4 mm long; sepals 5, free, lanceolate to ovate, acute at the apex, not glandular along the margin, to 3 mm long; petals 5, free, yellow, 3–7 mm long; stamens 5; ovary superior; capsules depressed-globose, glabrous, 1.8–2.0 mm in diameter, splitting into 10 valves at maturity, with numerous seeds. June–August.

Wet meadows, swampy woods, bogs.

IL, IN, MO (FACW–), KY, OH (FACW).

Ridged yellow flax; ridgestem yellow flax.

The preponderance of lower opposite leaves and the absence of glands along the margins of the sepals are distinctive for this species. It also has stems that are conspicuously striate-angular.

83. LYTHRACEAE—LOOSESTRIFE FAMILY

Annual or perennial herbs or shrubs; leaves simple, usually opposite, less often alternate or whorled, entire, without stipules; flowers perfect, actinomorphic or slightly zygomorphic, perigynous, sometimes dimorphic or trimorphic, axillary or in spikes or racemes; calyx campanulate to globose to tubular, 4- to 7-toothed, sometimes with appendages between the teeth; petals 0, 4, 5, 6, or 7, free from each other; stamens 4–14, attached near top of hypanthium; ovary superior, 1- to 5-locular; style 1; fruit a capsule, rarely indehiscent.

The Lythraceae consists of twenty-five genera and about five hundred species in the warmer climates of the world. In addition to the five genera enumerated below that sometimes occur in standing water, *Cuphea* is also found in the central Midwest in drier soils.

1. Plants shrubby; leaves, or some of them, whorled; petals at least 10 mm long 2.*Decodon*
1. Plants herbaceous; leaves opposite or occasionally alternate; petals less than 10 mm long, or absent.
   2. Petals absent; calyx teeth without appendages between them; fruit indehiscent ....
   ....3. *Didiplis*
   2. Petals present; calyx teeth with an appendage between them; fruit dehiscent.
      3. Calyx tubular, with usually (4–) 6 teeth; petals 6; flowers showy (except in the annual *L. hyssopifolia*) .... 4. *Lythrum*
      3. Calyx campanulate to globose, about as broad as long, with 4 teeth; petals 4; flowers not showy.
         4. Leaves tapering to the base; appendages between the calyx teeth triangular; capsule splitting into 4 valves .... 5. *Rotala*
         4. At least some of the leaves dilated at the base and often slightly clasping; appendages between the calyx teeth subulate; capsule splitting irregularly .... 1. *Ammannia*

Fig. 307. *Linum striatum* (Ridged yellow flax). Habit (center).
Leaves (left). Fruit (right).

1. **Ammannia** L.—Tooth-cup

Glabrous annual herbs; leaves opposite, entire, at least some of them dilated at the base and somewhat auriculate; flowers axillary, usually 3 or more in each axil; calyx campanulate to globose, with 4 teeth, with a subulate appendage between the teeth; petals 4, early deciduous, rarely absent, purplish, up to 2.5 mm long; stamens 4, rarely 5–12; capsule irregularly dehiscent, with many seeds.

Ammannia consists of about twenty-five species, most of them in the tropics. This genus looks very much like *Rotala*, but the leaves are dilated and often auriculate at the base, the appendages between the calyx teeth are subulate, and the capsule splits irregularly. The following wetland species may on occasion be in standing water for a short period in the central Midwest. All of them tend to intergrade, and some of them may be found in the same colony.

1. Stems slender, unbranched except sometimes near the upper end; peduncles very slender, 3–9 mm long; flowers 7–15 per leaf axil; capsules up to 3 mm in diameter ....1. *A. auriculata*
1. Stems robust, often branched from near the base; peduncles stout, rarely more than 4 mm long, or absent; flowers 3–5 (–14) per leaf axil; capsules at least 3.5 mm in diameter.
   2. Flowers sessile, usually 3 per leaf axil; petals pale pink; capsules 4–6 mm in diameter .... 3. *A. robusta*
   2. Flowers short-pedicellate, 3–5 per leaf axil; petals deep rose-purple; capsules 3.5– 5.0 mm in diameter .... 2. *A. coccinea*

1. **Ammannia auriculata** Willd. Hort. Berol. 7, pl. 7. 1806. Fig. 308.

Glabrous annuals with slender, often reddish stems, unbranched or branched in the upper part, up to 80 cm tall; leaves simple, opposite, linear to linear-lanceolate, all except sometimes the lowermost auriculate and clasping, entire, to 6 cm long, to 1 cm wide, thin in texture; flowers 7–15 in axils of the leaves, usually on filiform pedicels 3–9 mm long, the pedicels 1–6 mm long; calyx 1.5–2.0 mm long, with 4 triangular teeth; petals 4, deep rose-purple, falling away early, about 1.5 mm long and about as wide; stamens 4 or less commonly 8, long-exserted; capsules more or less globose, 1.5–3.0 mm in diameter. July–October.

Around ponds and lakes, in wet ditches, occasionally in swamps.

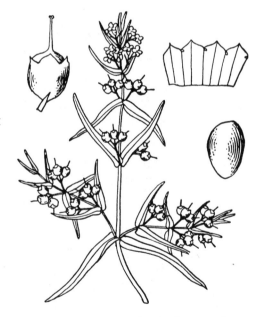

Fig. 308. *Ammannia auriculata* (Red-stemmed tooth-cup). Habit. Capsule (upper left). Corolla open (upper right). Seed (lower right).

KS, NE (OBL).

Red-stemmed tooth-cup.

The combination of slender stems branched only above, slender peduncles up to 9 mm long, 7–15 flowers per leaf axil, and capsules up to 3 mm in diameter distinguishes this species form the other species of *Ammannia* in the central Midwest.

2. Ammannia coccinea Rottb. Pl. Hort. Havn. Descr. 7. 1773. Fig. 309.

Glabrous annuals with rather robust stems often branched from near the base, up to 1 m long, with some of the lowest branches often decumbent; leaves simple, opposite, linear-lanceolate to elliptic, all except the lowermost auriculate and cordate, to 8 cm long, to 1.5 cm wide, somewhat thick; flowers 3–5 (–14) per leaf axil, sessile or more commonly on stout pedicels up to 4 mm long or occasionally longer, sessile or on pedicels up to 2 mm long; calyx to 5 mm long, with 4 triangular teeth; petals 4, deep rose-purple, falling away early, about 2 mm long and nearly as wide; stamens 4, rarely more, exserted; capsules more or less globose, 3.5–5.0 mm in diameter. July–October.

Around ponds and lakes, along streams, in wet ditches, in swamps.

IA, IL, IN, KS, KY, MO, NE, OH (OBL).

Common tooth-cup.

This is usually the most common species of *Ammannia* in the central Midwest. Colonies of this species sometimes may have specimens of *A. auriculata* and/or *A. robusta* mixed with it. *Ammannia coccinea* differs from *A. auriculata* by the more robust stems that often branch at the base, by its shorter, stouter peduncles, by its fewer flowers per leaf axil, and by its smaller capsules. It differs from *A. robusta* by its deeper rose-purple flowers, its short-pedunculate inflorescences, and its having 3–5 flowers in each leaf axil.

3. Ammannia robusta Heer & Regel, Index Sem. Zurich 1. 1842. Fig. 310.

Glabrous annuals with robust stems often branched from near the base, up to 1 m long, with some of the lowest branches often decumbent; leaves opposite, simple, linear-lanceolate, all except sometimes the lowermost auriculate to cordate and clasping, to 8 cm long, to 1.5 cm wide, somewhat thick; flowers (1–) 3 per leaf axil, sessile; calyx to 3.5 mm long, with 4 triangular teeth; petals 4, pale pink, falling away early, about 2.5 mm long, usually a little wider; stamens 4, rarely more, exserted; capsules more or less globose, 4–6 mm in diameter. July–October.

Around ponds and lakes, in wet ditches, in swamps.

IA, IL, IN, KS, KY, MO, NE, OH (not listed by the U.S. Fish and Wildlife Service).

Sessile tooth-cup.

During most of the twentieth century, this species was not recognized as distinct from *A. coccinea*. It differs by its sessile flowers that are usually fewer per leaf axis and that are pale pink in color.

2. Decodon J. F. Gmel.—Swamp Loosestrife; Water Willow

Shrubs with arching branches; leaves mostly whorled, simple, entire; flowers several in axillary clusters, trimorphic, with flowers having stamens of three different

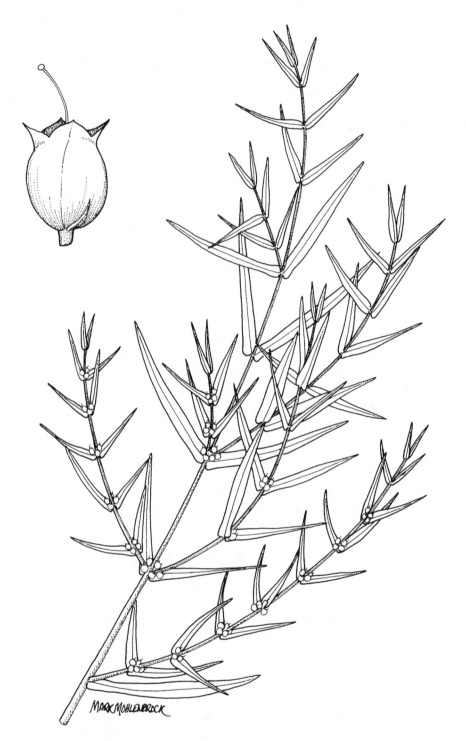

Fig. 309. *Ammannia coccinea* (Common tooth-cup). Habit. Fruit (upper left).

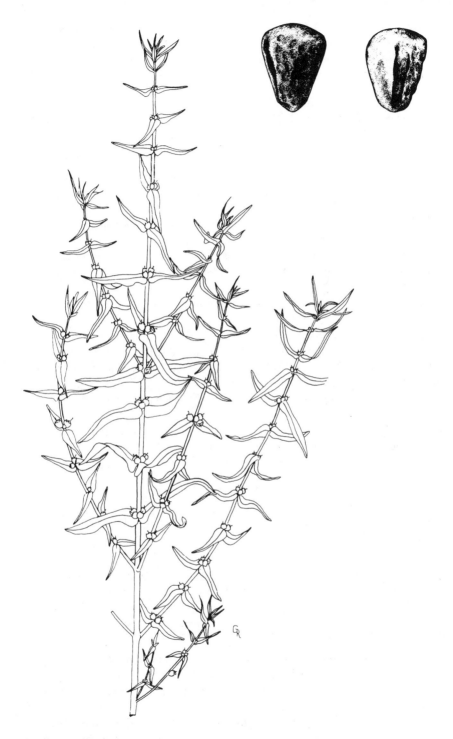

Fig. 310. *Ammannia robusta* (Sessile tooth-cup). Habit. Fruits (right).

lengths; calyx campanulate to globose, with 5–7 teeth and as many hornlike appendages between the teeth; petals 5, free from each other, fringed; stamens 8 or 10; capsule globose, usually 3-locular, with many seeds.

Decodon verticillatus is the only species in the genus.

1. **Decodon verticillatus** (L.) Ell. Bot. S. C. & Ga. 1:544. 1821. Fig. 311.
Lythrum verticillatum L. Sp. Pl. 446. 1753.
Decodon verticillatus (L.) Ell. var. *laevigatus* Torr.& Gray, Fl. N. Am. 2:483. 1843.

Shrubs with arching stems often rooting at the tip, to 2.5 m long, glabrous or softly pubescent; leaves, or at least some of them, whorled, lanceolate, acute at the apex, tapering to the sessile or subsessile base, entire, to 15 cm long, to 4 cm wide, glabrous or softly pubescent on the lower surface; flowers clustered in the axils of the uppermost leaves, on glabrous or softly pubescent peduncles up to 5 mm long; calyx more or less campanulate, with 5–7 triangular teeth and an equal number of hornlike appendages between the teeth; petals 5, free from each other, bright pink, 10–15 mm long; stamens 8 or 10, of three different lengths; capsules globose, about 5 mm in diameter, with many seeds. July–September.

Swamps, edge of lakes, often in standing water.

IA, IL, IN, KY, MO, OH (OBL).

Swamp loosestrife; water willow.

Decodon verticillatus and *Cephalanthus occidentalis* are the only shrubs in the central Midwest with whorled leaves. The branches of *Decodon verticillatus* are arching, while those of *Cephalanthus* are ascending or erect.

When *Decodon verticillatus* grows in standing water, its submerged stems are spongy.

Most plants in the central Midwest have softly hairy stems and lower leaf surfaces. A few plants from Kentucky and Ohio have glabrous stems and lower leaf surfaces and may be known as var. *laevigatus*.

3. Didiplis Raf.—Water Purslane

Submersed or rooted annuals; leaves simple, opposite, entire, the submersed ones different from the emersed ones; flower axillary, solitary; calyx 4-parted, without appendages between the teeth; petals 0; stamens (2–) 4; capsule indehiscent, with many seeds.

There is only one species in the genus, found in North America, Europe, and Asia. Some botanists place this species in the genus *Peplis*.

1. **Didiplis diandra** (Nutt.) Wood, Bot. & Fl. 124. 1870. Fig. 312.
Peplis diandra Nutt. in DC. Prodr. 2:77. 1828.

Dwarf annuals, submersed in water or rooting in mud; stems very slender, weakly branched, glabrous; leaves simple, opposite, rarely alternate, entire, glabrous, the submersed ones membranous, linear, acute at the apex, truncate at the sessile base, to 3 cm long, to 4 mm wide, the emersed ones thicker, elliptic, acute at the apex, tapering to the narrow base, to 3 cm long, to 4 mm wide; flower solitary in the leaf axils, greenish; calyx 2–3 mm long, with 4 lobes, but lacking appendages

Fig. 311. *Decodon verticillatus* (Swamp loosestrife). Habit. Flower (lower left).

between the lobes; petals 0; stamens 2 or 4, not exserted; fruit globose, about 3 mm in diameter, indehiscent. June–August.

Shallow water and mud around lakes and ponds.

IA, IL, IN, KS, KY, MO, OH (OBL).

Water purslane.

This easily overlooked diminutive species usually occurs in shallow standing water at the edges of lakes and ponds, although it may be stranded on mud for considerable periods of time. The somewhat similar appearing *Ludwigia palustris* has larger, petiolate leaves and larger fruits.

Fig. 312. *Didiplis diandra* (Water purslane). Habit. Fruit (below).

4. Lythrum L.—Loosestrife

Annual or more commonly perennial herbs; stems usually 4-angled; leaves simple, usually opposite or subopposite or sometimes whorled, entire; flowers axillary, solitary or in clusters, sometimes forming spikes, actinomorphic; calyx tubular, with usually 6 lobes and 6 appendages that are usually longer than the lobes; petals usually 6, less commonly 4, free from each other, purple or sometimes white; stamens (4–) 6 or 12; capsules enclosed by the persistent hypanthium, with many seeds.

Thirty species, found worldwide, comprise this genus.

1. Plants annual; stamens 4 or 6; petals 4 or 6, 2–3 mm long, pale purple to white; most of the leaves alternate ....3. *L. hyssopifolia*
1. Plants perennial; stamens 6 or 12; petals 6, 3–12 mm long, purple; leaves opposite or occasionally whorled.
    2. Stamens 12; petals 7–12 mm long; flowers in showy spikes; plants often more or less pubescent.. 4. *L. salicaria*
    2. Stamens 6; petals up to 6 mm long; flowers 1–2 in the leaf axils, not showy; plants usually glabrous.
        3. Leaves membranous, deep green, some of them ovate ....1. *L. alatum*
        3. Leaves more of less fleshy, gray-green, all of them linear-oblong to linear-lanceolate ....
        .... 2. *L. californicum*

1. **Lythrum alatum** Pursh, Fl. Am. Sept. 334. 1814. Fig. 313.
Lythrum dacotanum Nieuw. Am. Midl. Nat. 3:265. 1914.

Glabrous perennials to 75 cm tall; stems stiffly branched, 4-angled and slightly winged on the angles; leaves simple, opposite except for the uppermost alternate ones, ovate to oblong to linear-lanceolate, acute at the apex, rounded to subcordate to tapering to the base, sessile or on petioles to 3 mm long, dark green on the upper surface, pale below, to 6 cm long, 5–15 mm wide; flowers 1–2 in the axils of the leaves, dimorphic, with either the stamens or the styles exserted; calyx tubular, 4–7 mm long, 12-nerved, with 6 teeth about half as long as the 6 appendages; petals 6, free, 3–6 mm long, purple; stamens 6; ovary surrounded by a fleshy disk; capsules spherical. June–September.

Wet ditches, low depressions in floodplains, swamps, occasionally in standing water for a short period.

IA, IL, IN, KS, MO, NE (OBL), KY, OH (FACW+).

Winged loosestrife.

The slightly winged, 4-angled stems are distinctive for this species. The narrow uppermost leaves are usually alternate, while the broader leaves lower down the stem are opposite. The dark green upper surface of the leaves and the generally broader leaves distinguish this species from the more western *L. californicum*.

2. **Lythrum californicum** Torr. & Gray, Fl. N. Am. 1:482. 1840. Fig. 314.

Glabrous perennials to 60 cm tall; stems slender, branched, the branches stiff or lax, sometimes 4-angled but the angles not winged; leaves simple, opposite except for the uppermost alternate ones, narrowly oblong to linear-lanceolate to linear, acute at the apex, rounded at the base, sessile or nearly so, somewhat fleshy, gray-green or sometimes glaucous, to 2.5 cm long, 3–8 mm wide; flowers 1–2 in the

Fig. 313. *Lythrum alatum* (Winged loosestrife). Habit. Flower (lower left).

Fig. 314. *Lythrum californicum* (Western loosestrife). Habit. Upper part of plant (center). Flower (right).

axils of the leaves, dimorphic, with either the stamens or the styles exserted; calyx tubular, 4–7 mm long, 12-nerved, with 6 teeth about half as long as the 6 appendages; petals 6, free, 4–6 mm long, purple; stamens 6; ovary surrounded by a fleshy disk; capsules more or less spherical. June–September.

Wet ditches, around lakes, sometimes in shallow water for a short time. KS (OBL).

Western loosestrife.

This species resembles *L. alatum* but lacks the wings on the stems, has a smaller stature, and possesses narrower gray-green leaves.

3. Lythrum hyssopifolium L. Sp. Pl. 447. 1753. Fig. 315.

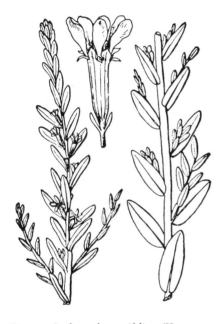

Glabrous annuals; stems 4-angled but not winged, branched or unbranched, to 60 cm tall; leaves simple, alternate, linear to linear-oblong, obtuse at the apex, tapering to the base, pale green, to 3 cm long, 1–7 mm wide; calyx tubular, 3–5 mm long, with 6 teeth slightly shorter than the 6 appendages; petals (4–) 6, free, pale purple to white, 2–3 mm long; stamens (4–) 6, never exserted; ovary not surrounded by a disk; capsules more or less spherical. June–September.

Marshes, rarely in standing water. OH (OBL).

Hyssop-leaved loosestrife.

This is primarily a species of salt marshes on both the Atlantic and Pacific coasts. The pale color of the leaves resembles the leaves of *L. californicum*, but *L. hyssopifolia* is a more slender plant with narrower leaves and smaller, paler flowers. The leaves are alternate, as well.

Fig. 315. *Lythrum hyssopifolium* (Hyssop-leaved loosestrife). Habit (left). Flower (center). Lower part of plant (right).

4. Lythrum salicaria L. Sp. Pl. 446. 1753. Fig. 316.

Robust perennials; stems to 2.5 m tall, more or less softly pubescent; leaves simple, opposite or occasionally whorled, lanceolate, acute at the apex, rounded to cordate at the base, sessile, glabrous or softly pubescent, to 10 cm long, 5–15 mm wide; flowers in terminal spikes, showy, trimorphic, with stamens and styles of different lengths, each flower subtended by foliaceous bracts; calyx tubular, 4–6 mm long, more or less pubescent, with 6 appendages about twice as long as the 6 calyx lobes; petals 6, free, rose-purple, 7–12 mm long; stamens usually 12; ovary not surrounded by a disk; capsules more or less spherical. June–September.

Marshes, wet ditches, wet meadows, floodplains.

IA, IL, IN, MO (OBL), KS, NE (NI), KY, OH (FACW+).

Purple loosestrife.

This handsome species is native to Europe, Asia, and Africa and is popular in gardens because of its profusion of rose-purple flowers. However, it readily escapes from cultivation and is an aggressive weed in wetlands, often driving out native wetland species.

The rose-purple flowers borne in terminal spikes and the sessile, often whorled leaves are distinctive for this species.

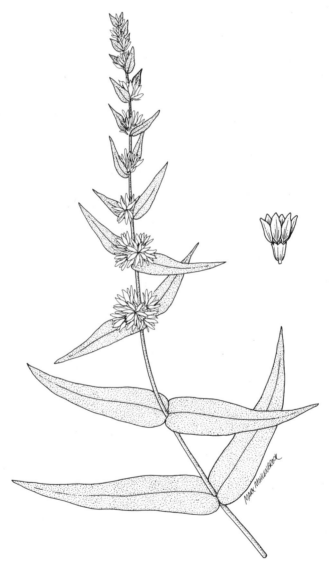

Fig. 316. *Lythrum salicaria* (Purple loosestrife). Habit (left). Flower (right).

5. Rotala L—Tooth-cup

Glabrous annuals; leaves simple, opposite, entire, narrowed to the base; flower solitary in the axils of the leaves, actinomorphic; calyx campanulate to globose, with 4 teeth and 4 appendages; petals 4, free from each other, rarely absent; stamens 4 (–6); ovary not surrounded by a disk; capsules dehiscent by (3–) 4 valves; seeds numerous.

Rotala is a genus of approximately forty-five species, most of them in Europe, Asia, and Africa. Only the following species occurs in the central Midwest.

1. **Rotala ramosior** (L.) Koehne in Mart. Fl. Bras. 13 (2):194. 1875. Fig. 317.
Ammannia ramosior L. Sp. Pl. 120. 1753.
Rotala ramosior (L.) Koehne var. *interior* Fern. & Grisc. Rhodora 37:169. 1935.

Glabrous annuals; stems erect or prostrate, sometimes branched, to 45 cm long; leaves simple, opposite, linear to oblanceolate, obtuse to subacute at the apex, distinctly tapering to the short-petiolate or sessile base, to 4 (–5) cm long, 1–12 mm wide; flower solitary in the axils of the leaves, sessile, actinomorphic; calyx campanulate to globose, 2–5 mm long, with 4 teeth about as long as the 4

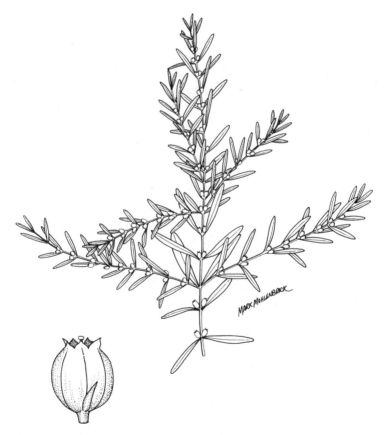

Fig. 317. *Rotala ramosior* (Tooth-cup). Habit. Fruit (lower left).

appendages; petals 4, rarely absent, pink to purplish to whitish, 2–3 mm long; stamens 4 (–6); capsules 2–5 mm long, nearly as wide, transversely striate, dehiscing by (3–) 4 valves. July–October.

Around ponds and lakes, marshes, wet ditches, depressions in floodplains.

IA, IL, IN, KS, KY, MO, NE, OH (OBL).

Tooth-cup.

This species resembles species of *Ammannia*, but the leaves of *Rotala* taper to the base rather than becoming dilated at the base. There is only one flower per leaf axil, and the capsules dehisce by means of 3 or 4 valves.

84. MALVACEAE—MALLOW FAMILY

Herbs or shrubs; leaves alternate, simple, toothed or lobed, with stipules; flowers usually perfect, actinomorphic; sepals 5, united at base, often subtended by an involucre of bracts that resembles a second calyx; petals 5, free from each other; stamens numerous, attached to a column that surrounds the styles; pistils several, the ovaries united in a ring at the base of the column, the stigmas exserted beyond the tip of the column; capsules several-locular, often forming a ring.

This family consists of about seventy-five genera and more than one thousand species found worldwide. Only the following genus sometimes occurs in standing water in the central Midwest.

1. Hibiscus L.—Rose Mallow

Annual or perennial, often robust herbs or shrubs; leaves simple, alternate, toothed, sometimes lobed; flowers actinomorphic, large (in the aquatic and wetland species), solitary, axillary, each subtended by usually 12 bracts forming an involucre; sepals 5, united at the base; petals 5, free, showy, white or pink or rose; stamens numerous on a central column; stigmas 5, protruding from the tip of the column; carpels 5, forming a ring around the base of the column; fruit a capsule, enclosed by or subtended by the persistent calyx.

Hibiscus consists of about two hundred species found in the warmer regions of the world.

Hibiscus syriaca, Rose-of-Sharon, was a popular ornamental shrub at one time. Three species occur as aquatic species or in wetlands in the central Midwest.

1. Stems and leaves glabrous; at least some of the leaves hastately lobed .... 1. *H. laevis*
1. Stems and at least the lower surface of the leaves pubescent; leaves unlobed or, if lobed, not hastately lobed.
    2. Leaves soft-hairy on both surfaces; none of the leaves lobed; capsules densely tomentose.
    .... 2. *H. lasiocarpos*
    2. Leaves glabrous on the upper surface, somewhat pubescent on the lower surface; leaves unlobed, or occasionally some of the leaves obtusely lobed; capsules glabrous ....
    .... 3. *H. moscheutos*

1. **Hibiscus laevis** Scop. Del. Insub. 3:35. 1777. Fig. 318.
Hibiscus militaris Cav. Diss. 3:352, pl. 198. 1787.

Perennial herbs; stems robust, erect, glabrous, to 2.5 m tall; leaves simple, alternate, triangular in outline, ovate, acute to acuminate at the apex, broadly rounded

Fig. 318. *Hibiscus laevis*
(Halberd-leaved rose mallow).

a. Habit.
b. Fruit .

c. Seed.

to cordate to truncate at the base, often hastately 3-lobed, serrate, glabrous, to 15 cm long, to 6 cm wide, the petioles up to 10 cm long, slender, glabrous; flowers large, showy, subtended by an involucre of filiform bracts up to 3 cm long; sepals 5, ovate, glabrous, united at the base, up to 2 cm long; petals 5, free, obovate, to 8 cm long, pink or white, with a dark purple center; capsules ovoid, glabrous to pubescent, 1.5–3.0 cm long, loosely enclosed by the calyx; seeds pubescent. July–September.

Marshes, along streams, wet ditches, wooded swamps, sometimes in standing water.

IA, IL, IN, KS, KY, MO, NE, OH (OBL).

Halberd-leaved rose mallow.

This handsome species is readily distinguished from the other species of *Hibiscus* in the central Midwest by its leaves which are often hastately lobed.

In the past, many botanists used *H. militaris* as the binomial for this species.

2. **Hibiscus lasiocarpos** Cav. Diss. 3:159, pl. 70. 1787. Fig. 319.

Perennial herbs; stems robust, erect, tomentose, to 2 m tall; leaves simple, alternate, ovate, acute to acuminate at the apex, cordate to subcordate at the base, rarely shallowly lobed but never hastately lobed, crenate to dentate, densely tomentose on both surfaces with stellate hairs, to 20 cm long, to 15 cm wide, on tomentose petioles up to 10 cm long; flowers large, showy, subtended by an involucre of linear bracts up to 2.5 cm long; sepals 5, ovate, tomentose, united at the base, up to 2.5 cm long; petals 5, free, obovate, to 10 cm long, white, often with a dark purple base, rarely pink throughout; capsules ovoid, tomentose to villous-hirsute, 2–3 cm long, closely enclosed by the calyx; seeds glabrous. June–September.

Marshes, along streams, swampy woods, wet ditches.

IL, IN, KS, KY, MO (considered the same as *H. moscheutos* by the U.S. Fish and Wildlife Service).

Hairy rose mallow.

This mostly southern species is readily distinguished by its densely tomentose leaves and very pubescent capsules. Some botanists reduce this plant to a variety of *H. moscheutos*, but I reject this view.

3. **Hibiscus moscheutos** L. Sp. Pl. 693. 1753. Fig. 320.
Hibiscus palustris L. Sp. Pl. 693. 1753.

Perennial herbs; stems robust, erect, glabrous, to 2.5 m tall; leaves simple, alternate, lanceolate to ovate, acute to acuminate at the apex, tapering to rounded to subcordate at the base, sometimes shallowly obtusely lobed, dentate to crenate, glabrous on the upper surface, somewhat gray-pubescent on the lower surface, to 20 cm long, to 10 cm wide, on slender, usually glabrous petioles up to 5 cm long; flowers large, showy, subtended by an involucre of linear bracts up to 3 cm long; sepals 5, ovate, somewhat gray-canescent, united at the base, up to 2.5 cm long; petals 5, free, obovate, up to 10 cm long, white, sometimes with a purple center, or pink throughout; capsules ovoid, glabrous except along the sutures, 2.5–3.0 cm long, closely enclosed by the calyx; seeds glabrous. July–September.

Marshes, along streams, swampy woods, wet ditches, wet meadows.

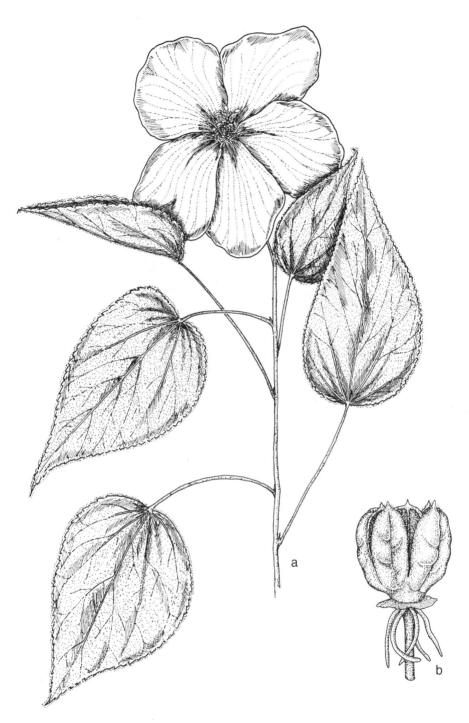

Fig. 319. *Hibiscus lasiocarpos* (Hairy rose mallow).   a. Habit.   b. Fruit.

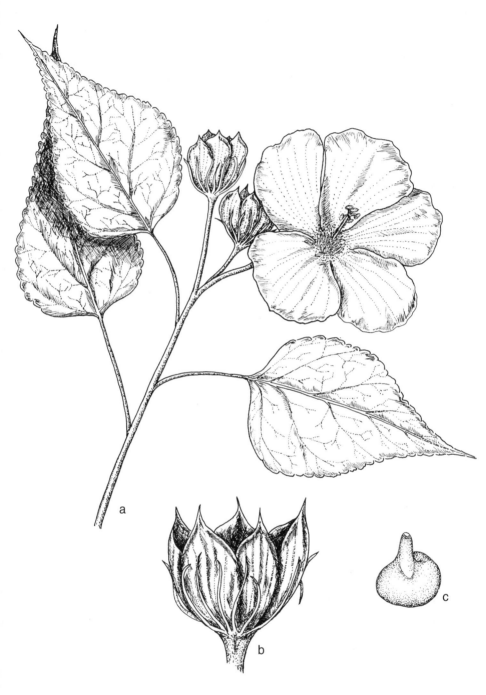

Fig. 320. *Hibiscus moscheutos*   a. Habit.   c. Seed.
(Smooth rose mallow).   b. Capsule.

IL, IN, KS, KY, OH (OBL).

Smooth rose mallow.

Plants with some of the leaves 3-lobed and with pink petals throughout have been segregated as *H. palustris* by some botanists

85. MELASTOMATACEAE—MEADOW BEAUTY FAMILY

Herbs (in our area) or woody plants; leaves opposite, simple, entire, with 3–9 prominent arching veins, usually without bracts; flowers showy, usually in cymes, actinomorphic, perfect; sepals 4–5, united below; petals 4–5, free; stamens 4, 5, 8, or 10; ovary 3- to 5- locular; fruit a capsule or berry, usually with coiled seeds.

This mostly tropical family consists of two hundred genera and four thousand species, many of them woody.

1. **Rhexia** L.—Meadow Beauty

Perennial herbs; leaves opposite, simple, entire, with 3–9 prominent arching veins; flowers showy, in cymes, actinomorphic, perfect; sepals 4, united below; petals 4, free; stamens 8; ovary 4-locular; seeds coiled.

Rhexia consists of about twelve species, most of them in the southeastern United States.

1. All four sides of the stems equal, sharply angled or winged.
   2. Angles of stems winged ....3. *R. virginica*
   2. Angles of stems unwinged.. 1. *R. interior*
1. Two of the sides of the stems broader than the other two, neither sharply angled nor winged.. 2. *R. mariana*

1. **Rhexia interior** Pennell, Bull. Torrey Club 45:480. 1918. Fig. 321.

Perennial herbs from slender rhizomes; stems erect, branched or unbranched, hirsute, more or less 4-sided, to 1 m tall, with narrow wings up to 0.5 mm wide; leaves opposite, simple, lanceolate to elliptic, acute at the apex, rounded at the sessile base, glabrous to somewhat hirsute on both surfaces, with 3 prominent arching veins, finely serrate, to 4 cm long, to 2 cm wide; flowers showy, in cymes, actinomorphic, perfect, on hirsute pedicels; sepals 4, the tube hirsute, to 10 mm long; petals 4, free, glandular-hairy, rose-pink, 12–18 mm long; stamens 8; fruit flask-shaped, hirsute, 6–10 mm long; seeds 0.5–0.7 mm long, tuberculate or papillose. June–July.

Wet ground in prairies, wet ditches.

KY, MO (OBL), KS (FACW+). The U.S. Fish and Wildlife Service does not distinguish this plant from *R. mariana*.

Showy meadow beauty.

The slightly winged stems and lanceolate to elliptic leaves distinguish this species which is often considered to be a variety of *R. mariana*.

2. **Rhexia mariana** L. Sp. Pl. 346, 1753. Fig. 322.

Rhexia mariana L. var. *leiosperma* Ferno & Grisc. Rhodora 37:171. 1935.

Perennial herbs from slender rhizomes; stems erect, branched or unbranched, hirsute, more or less 4-sided, unwinged, to 1 m tall; leaves simple, opposite, lanceolate to elliptic, glabrous to somewhat hirsute on both surfaces, with 3 prominent

Fig. 321. *Rhexia interior* (Showy
meadow beauty). Habit.

Fig. 322. *Rhexia mariana* (Dull meadow beauty).
Habit. Capsule (left).

arching veins, finely serrate, to 4 cm long, to 2 cm wide; flowers showy, in cymes, actinomorphic, perfect, on hirsute pedicels; sepals 4, the tube glandular-hirsute, to 10 mm long; petals 4, free, glabrous, rose-pink, 12–18 mm long; stamens 8; fruit flask-shaped, hirsute, 6–10 mm long; seeds 0.5–0.7 mm long, tuberulate or papillose or smooth. June–September.

Swampy woods, depressions in prairies, wet ditches.

IL, IN, KY, MO, OH (OBL), KS (FACW+).

Dull meadow beauty.

This beautiful species does not have any wings on the stems. The petals are glabrous.

Plants without papillate seeds have been called var. *leiosperma*.

3. **Rhexia virginica** L. Sp. Pl. 346. 1753. Fig. 323.

Perennial herbs from slender rhizomes and tuberous-thickened roots; stems erect, usually unbranched, 4-angled, glabrous or glandular-hairy, often with reddish nodes, with wings up to 2 mm wide, sometimes with stiff red pubescence, to 1 m tall; leaves simple, opposite, lanceolate to ovate, acute at the apex, rounded at the sessile base, hirsute on the upper surface and sometimes on the lower, with 3 prominent arching veins, serrate, to 7 cm long, to 3 cm wide; flowers showy, in cymes, actinomorphic, perfect, on hirsute pedicels; sepals 4, the tube glandular-hirsute, to 10 mm long; petals 4, free, rose-pink, 15–20 mm long; stamens 8; fruit flask-shaped, hirsute, to 10 mm long; seeds up to 1 mm long, strongly papillose. July–September.

Swampy woods, wet meadows, wet ditches, around sinkhole ponds.

IA, IL, IN, KY, MO, OH (OBL).

Wingstem meadow beauty.

This meadow beauty has a winged stem with the wing 0.5–2.0 mm wide.

86. MENYANTHACEAE—BOGBEAN FAMILY

Aquatic or bog-inhabiting perennial herbs; leaves basal or alternate or opposite, simple or compound, entire, without stipules; flowers usually perfect, actinomorphic, in racemes or umbels; sepals 5, united at the base; petals 5, united below; stamens 5; ovary superior to half-inferior, 1-locular; fruit a capsule (in our species), with few to many seeds.

This family consists of five genera and about thirty-five species, found throughout the world. The members of this family are sometimes placed in the Gentianaceae. Two genera and two species occur in the central Midwest, and all are found in standing water.

1. Leaves compound, trifoliolate; petals bearded on the upper surface; flowers whitish or pinkish .... 1. *Menyanthes*
1. Leaves simple; petals fringed; flowers yellow .... 2. *Nymphoides*

1. **Menyanthes** L.—Bogbean; Buckbean

Only the following species comprises the genus.

1. **Menyanthes trifoliata** L. Sp. Pl. 145. 1753. Fig. 324.
Menyanthes trifoliata L. var. *minor* Mill. ex Raf. Med. Fl. 2:34. 1830.

Fig. 323. *Rhexia virginica* (Wingstem meadow beauty). Habit. Flower (lower left).

Fig. 324. *Menyanthes trifoliata* (Bogbean). Habit (center). Fruiting branch (upper right). Corolla cut open (lower right).

Perennials with a thick, creeping rhizome; leaves basal, emergent, trifoliolate, on petioles up to 30 cm long, the leaflets oval to oblong to elliptic, entire or undulate, to 10 cm long, to 5 cm wide, glabrous, sessile or subsessile; flowering scape to 35 cm long, glabrous, bearing a raceme of flowers at the tip; flowers dimorphic, some with exserted stamens and included styles, and some with included stamens and exserted styles, on peduncles to 2 cm long, subtended by bracts; calyx 5-parted, united near the base, the lobes 1.5–3.0 mm long; corolla salverform, whitish or pinkish, united at the base for about half its length, the 5 lobes 8–12 mm long, recurved, bearded on the upper surface; stamens 5, attached to the base of the petals; ovary half-inferior; capsules globose, up to 1 cm in diameter, splitting irregularly; seeds yellow-brown, shiny. April–September.

Shallow water, bogs.

IA, IL, IN, MO, NE, OH (OBL).

Bogbean, buckbean.

This somewhat fleshy species is unique because of its large, entire, trifoliolate leaves and its scapose inflorescence. Since plants of the eastern, central, and southern United States have flowers slightly smaller and with shorter beards on the petals than plants on the Pacific coast and in Europe and Asia, our plants are sometimes designated as var. *minor*.

2. **Nymphoides** Hill—Floating Hearts

Aquatic herbs with rhizomes; leaves opposite (in our species), simple, floating, ovate to orbicular, cordate, more or less entire, on long petioles; flowers yellow (in our species) in whorls at the base of the leaf petioles, actinomorphic, perfect; calyx 5 parted; corolla 5-parted, rotate, the petals united at the base, erose or fringed at the apex; stamens 5, attached to the base of the petals; ovary inferior, 1-locular, with a short style, or style absent; capsules indehiscent or splitting irregularly; seeds numerous.

About twenty species are in this genus in both temperate and tropical regions of the world.

1. **Nymphoides peltata** (Gmel.) Ktze, Rev. Gen. Pl. 2:429. 1891. Fig. 325.
Limnanthes peltata Gmel. Nov. Comm. Acad. Sci. Imp. Petrop. 14:257. 1769.

Aquatic perennial herbs; stems emersed or creeping on mud, branched, glabrous; leaves floating, broadly ovate to orbicular, obtuse at the apex, cordate at the base, undulate along the margin, glabrous, to 10 cm wide, on long, stout petioles; flowers umbellate, on stout pedicels up to 15 cm long, actinomorphic, up to 2.2 cm across; calyx 10–14 mm long; corolla deeply 5-cleft, erose at the tip of each petal, bright yellow, 1.0–1.5 cm long; stamens 5; ovary inferior; capsules ovoid, beaked at the apex, 15–25 mm long, with numerous fringed seeds. June–September.

Ponds.

IL, IN, KY, MO, OH (OBL).

Yellow floating hearts.

This native of Europe is distinguished by its cordate floating leaves, its erose bright yellow petals, and its fringed seeds.

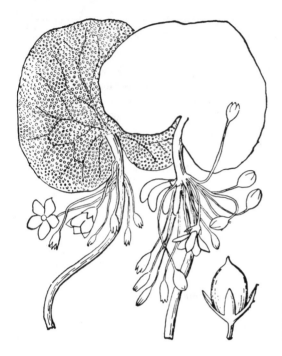

Fig. 325. *Nymphoides peltata* (Yellow floating hearts). Habit. Capsule (lower right).

87. MOLLUGINACEAE—CARPETWEED FAMILY

Herbs; leaves simple, alternate, opposite, or whorled, entire; flowers in cymes or solitary in the leaf axils, usually perfect, actinomorphic; sepals 4–5, free or nearly so; petals 0, 4, 5, minute; stamens 2–10; ovary superior, 2- to 5-locular, with 1–many seeds.

This mostly tropical and subtropical family consists of about twelve genera and one hundred species.

1. **Glinus** L.—Glinus

Annual herbs; leaves simple, whorled, entire; flowers axillary, actinomorphic, perfect; sepals 5, united below; petals absent; stamens 3–5; ovary superior, 3- to 5-locular; capsules with few seeds.

This genus consists of approximately ten species native to the tropics and subtropics.

1. **Glinus lotoides** L. Sp. Pl. 1:463. 1753. Fig. 326.

Annual herbs from fibrous roots; stems prostrate to ascending, branched, stellate-pubescent, to 30 cm long; leaves simple, whorled, obovate to suborbicular, obtuse at the apex, rounded or tapering at the base, to 2 cm long, to 1.5 cm wide, stellate-pubescent, entire, short-petiolate; flowers in the leaf axils, sessile, actinomorphic, perfect; sepals 5, united below, 3–5 mm long; petals absent; stamens 3–5; ovary superior, 3- to 5-locular; seeds several. July–September.

Wet soil around lakes and along streams, depressions in old fields.

Native of Africa; KS (FACW–), MO (OBL).

Glinus.

This small non-native is distinguished by its prostrate habit, rounded, whorled leaves, and apetalous flowers.

88. MYRICACEAE—MYRTLE FAMILY

Shrubs (in our area) or trees; leaves alternate, simple, aromatic, with resin glands; flowers unisexual, the plants monoecious or dioecious, arranged in catkins, subtended by bracteoles; sepals 0; petals 0; stamens 2–8 (–12); ovary inferior, 1-celled, with 1 basal ovule and 2 stigmas; nutlets sometimes covered with a whitish wax.

Depending upon one's point of view, there are 1 to 3 genera in this family, consisting of about fifty species in most parts of the world. I recognize three genera, one of which barely enters our range in wetlands.

Fig. 326. *Glinus lotoides* (Glinus). Habit (above). Flower (below habit). Pistil (right of flower). Sepal (lower left). Capsules (lower right). Stamen (between sepal and capsules).

1. **Gale** Duhamel—Sweet Gale

Only the following species comprises this genus.

1. **Gale palustris** (Lam.) A. Chev. Rev. Bot. Appl. Agric. Trop. 15:946. 1936.
   Fig. 327.
Myrica gale L. Sp. Pl. 2:1024. 1753.
Myrica palustris Lam. Fl. Franc. 2:236. 1779.

   Shrubs, dioecious, much branched, to 2 m tall; leaves simple, alternate, oblanceolate, more or less acute at the apex, tapering to the sessile base, several-toothed
near the apex, glabrous or pubescent on both surfaces, aromatic when crushed
because of the numerous resin glands, without stipules; flowers appearing before
the leaves unfold, the staminate in cylindrical catkins to 2 cm long, the pistillate in
ovoid catkins to 1 cm long; staminate flowers without a perianth, with 2–8 stamens, subtended by a pair of small triangular bracteoles; pistillate flowers without
a perianth, the ovary subtended by a pair of nearly orbicular bracteoles; nutlets flat,
ovate, beaked at the summit, 2.5–3.0 mm long, in catkins expanding in length to
1.2 cm. April–May.
   Bogs.
   OH (OBL).
   Sweet gale.
   This shrub is readily distinguished by its aromatic oblanceolate leaves that are
toothed only near the apex.

Fig. 327. *Gale palustris* (Sweet gale). Habit. Staminate
flower (center below). Pistillate flower (lower left).
Catkins (upper right).

Glossary

Acaulescent. Seemingly without aerial stems.

Achene. A type of one-seeded, dry, indehiscent fruit with the seed coat not attached to the mature ovary wall.

Acicular. Awl-shaped; tapering to an extremely narrow point.

Actinomorphic. Having radial symmetry; regular, in reference to a flower.

Acuminate. Gradually tapering to a long point.

Acute. Sharply tapering to a point.

Adaxial. Toward the axis; when referring to a leaf, the upper surface.

Ament. A spike of unisexual, apetalous flowers; a catkin.

Anther. The terminal part of a stamen which bears pollen.

Anthesis. Flowering time.

Antrorse. Projecting forward.

Apical. At the apex.

Apiculate. Abruptly short-pointed at the tip.

Appressed. Lying flat against the surface.

Arcuate. Curving.

Areole. A small area between leaf veins.

Aristate. Bearing an awn.

Articulated. Jointed.

Attenuate. Gradually becoming narrowed.

Auricle. An earlike lobe.

Auriculate. Bearing an earlike process.

Awn. A bristle usually terminating a structure.

Axillary. Borne from an axil.

Beard. A tuft of stiff hairs.

Berry. A type of fruit where the seeds are surrounded only by fleshy material.

Bicostate. With two veins.

Bidentate. Having two teeth.

Bifid. Two-cleft.

Bifoliolate. Bearing two leaflets.

Bifurcate. Forked.

Biglandular. Bearing two glands.

Bilabiate. Two-lipped.

Bipinnate. Divided once into distinct segments. with each segment in turn divided into distinct segments.

Bipinnatifid. Divided partway to the center, with each lobe again divided partway to the center.

Bisexual. Referring to a flower that contains both stamens and pistils.

Biternate. Divided into three segments two times.

Bract. An accessory structure at the base of many flowers, usually appearing leaflike.

Bracteate. Bearing one or more bracts.

Bracteole. A secondary bract.

Bractlet. A small bract.

Bristle. A stiff hair or hairlike growth; a seta.

Bulblet. A small bulb.

Callosity. Any hardened thickening.

Calyx. The outermost segments of the perianth of a flower, composed of sepals.

Campanulate. Bell-shaped.

Canescent. Grayish-hairy.

Capillary. Threadlike.

Capitate. Forming a head.

Capsule. A dry, dehiscent fruit composed of more than one carpel.

Carpel. A simple pistil, or one member of a compound pistil.

Cartilaginous. Firm but flexible.

Catkin. A spike of unisexual, apetalous flowers; an ament.

Caudate. With a taillike appendage.

Caudex. The woody base of a perennial plant.

Cauline. Belonging to a stem.

Cespitose. Growing in tufts.

Chaffy. Covered with scales.

Chartaceous. Papery.

Cilia. Marginal hairs.

Ciliate. Bearing cilia.

Ciliolate. Bearing small cilia.

Circumscissile. Usually referring to a fruit which dehisces by a horizontal, circular line.

Claw. A narrow, basal stalk, particularly of a petal.

Cleistogamous. Hidden.

Coherent. The growing together of like parts.

Columella. A small column to which stamens and pistils are usually attached.

Coma. A tuft of hairs at the end of a seed.

Compressed. Flattened.

Concave. Curved on the inner surface; opposed to convex.

Connate. Having a union of like parts.

Convex. Curved on the outer surface.

Convolute. Rolled lengthwise.

Cordate. Heart-shaped.

Coriaceous. Leathery.

Corm. An underground, vertical stem with scaly leaves, differing from a bulb by lacking fleshy leaves.

Corolla. The segments of a flower just within the calyx, composed of petals.

Corona. A crown of petal-like structures.

Corrugated. Folded or wrinkled.

Corymb. A type of inflorescence where the pedicellate flowers are arranged along an elongated axis but with the flowers all attaining about the same height.

Corymbiform. Shaped like a corymb.

Cotyledon. A seed leaf.

Crateriform. Cone-shaped but sunken in the center at the top.

Crenate. With round teeth.

Crenulate. With small, round teeth.

Crest. A ridge.

Crisped. Curled.

Cucullate. Hood-shaped.

Culm. The stem that terminates in an inflorescence.

Cuneate. Wedge-shaped; tapering to the base.

Cupular. Shaped like a small cup.

Cuspidate. Terminating in a very short point.

Cyathium. A cuplike structure enclosing flowers.

Cyme. A type of broad and flattened inflorescence in which the central flowers bloom first.

Cymose. Bearing a cyme.

Deciduous. Falling away.

Decumbent. Lying flat, but with the tip ascending.

Decurrent. Adnate to the petiole or stem and then extending beyond the point of attachment.

Deflexed. Turned downward.

Dehiscent. Splitting at maturity.

Deltoid. Triangular.

Dentate. With sharp teeth, the tips of which project outward.

Denticulate. With small, sharp teeth, the tips of which project outward.

Diffuse. Loosely spreading.

Digitate. Radiating from a common point, like the fingers from a hand.

Dilated. Swollen; expanded.

Dimorphic. Having two forms.

Dioecious. With staminate flowers on one plant, pistillate flowers on another.

Disarticulate. To come apart; to become disjointed.

Disk. An enlarged outgrowth of the receptacle.

Divergent. Spreading apart.

Drupe. A type of fruit in which the seed is surrounded by a hard, dry covering, which in turn is surrounded by fleshy material.

Ebracteate. Without bracts.

Echinate. Spiny.

Eciliate. Without cilia.

Eglandular. Without glands.

Ellipsoid. Referring to a solid object that is broadest at the middle, gradually tapering to both ends.

Elliptic. Broadest at the middle, gradually tapering to both ends.

Emarginate. Having a shallow notch at the extremity.

Epidermis. The outermost layer of cells.

Epunctate. Without dots.

Erose. With an irregularly notched margin.

Exudate. Secreted material.

Falcate. Sickle-shaped.

Fascicle. Cluster.

Ferruginous. Rust-colored.

Fetid. Foul-smelling.

Fibrous. Referring to roots borne in tufts.

Filament. That part of the stamen supporting the anther.

Filiform. Threadlike.

Fimbriate. Fringed.

Flaccid. Weak; flabby.

Flexible. Able to be bent readily.

Flexuous. Zigzag.

Floret. A small flower.

Follicle. A type of dry, dehiscent fruit that splits along one side at maturity.

Friable. Breaking easily into small particles.

Funnelform. Shaped like a funnel.

Geniculate. Bent.

Gibbous. Swollen on one side.

Glabrate. Becoming smooth.

Glabrous. Without pubescence or hairs.

Gland. An enlarged, spherical body functioning as a secretory organ.

Glandular. Bearing glands.

Glaucous. With a whitish covering that can be rubbed off.

Globose. Round; globular.

Glomerule. A small compact cluster.

Glutinous. Covered with a sticky secretion.

Hastate. Spear-shaped; said of a leaf that is triangular with spreading basal lobes.

Haustoria. Structures of a parasitic plant that attach it to the host plant.

Hirsute. With stiff hairs.

Hirsutulous. With minute stiff hairs.

Hirtellous. Finely hirsute.

Hispid. With rigid hairs.

Hispidulous. With minute rigid hairs.

Hoary. Grayish-white, usually referring to pubescence.

Hood. That part of an organ, usually of a flower, that is strongly concave and arching.

Horn. An accessory structure found in certain flowers.

Hyaline. Transparent.

Hypanthium. A development of the receptacle beneath the calyx.

Imbricate. Overlapping.

Indehiscent. Not splitting open at maturity.

Indurate. Hard.

Inferior. Referring to the position of the ovary when it is surrounded by the adnate portion of the floral tube or is embedded in the receptacle.

Inflexed. Turned inward.

Inflorescence. A cluster of flowers.

Internode. The area between two adjacent nodes.

Involucel. A cluster of bracteoles that subtends a secondary flower cluster.

Involucre. A circle of bracts that subtends a flower cluster.

Involute. Rolled inward.

Keel. A ridgelike process.

Laciniate. Divided into narrow, pointed divisions.

Lanceolate. Lance-shaped; broadest near base, gradually tapering to the narrower apex.

Lanceoloid. Referring to a solid object that is broadest near base, gradually tapering to the narrower apex.

Latex. Milky juice.

Leaflet. An individual unit of a compound leaf.

Legume. A dry fruit usually dehiscing along two sides at maturity.

Lenticel. Corky opening on the bark of twigs and branches.

Lenticular. Lens-shaped.

Lepidote. Scaly.

Linear. Elongated and uniform in width throughout.

Lobulate. With small lobes.

Locular. Referring to the locule, or cavity of the ovary or the anther.

Loculicidal. Said of a capsule that splits down the dorsal suture of each cell.

Lunate. Crescent-shaped.

Lustrous. Shiny.

Lyrate. Pinnatifid, with the terminal lobe much larger than the lower ones.

Moniliform. Constricted at regular intervals to resemble a string of beads.

Monoecious. Bearing both sexes in separate flowers on the same plant.

Monomorphic. Having but one form.

Mucilaginous. Slimy.

Mucro. A short, abrupt tip.

Mucronate. Possessing a short, abrupt tip.

Mucronulate. Possessing a very short, abrupt tip.

Muricate. Minutely spiny.

Nectariferous. Producing nectar.

Nigrescent. Blackish.

Node. That place on the stem from which leaves and branchlets arise.

Nutlet. A small nut.

Oblanceolate. Reverse lance-shaped; broadest at apex, gradually tapering to narrow base.

Oblong. Broadest at the middle, and tapering to both ends, but broader than elliptic.

Oblongoid. Referring to a solid object that, in side view, is nearly the same width throughout.

Obovoid. Referring to a solid object that is broadly rounded at the apex, becoming narrowed below.

Obpyramidal. Reverse pyramid-shaped.

Obtuse. Rounded at the apex.

Ocrea. A sheathing stipule, often tubular.

Ocreola. A secondary, usually tubular, sheath.

Opaque. Referring to an object that cannot be seen through.

Orbicular. Round.

Oval. Broadly elliptic.

Ovary. The lower swollen part of the pistil that produces the ovules.

Ovoid. Referring to a solid object that is broadly rounded at the base, becoming narrowed above.

Palmate. Divided radiately, like the fingers of a hand.

Pandurate. Fiddle-shaped.

Panduriform. Fiddle-shaped.

Panicle. A type of inflorescence composed of several racemes.

Papilla. A small wart.

Papillate. Bearing small warts, or papillae.

Papillose. Bearing pimplelike processes.

Pappus. The modified calyx in the Asteraceae.

Papule. A pimplelike projection.

Pectinate. Pinnatifid into close, narrow segments; comblike.

Pedicel. The stalk of a flower of an inflorescence.

Pedicellate. Bearing a pedicel.

Peduncle. The stalk of an inflorescence.

Peltate. Attached away from the margin, in reference to a leaf.

Perennial. Living more than two years.

Perfect. Bearing both stamens and pistils in the same flower.

Perfoliate. Referring to a leaf that appears to have the stem pass through it.

Perianth. Those parts of a flower including both the calyx and corolla.

Pericarp. The ripened ovary wall.

Petal. One segment of the corolla.

Petaloid. Resembling a petal in texture and appearance.

Petiolate. Bearing a petiole, or leafstalk.

Petiole. The stalk of a leaf.

Petiolulate. Bearing a petiolule, or leaflet-stalk.

Petiolule. The stalk of a leaflet.

Pilose. Bearing soft hairs.

Pinna. A primary division of a compound blade.

Pinnate. Divided once into distinct segments.

Pinnatifid. Said of a simple leaf or leaf-part that is cleft or lobed only partway to its axis.

Pinnule. The secondary segments of a compound blade.

Pistil. The ovule-producing organ of a flower normally composed of an ovary, a style, and a stigma.

Pistillate. Bearing pistils but not stamens.

Plicate. Folded.

Plumose. Bearing fine hairs, like the plume of a feather.

Procumbent. Lying on the ground.

Pruinose. Having a waxy covering.

Puberulent. With minute hairs.

Pubescent. Bearing some kind of hairs.

Punctate. Dotted.

Pyramidal. Shaped like a pyramid.

Pyriform. Pear-shaped.

Quadrate. Four-sided.

Raceme. A type of inflorescence where pedicellate flowers are arranged along an elongated axis.

Rachis. A primary axis.

Receptacle. That part of the flower to which the perianth, stamens, and pistils are usually attached.

Reflexed. Turned downward.

Reniform. Kidney-shaped.

Repand. Wavy along the margin.

Repent. Creeping.

Resinous. Producing a sticky secretion, or resin.

Reticulate. Resembling a network.

Reticulum. A network.

Retrorse. Pointing downward.

Retuse. Shallowly notched at a rounded apex.

Revolute. Rolled under from the margin.

Rhizome. An underground horizontal stem, bearing nodes, buds, and roots.

Rhombic. Becoming quadrangular.

Rosette. A cluster of leaves in a circular arrangement at the base of a plant.

Rotate. Flat and circular.

Rufescent. Reddish brown.

Rufous. Red-brown.

Rugose. Wrinkled.

Rugulose. With small wrinkles.

Saccate. Sac-shaped.

Sagittate. Shaped like an arrowhead.

Salverform. Referring to a tubular corolla that abruptly expands into a flat limb.

Samara. An indehiscent winged fruit.

Scaberulous. Slightly rough to the touch.

Scabrous. Rough to the touch.

Scape. A leafless stalk bearing a flower or inflorescence.

Scarious. Thin and membranous.

Scorpioid. Strongly curved, resembling the tail of a scorpion.

Scurfy. Bearing scaly particles.

Secund. Borne on one side.

Sepaloid. Resembling a sepal in texture.

Septate. With dividing walls.

Septicidal. Said of a capsule that splits between the locules.

Sericeous. Silky; bearing soft, appressed hairs.

Serrate. With teeth that project forward.

Serrulate. With very small teeth that project forward.

Sessile. Without a stalk.

Seta. Bristle.

Setaceous. Bearing bristles, or setae.

Setiform. Bristle-shaped.

Setose. Bearing setae.

Silicle. A short silique.

Silique. An elongated capsule with a central partition separating the valves.

Sinuate. Wavy along the margins.

Sinus. The cleft between two lobes or teeth.

Spatulate. Oblong, but with the basal end elongated.

Spicate. Bearing a spike.

Spike. A type of inflorescence where sessile flowers are arranged along an elongated axis.

Spikelet. A small spike.

Spinescent. Becoming spiny.

Spinose. Bearing spines.

Spinule. A small spine.

Spinulose. Bearing small spines.

Spur. A saclike extension of the flower.

Stamen. The pollen-producing organ of a flower composed of a filament and an anther.

Staminate. Bearing stamens but not pistils.

Staminodium. A sterile stamen.

Standard. The upper, usually enlarged, petal of a pea-shaped flower.

Stellate. Star-shaped.

Stipe. A stalk.

Stipel. A small stipe.

Stipitate. Possessing a stipe.

Stipular. Pertaining to a stipule.

Stipule. A leaflike or scaly structure found at the point of attachment of a leaf to the stem.

Stolon. A slender, horizontal stem on the surface of the ground.

Stoloniferous. Bearing stolons.

Stramineous. Straw-colored.

Striate. Marked with grooves.

Strigillose. With short, appressed, straight hairs.

Strigose. With appressed, straight hairs.

Strigulose. With short, appressed, straight hairs.

Style. That elongated part of the pistil between the ovary and the stigma.

Subcuneate. Nearly wedge-shaped.

Suborbicular. Nearly spherical.

Subulate. With a very short, narrow point.

Succulent. Fleshy.

Suffused. Spread throughout; flushed.

Superior. Referring to the position of the ovary when the free floral parts arise below the ovary.

Tendril. A spiraling, coiling structure that enables a climbing plant to attach itself to a supporting body.

Terete. Round, in cross-section.

Ternate. Divided three times.

Tomentose. Pubescent with matted wool.

Tomentulose. Finely pubescent with matted wool.

Tomentum. Woolly hair.

Torulose. With small contractions.

Translucent. Partly transparent.

Trifoliolate. Divided into three leaflets.

Trigonous. Triangular in cross-section.

Truncate. Abruptly cut across.

Tuber. An underground fleshy stem formed as a storage organ at the end of a rhizome.

Tubercle. A small, wartlike process.

Tuberculate. Warty.

Turgid. Swollen to the point of bursting.

Turion. A swollen asexual structure.

Umbel. A type of inflorescence in which the flower stalks arise from the same level.

Undulate. Wavy.

Unisexual. Bearing either stamens or pistils in one flower, but not both.

Urceolate. Urn-shaped.

Utricle. A small, one-seeded, indehiscent fruit with a thin covering.

Verrucose. Warty.

Verticil. A whorl.

Verticillate. Whorled.

Villous. With long, soft, slender, unmatted hairs.

Virgate. Wandlike.

Viscid. Sticky.

Whorl. An arrangement of three or more structures at a point on the stem.

Zygomorphic. Bilaterally symmetrical.

Illustration Credits

Illustrations 3, 4, 5, 6, 11, 12, 16, 24, 28, 31, 32, 35, 36, 37, 39, 42, 43, 48, 49, 50, 55, 60, 65, 68, 69, 72, 74, 78, 79, 80, 82, 83, 87, 90, 91, 92, 95, 96, 97, 98, 99, 100, 101, 102, 105, 106, 107, 111, 112, 115, 116, 117, 124, 125, 126, 147, 151, 152, 153, 154, 155, 156, 157, 162, 165, 167, 175, 178, 185, 192, 195, 196, 200, 202, 205, 209, 221, 223, 242, 260, 263, 266, 267, 269, 270, 276, 278, 281, 282, 286, 288, 289, 290, 291, 309, 311, 313, 316, 317, 318, 319, 320, 322, and 323 were prepared by Mark W. Mohlenbrock.

Illustrations 7, 8, 9, 10, 27, 66, 119, 120, 121, 122, 123, 130, 131, 133, 135, 136, 137, 138, 139, 141, 142, 143, 144, 168, 169, 171, 174, 179, 180, 181, 182, 183, 201, 203, 210, 211, 212, 214, 215, 216, 240, 264, 265, 287, and 300 were prepared by Paul W. Nelson.

Illustration 259 was prepared by Fredda J. Burton.

Illustrations 232, 233, 235, 236, 238, 239, and 241 were prepared by Miriam Wysong Meyer and are reprinted from *Contributions to Transactions of the Illinois State Academy of Science I, 1965–1999*.

Illustrations 173, 188, and 190 were prepared by Peggy Kessler Duke and are reprinted from *Herbaceous Plants of Maryland* by Melvin L. Brown and Russell G. Brown.

Illustration 254 was prepared by Melanie Darst and was first published in *Trees, Shrubs, and Woody Vines of Northern Florida and Adjacent Georgia and Alabama* by Robert K. Godfrey and Jean W. Wooten. Reprinted from *Florida Wetland Plants: An Identification Manual*. Copyright 1988 by the University of Georgia Press.

Illustrations 213 and 292 were prepared by Melanie Darst and are reprinted from *Trees, Shrubs, and Woody Vines of Northern Florida and Adjacent Georgia and Alabama* by Robert K. Godfrey and Jean W. Wooten. Copyright 1988 by the University of Georgia Press.

Illustrations 19, 86, and 207 were prepared by Amy V. McIntosh and are reprinted from *Plant Life of Kentucky: An Illustrated Guide to the Vascular Flora* by Ronald L. Jones. Lexington: University of Kentucky Press, 2005.

Illustration 231 is copyright APIRS, Center for Aquatic and Invasive Plants and is reprinted from *Plant Life of Kentucky: An Illustrated Guide to the Vascular Flora* by Ronald L. Jones. Lexington: University of Kentucky Press, 2005.

Illustration 33 is reprinted from *Plant Life of Kentucky: An Illustrated Guide to the Vascular Flora* by Ronald L. Jones. Lexington: University of Kentucky Press, 2005.

Illustrations 18, 30, 184, 310, and 321 are reprinted from *Aquatic and Wetland Plants of the Southeastern United States: Dicotyledons* by Robert K. Godfrey and Jean W. Wooten. Athens: University of Georgia Press.

Illustration 326 is reprinted from *Steyermark's Flora of Missouri*, vol. 2, by G. Yatskievych. St. Louis: Missouri Botanical Garden Press. With Permission from the Missouri Botanical Garden Press.

Illustrations 38, 40, 191, 194, 197, 198, 244, 245, 246, 247, 248, 249, 250, 251, 252, 253, 255, 256, 257, and 258 are reprinted from *Flowering Plants: Hollies to Loasa* by Robert H. Mohlenbrock, from the Illustrated Flora of Illinois series, Southern Illinois University Press.

Illustrations 145 and 146 are reprinted from *Flowering Plants: Magnolias to Pitcher Plants* by Robert H. Mohlenbrock, from the Illustrated Flora of Illinois series, Southern Illinois University Press.

Illustrations 132 and 140 are reprinted from *Flowering Plants: Willows to Mustards* by Robert H. Mohlenbrock, from the Illustrated Flora of Illinois series, Southern Illinois University Press.

Illustration 327 is reprinted from *An Illustrated Flora of the Northern United States and Canada*, vol. 1, by Nathaniel Britton and Addison Brown. Charles Scribner's Sons, 1913; reprinted 1970 by Dover Publications, New York.

Illustrations 15, 17, 20, 21, 23, 25, 26, 29, 34, 41, 143, 170, 176, 177, 189, 199, 204, 206, 208, 225, 226, 227, 228, 229, 230, 234, 237, 305, 306, 308, and 315 are reprinted from *An Illustrated Flora of the Northern United States and Canada*, vol. 2, by Nathaniel Britton and Addison Brown. Charles Scribner's Sons, 1913; reprinted 1970 by Dover Publications, New York.

Illustrations 1, 2, 45, 46, 47, 59, 61, 62, 63, 64, 67, 71, 73, 75, 76, 77, 81, 84, 85, 88, 89, 93, 94, 103, 108, 113, 114, 127, 128, 158, 159, 160, 161, 163, 164, 166, 217, 218, 220, 222, 224, 243, 261, 262, 274, 275, 277, 279, 283, 284, 285, 293, 294, 295, 296, 297, 298, 301, 302, 303, 304, 324, and 325 are reprinted from *An Illustrated Flora of the Northern United States and Canada*, vol. 3, by Nathaniel Britton and Addison Brown. Charles Scribner's Sons, 1913; reprinted 1970 by Dover Publications, New York.

Illustration 273 was first published in *Flora of Texas* by C. L. Lundell. Renner: Texas Research Foundation, 1961–1969, and is reprinted from *Shinners and Mahler's Illustrated Flora of North Central Texas* by George M. Diggs Jr., Barney L. Lipscomb, and Robert J. O'Kennon.

Illustration 314 was originally published in *Aquatic and Wetland Plants of the Southwestern United States* by D. S. Correll and H. B. Correll. Washington, D.C.: Environmental Protection Agency, 1972, and is reprinted from *Shinners and Mahler's Illustrated Flora of North Central Texas* by George M. Diggs Jr., Barney L. Lipscomb, and Robert J. O'Kennon.

Illustrations 13, 22, 44, 51, 52, 53, 54, 56, 57, 58, 70, 104, 109, 110, 129, 149, 172, 186, 193, 219, 271, 272, 280, 307, 312 are reprinted from *Northeast Wetland Flora: Field Office Guide to Plant Species*.

Illustration 118 is reprinted from *Southern Field Office Guide to Wetland Plants of the South*.

Illustrations 14, 148, 150, 268, are reprinted from *Western Wetland Flora: Field Office Guide to Plant Species*.

Index to Families, Genera, and Species

Names in roman type are accepted, while those in italics are synonyms and are not considered valid. Page numbers in bold refer to pages that have illustrations.

Index to Common Names

Robert H. Mohlenbrock is in his sixtieth year of plant study. After receiving his doctorate from Washington University in St. Louis, Mohlenbrock taught botany and plant taxonomy at Southern Illinois University for thirty-four years, sixteen of which he served as chairman of the department. During his career at SIU Carbondale, he was major professor for ninety graduate students and carried out a vigorous program of research. To date, his research has resulted in the publication of fifty-eight books and more than five hundred other publications. Since 1984, he has written "This Land," a monthly column for *Natural History*, published by the American Museum of Natural History in New York. Since retiring from SIU in 1990, Mohlenbrock has been a senior scientist for Biotic Consultants Inc., where he teaches weeklong wetland plant identification classes for government employees and consultants throughout the country.